ILLUSTRATED
ENCYCLOPEDIC
DICTIONARY OF
ELECTRONICS

ACKNOWLEDGMENT

"Speed was of the essence" in writing this book, and I owe a great debt of gratitude to Leatrice, my wife, who prepared the manuscript. Since this involved deciphering thousands of index cards containing innumerable unfamiliar terms and symbols, all written in my crabbed script, and without detriment to her other vocations of wife, mother, housekeeper, farmer, gardener, chauffeur, etc., etc., I can only say, rather tritely I'm afraid, that I can never adequately repay her for this "labor of love."

ILLUSTRATED ENCYCLOPEDIC DICTIONARY OF ELECTRONICS

John Douglas-Young

PARKER PUBLISHING COMPANY, INC.
West Nyack, N.Y.

Library of Congress Cataloging in Publication Data

Douglas-Young, John.
    Illustrated encyclopedic dictionary of electronics.

    1. Electronics—Dictionaries.   I.   Title.
TK7804.D68   621.381'03'21   80-23639
ISBN 0-13-450791-6

Printed in the United States of America

The Unique Practical Value
This Book Offers...

The need for this reference work brings to mind remarks made by the Red Queen to Alice during her visit through the looking glass. In essence, the message was that it takes all the running we can do to stay in the same place—and if we want to get somewhere else, it's necessary to run at least twice as fast.

All during the time this extensive manuscript was in preparation, we were aware of the accelerating expansion of the electronics universe and the necessity of keeping up with it if the "state-of-the-art" was not going to leave us hopelessly behind. A new and more comprehensive approach to electronics reference data was clearly needed.

This book, therefore, has been prepared expressly for technicians, engineers, and experimenters in a carefully organized manner in order to present a broad, up-to-the-minute overview of the electronics profession. For the most part, the material is presented in straightforward, nonmathematical language, while at the same time it provides all the formulas, tables, and other essential information the electronics technologist needs in his daily work.

Most electronic reference books are presented in one of two forms: *alphabetical*, i.e., dictionary form; or *topical*, with the information arranged in sections by subject. Both styles have their advantages and disadvantages. In an alphabetical listing each entry is necessarily condensed, and the general subject of which it is a part is fragmented, so that it tends to be illogically distributed throughout the work; but it is easy to find individual items. In a topical format the information is logically organized by subject so that related data are grouped together ... but you need an index to find the individual items.

The *Illustrated Encyclopedic Dictionary of Electronics* gets the benefit of the advantages and overcomes the disadvantages of both methods by using a new, practical approach. It combines ready-reference data with essential information in depth. The alphabetical listings include thousands of electronic terms, with

5

illustrations and brief definitions, that are meaningful and intelligible in themselves. However, where the individual item is part of a larger subject covered by a major reference, the title of the latter is given in the minor entry; so when you need more detailed data, or related background information, you can readily find it—because the major references are also arranged alphabetically. These major subjects are treated with expanded presentations that give comprehensive information on electronic topics of paramount importance; e.g., "Antennas," "Broadcasting," "Computer Hardware," "Electromagnetic-Wave Propagation," "Microelectronics," "Radar," "Recording," "Semiconductors," "Waveguides and Resonators," to mention only a few of the fifty subjects covered in depth. These are practical, more detailed reviews for readers who have a basic grasp of the subject, but need additional information in these key areas.

For example, to illustrate how you will use this book, take the entry for the term "filter," which reads as follows:

> **filter**—Arrangement of electronic components that allows some frequencies to pass and blocks others. *Active Filters, Passive Filters.*

The entry tells in general terms what a filter is. It then gives in italics two major references to turn to for more information. For instance, if you want to design a bandpass filter for use in a modern solid-state circuit you proceed to one of these major references, *Active Filters*, and there you may find not only the circuit you need but also the easy way to calculate the values of its components for the frequency band you want it to pass.

From this example you can see that the *Illustrated Encyclopedic Dictionary of Electronics* has an unusually helpful format, written in understandable language, abundantly illustrated with functional drawings and schematic diagrams, and packed with mathematical tables, formulas, conversion factors, graphic symbols and much more, far more than there is room to mention here. You will find it offers extraordinary value to technicians, engineers, and experimenters alike. For practical utility, fast reference, and dependable data it is a "must" for all those who are active in the field of electronics.

John Douglas-Young

A—1. Symbol for ampere (q.v.). 2. Symbol for angstrom unit (q.v.); also written Å.

a—Symbol for prefix atto (q.v.).

A− (minus)—Negative terminal of an A battery or other source of filament voltage (also called F−).

A+ (plus)—Positive terminal of an A battery or other source of filament voltage (also called F+).

abc—Automatic bass compensation (q.v.).

aberration—In lenses, the failure of rays of light or electrons in a beam to converge to a single focus.

abnormal propagation—The phenomenon when electromagnetic waves do not follow their normal path through space because of unstable or changing atmospheric or ionospheric conditions. *Electromagnetic-Wave Propagation.*

abort—To halt an operation in progress. Said of a missile launch countdown, computer run, or the like.

absolute altimeter—An aircraft altimeter that uses radio or radar reflections from the ground to determine the actual height above the surface beneath; as distinct from altitude above sea level given by an aneroid altimeter.

absolute temperature—Temperature expressed in kelvins (q.v.).

absolute value—If r is a number, the absolute value of r—denoted by |r|—has no sign, because it only answers the question "How many?"

absolute zero—Temperature of zero thermal energy, corresponding to −273.16 degrees Celsius. (See also kelvin.)

absorbed dose—Energy absorbed per mass of object irradiated, expressed in grays. *Nuclear Physics.*

absorption—The dissipation of the energy of a wave as it passes through matter. *Electromagnetic-Wave Propagation, Electroacoustics.*

absorption circuit—A tuned circuit that dissipates energy taken from another circuit. (See also trap.) (See Figure A-1.)

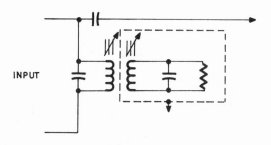

Figure A-1
Absorption Circuit

absorption coefficient—Measure of the attenuation of radiant or sound energy that results from its passage through a substance or reflection from a surface. *Electromagnetic-Wave Propagation, Electroacoustics.*

absorption current—After the plates of a capacitor have been fully charged, a small absorption current continues to flow as the electrical stress between them gradually forces electric charges into the dielectric. A similar current flows after the initial discharge as the charges emerge from the dielectric. *Capacitors.*

absorption fading—Slow type of fading, primarily caused by variations in the rate of absorption (q.v.) along the radio path.

absorption loss—That part of the transmission loss due to dissipation or conversion of either electromagnetic or sound energy into other forms of energy resulting from its passage through a substance or reflection from a surface. *Electromagnetic-Wave Propagation, Electroacoustics.*

absorption marker—Type of marker consisting of a sharp dip on the frequency-response curve of a circuit displayed on an oscilloscope screen. It is produced by absorption of energy by a circuit sharply tuned to the marker frequency.

absorption wavemeter—Device for measuring the frequency of a radio wave by extracting power from the frequency source to actuate its indicating device (microammeter, lamp, or earphone). Maximum indication is given when the wavemeter's cavity or circuit is tuned to resonance with the frequency being measured. The frequency is read from the wavemeter's calibrated tuning dial or accompanying chart. (See also reaction wavemeter.) (See Figure A-2.)

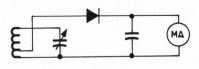

Figure A-2
Absorption Wavemeter

a.c.—Abbreviation for alternating current (q.v.). Also written A.C., AC, or ac.

a.c. bias—Alternating current in the 60-100 kilohertz range used to bias the recording head during recording. Its value affects frequency response, distortion, and signal-to-noise ratio. *Recording.*

accelerating electrode—In a cathode-ray tube or other electron tube, an electrode to which a potential is applied to accelerate the electron beam. (See also post-deflection acceleration.)

acceleration—Time rate at which velocity of a point is changing, measured with an accelerometer. The S.I. unit for acceleration is meter per second squared (m/s²). Acceleration is often expressed as a multiple of the acceleration due to gravity (g), which is 9.801 042 3 m/s² (32.15565ft/s²) at the U.S. National Gravity Base, Washington, D.C. *Navigation Aids.*

accelerator—Used to provide beams of various particles for study of nuclear structure and nuclear reactions. Accelerators with energies from 100 kiloelectron volts to 200 gigaelectron volts (proposed) accelerate stable charged particles from electrons to ^{238}U. *Nuclear Physics.*

accelerometer—Instrument that measures the rate at which the speed of a body is changing (i.e., its acceleration). *Navigation Aids.* (See Figure A-3.)

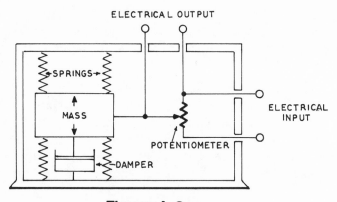

Figure A-3
Accelerometer

acceptor—Impurity element, such as boron from Group III of the periodic table (q.v.), which has three valence electrons in the outer shell of its atom, that increases the number of holes in a semiconductor, resulting in a p-type material in which holes are the majority carriers. *Semiconductors.*

access—Method of selecting a computer memory location. (See also random-access memory.) *Computer Hardware.*

access time—Time that elapses between initiation of memory-read operation by controller and appearance of data on bus; and, similarly, time taken by memory-write operation. *Computer Hardware.*

a.c. coupling—Coupling between stages or circuits via a capacitor or other device that passes the a.c. component but not the d.c. component of a signal.

accumulator—In a computer, a register that stores a number and arithmetically adds to it another number or numbers received in performing an arithmetic operation. *Computer Hardware.*

accuracy—The maximum difference between the true value and that indicated by an instrument is the measure of the instrument's accuracy. It is expressed as a percentage of the full-scale value or of the reading, according to the type of instrument.

a.c./d.c.—Receiver or other electronic equipment that can operate from either a.c. or d.c. power. Also written ac/dc, AC/DC.

acetate-base tape—A sound-recording tape made of smooth, transparent acetate, with a coating of finely divided ferrous-oxide particles. *Recording*.

a.c. generator—(See alternator.)

achromatic—Colorless, composed only of black, white, and shades of gray; also said of a lens that does not diffract white light into its spectral components.

acorn tube—Button or acorn-shaped electron tube with no base, used for ultra-high frequencies (330-3000 MHz).

acoustic—Having to do with hearing, heard sound, or the science of heard sound. *Electroacoustics*.

acoustic absorption loss—Energy lost by conversion to other forms when sound passes through or is reflected by a medium. *Electroacoustics*.

acoustic absorptivity—Ratio of sound energy absorbed by a surface to that arriving at the surface, equal to one minus the reflectivity of the surface. *Electroacoustics*.

acoustical reflectivity—(See sound-reflection coefficient.)

acoustic burglar alarm—Burglar alarm that is actuated by sounds made by an intruder when the sounds exceed a preset level.

acoustic capacitance—(See acoustic compliance.)

acoustic clarifier—System of cones loosely attached to the baffle of a speaker to absorb the excess energy of sudden loud sounds.

acoustic compliance—Parameter of acoustic energy storage analogous to capacitance in an electrical system. Acoustic compliance is expressed in centimeters^5/dyne, and is equal to the enclosed volume of gas in centimeters^3 divided by the square of the sound velocity in centimeters/second and the density of the gas in grams/centimeter^3

$(C_a=V_o/_c{}^2\rho)$.At audio frequencies acoustic compliance is also called adiabatic expansion. *Electroacoustics*.

acoustic impedance—Ratio of excess pressure to rate of volume flow in sound transmission. For a plane wave the acoustic impedance Z is real and equals the density times velocity divided by the wave-front area (V/S), hence it is taken to correspond to an electrical resistance. For a spherical wave there are both real and imaginary components corresponding to resistance and reactance in electrical impedance. *Electroacoustics*.

acoustic inertance—Parameter of acoustic energy storage analogous to inductance in an electrical system. Acoustic inertance is expressed in grams/centimeters^4 and is equal to the density of gas in grams/centimeters^3 times the length of the pipe in which the gas flows in centimeters, divided by the cross-sectional area of the pipe in centimeters^2(M=ρl/A). *Electroacoustics*.

acoustic labyrinth—Special baffle arrangement used with a speaker to prevent cavity resonance and to reinforce bass response.

acoustic line—Baffles, labyrinths, or resonators arranged at the rear of a loudspeaker to help reproduce very low audio frequencies.

acoustic reactance—That part of acoustic impedance which is due to the inertia and elasticity of the medium through which the sound travels. *Electroacoustics*.

acoustic reflectivity—Ratio of the rate of flow of sound energy reflected from the surface on the side of incidence to the incident rate of flow. *Electroacoustics*.

acoustic resistance—Parameter of acoustic energy dissipation analogous to resistance in an electrical system. Acoustic resistance is expressed in dyne-seconds/centimeter^5 and is equal to eight times the coefficient of viscosity in poises times the length of tube in which the gas flows, divided by the square of the cross-sectional area of the tube in centimeters^2 (R$_a$=8$\mu\pi$l/A^2). *Electroacoustics*.

acoustics—1. The science of heard sound.   2. Qualities of a room, theater, etc., that have to do with how clearly sounds can be heard or transmitted in it. *Electroacoustics*.

acquisition—In radar, the process that occurs between the location of a target in the search phase and the final alignment of the tracking equipment on the target. *Radar*.

acquisition and tracing radar—Radar set which locks on to a strong signal and tracks the object emitting the signal. *Radar*.

activator—A substance such as manganese, of which a small amount is added to the active material (q.v.) that is used to coat the screen of a cathode-ray tube.

active device—A component such as an electron tube or transistor that turns a current on or off, or controls its rate of flow, in accordance with an external input signal, to provide switching action or to obtain amplification. Also called active element. *Electron Tubes, Magnetic Amplifiers, Transistors.*

active filter—Arrangements of resistors and capacitors together with an active device (usually an operational amplifier) that passes some frequencies while blocking others, performs phase shifting, or delays a signal in time. *Active Filters.*

ACTIVE FILTERS

Filters are networks of electronic components that tailor signals to meet requirements of bandpass, phase shift, or time delay. There are passive filters (q.v.) and active filters. Because passive filters include inductors in their networks they are not readily adaptable to integrated-circuit techniques. Consequently there has been a rapid growth of the use of active filters in this area.

An active filter usually consists of an operational-amplifier IC, such as the popular bipolar 741 type or the FET 536 type, with an external network of resistors and capacitors. This network determines the function of the filter, which may be low pass, high pass, bandpass, bandstop, phase shift, or constant time delay.

Low-Pass Filter

A low-pass filter passes lower frequencies but not higher ones, as in Figure A-4, where OF represents increasing frequency and OA amplitude. The response of the filter is shown by the curve AF, which indicates that frequencies below f_c are passed readily, but those above f_c are greatly attenuated. The range of frequencies lower than f_c is

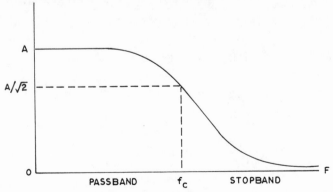

Figure A-4
Low-Pass Filter Response

therefore called the passband, and the range higher than f_c is the stopband. The cutoff frequency f_c is defined as the frequency at which the amplitude A has fallen to $A/\sqrt{2}$, or 0.707 times its full value. (This is also 3 dB below the maximum value of A.)

Figure A-5 gives the circuit of a commonly used active low-pass filter. The value of C1 is found by dividing the required f_c into 10, which gives C1's capacitance in

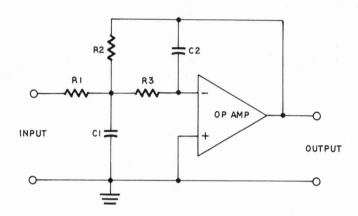

Figure A-5
Commonly Used Low-Pass Filter

microfarads. This capacitance is then multiplied by f_c, and their product divided into 159 to obtain the scale factor. Values for the other components can then be calculated from the following table:

Component	For Gain of 2	For Gain of 10
C2 (in μF)	0.150 × value of C1	0.033 × value of C1
R1 (in kΩ)	1.612 × scale factor	1.021 × scale factor
R2 (in kΩ)	3.223 × scale factor	10.211 × scale factor
R3 (in kΩ)	2.068 × scale factor	2.968 × scale factor

Example:

When f_c is required to be 1000 Hz and a gain of 2 is desired,

$$\text{C1} = \frac{10}{1000} = 0.01 \ \mu\text{F}$$

$$\text{C2} = 0.150 \times 0.01 = 0.0015 \ \mu\text{F}$$

$$\text{Scale factor} = \frac{159}{0.01 \times 1000} = 15.9$$

$$\text{R1} = 1.612 \times 15.9 = 25.631 \ \text{k}\Omega$$
$$\text{R2} = 3.223 \times 15.9 = 51.246 \ \text{k}\Omega$$
$$\text{R3} = 2.068 \times 15.9 = 32.881 \ \text{k}\Omega$$

Using standard values, the components required would be:

Capacitors (10% polystyrene or Mylar):
0.01 μF (C1), 0.0015 μF (C2)
Resistors (5%): 27 kΩ (R1), 51 kΩ (R2), 33 kΩ (R3)
Operational Amplifier: 741 or 536

High-Pass Filter

A high-pass filter passes higher frequencies but not lower ones, as in Figure A-6, where OF represents increasing frequency and OA amplitude. The response of the filter is

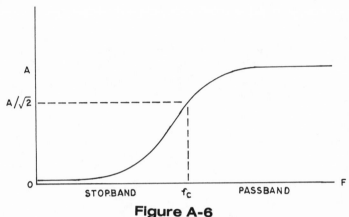

Figure A-6
High-Pass Filter Response

shown by the curve AF, which indicates that frequencies above f_c are passed readily, but those below f_c are greatly attenuated. The range of frequencies above f_c is therefore called the passband and the range below f_c is the stopband. The cutoff frequency f_c is defined as the frequency at which the amplitude of A has risen to $A/\sqrt{2}$, or 0.707 times its full value. (This is also 3 dB below the maximum value of A.)

Figure A-7 gives the circuit of a commonly used active high-pass filter. The value of C1 is found by dividing the required f_c into 10, which gives C1's capacitance in

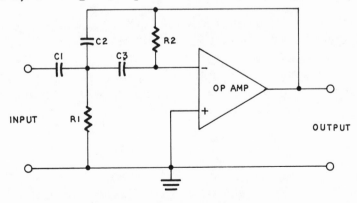

Figure A-7
Commonly Used High-Pass Filter

microfarads. This capacitance is then multiplied by f_c, and their product divided into 159 to obtain the scale factor. Values for the other components can then be calculated from the following table, as in the example given for a low-pass filter:

Component	For Gain of 2	For Gain of 10
C2 (in μF)	0.500 × value of C1	0.100 × value of C1
C3 (in μF)	1.000 × value of C1	1.000 × value of C1
R1 (in kΩ)	0.566 × scale factor	0.673 × scale factor
R2 (in kΩ)	3.536 × scale factor	14.849 × scale factor

Bandpass Filter

A bandpass filter passes a band of frequencies approximately symmetrical about a center frequency f_o, with a width inversely proportional to the circuit Q, as shown in Figure A-8. Frequencies f_{c1} and f_{c2} are the lower and upper cutoff frequencies at the points where the amplitude is $A/\sqrt{2}$.

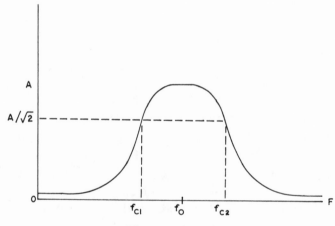

Figure A-8
Bandpass Filter Response

Figure A-9 gives the circuit of a commonly used active bandpass filter. The value of C1 is found by dividing the required f_o into 10, which gives C1's capacitance in microfarads. This capacitance is then multiplied by f_o, and their product divided into 159 to obtain the scale factor. Values for the other components to design a filter with a Q of 2 can then be calculated from the following table, as in the example for a low-pass filter:

Component	For Gain of 2	For Gain of 10
C2 (in μF)	1.000 × value of C1	2.000 × value of C1
R1 (in kΩ)	1.000 × scale factor	0.200 × scale factor
R2 (in kΩ)	0.333 × scale factor	1.000 × scale factor
R3 (in kΩ)	4.000 × scale factor	3.000 × scale factor

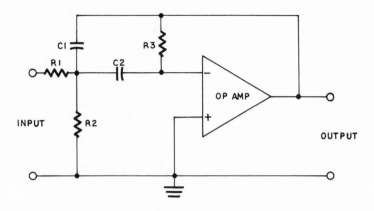

Figure A-9
Commonly Used Bandpass Filter

To obtain a narrower bandpass a higher value of Q is required, and the values for R1, R2, and R3 must be adjusted by multiplying them by the factors given in the following table:

Desired Value of	For Gain of 2		For Gain of 10	
Q	R1 and R3	R2	R1 and R3	R2
4	2	0.400	2	0.105
6	3	0.258	3	0.061
8	4	0.189	4	0.044
10	5	0.153	5	0.034

Bandstop Filter

A bandstop filter is also called a band-reject or notch filter. It rejects a band of frequencies approximately symmetrical about a center frequency f_o, with a width inversely proportional to the circuit Q, as shown in Figure A-10. Frequencies f_{c1} and f_{c2} are the lower and upper cutoff frequencies where the amplitude is $A/\sqrt{2}$.

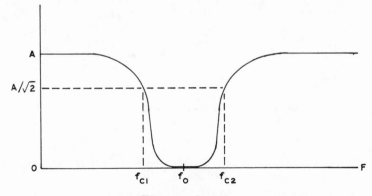

Figure A-10
Bandstop Filter Response

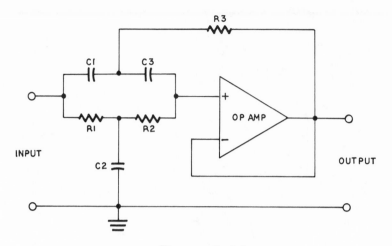

Figure A-11
Commonly Used Bandstop Filter

Figure A-11 gives the circuit of a commonly used active bandstop filter. The value of C1 is found by dividing f_o into 10, which gives C1's capacitance in microfarads. This capacitance is then multiplied by f_o, and their product divided into 159 to obtain the scale factor. C2's capacitance is twice that of C1. Values for the resistors in kilohms to design a filter with a gain of one can then be calculated for values of Q from one to ten from the following table by multiplying the scale factor by the values given:

Desired Value of Q	R1	R2	R3
1	0.500	2.000	0.400
2	0.250	4.000	0.235
3	0.167	6.000	0.162
4	0.125	8.000	0.123
5	0.100	10.000	0.099
6	0.083	12.00	0.083
7	0.071	14.000	0.071
8	0.063	16.000	0.062
9	0.056	18.000	0.055
10	0.050	20.000	0.050

Phase-Shifting Filter

A phase-shifting filter passes all frequencies with equal amplitude but shifts their phase in accordance with their frequency. It is therefore also called an all-pass filter. Figure A-12 gives the circuit of a commonly used active phase-shifting filter. The value of C1 is found by dividing the frequency of the signal to be phase-shifted into 10, which gives C1's capacitance in microfarads. This capacitance is then multiplied by the frequency (in hertz), and their product divided into 159 to obtain the scale factor. C2's capacitance is the same as that of C1. Values for the resistors in kilohms to design

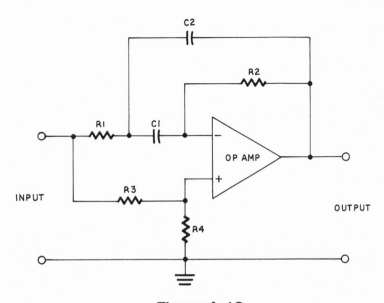

Figure A-12
Commonly Used Phase-Shifting Filter

a filter with a gain of 0.75 can then be calculated for various phase-shift angles by multiplying the scale factor by the values given in the following table:

Phase Shift	R1	R2	R3	R4
10°	1.948	23.374	31.165	93.495
20	1.026	12.317	16.423	49.268
30	0.735	8.824	11.765	35.296
40	0.597	7.169	9.558	28.675
50	0.518	6.219	8.292	24.875
60	0.467	5.605	7.473	22.420
70	0.431	5.175	6.900	20.700
80	0.405	4.855	6.474	19.421
90	0.384	4.606	6.141	18.422
−10	0.043	0.513	0.685	2.054
−20	0.081	0.974	1.299	3.897
−30	0.113	1.360	1.813	5.440
−40	0.139	1.674	2.232	6.696
−50	0.161	1.930	2.573	7.719
−60	0.178	2.141	2.855	8.564
−70	0.193	2.319	3.092	9.275
−80	0.206	2.472	3.295	9.886
−90	0.217	2.606	3.474	10.422

Constant-Time-Delay Filter

A constant-time-delay filter delays a signal by a time interval that is proportional to the frequency of the signal. If the time interval T_o (in seconds) is specified, the required frequency f_o (in hertz) is given by:

$$f_o = \frac{6}{13\pi T_o}$$

Conversely, the time delay T_o (in seconds) for a certain frequency f_o (in hertz) is given by:

$$T_o = \frac{6}{13\pi f_o}$$

The circuit of the filter is the same as that in Figure A-5. The value of C1 is found by dividing the required f_o into 10, which gives C1's capacitance in microfarads. This capacitance is then multiplied by f_o, and their product is divided into 159 to obtain the scale factor. Values of the other components can then be calculated from the following table:

Component	For Gain of 2	For Gain of 10
C2 (in µF)	0.200 × value of C1	0.047 × value of C1
R1 (in kΩ)	0.691 × scale factor	0.471 × scale factor
R2 (in kΩ)	1.382 × scale factor	4.709 × scale factor
R3 (in kΩ)	1.206 × scale factor	1.506 × scale factor

active line—A horizontal scanning line that carri   TV picture information. *Broadcasting*.

active material—1. Zinc orthosilicate or other luminescent material used to coat a cathode-ray tube screen. It usually requires, in addition, an activator such as a small amount of manganese.   2. Lead oxide or other active substance used in the plates of a storage battery (q.v.).

active network—A network with active as well as passive components.

active substrate—(See substrate.)

Active Swept Frequency Interferometer Radar—Dual radar air surveillance system that gives high-precision angle and range information for pinpointing target locations by trigonometric techniques. *Radar*.

active systems—In radio and radar, systems that require transmitting equipment, such as a beacon or transponder.

actual height—Highest altitude at which refraction of radio waves actually occurs. *Electromagnetic-Wave Propagation.*

actual power—Average of values of instantaneous power taken over one cycle.

actuating system—System in a device or vehicle that supplies and transmits energy for the operation of mechanisms or other devices.

actuator—1. The device that moves the load in a servo system.   2. The part of a relay that converts electrical energy into mechanical motion.

a.c. voltage—(See alternating voltage.)

adapter—Any means that enables different types of connectors to mate. (See Figure A-13.)

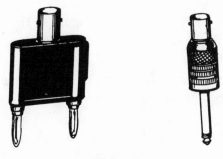

Figure A-13
Adapters

Adcock antenna—Pair of vertical antennas spaced at a half-wavelength or less and connected in phase opposition to produce a figure-eight directional radiation pattern. *Antennas.*

Adcock radio range—Type of high-frequency direction finder with four vertical antennas at the corners of a square (Adcock antennas), and a fifth antenna at the center of the square. *Antennas.*

a-d converter—(See analog-to-digital converter.)

adder—A device which accepts two words and produces a third word that is the sum of the two inputs. *Computer Hardware.*

additive primary colors—Primary colors that can be mixed to form other colors, such as the red, green, and blue in color television.

address—1. Location in memory where data are stored.   2. Address in memory from which the operand is to be read and then used to perform the specified operation. *Computer Hardware.*

A-display—Type of presentation on a cathode-ray indicator in which time is the horizontal coordinate and signals appear as vertical deflections. Also called A-scan and A-scope. *Radar.*

adjacent channel—Frequency band immediately above or below the channel of interest. *Broadcasting.*

adjacent sound channel—The carrier frequency containing the sound modulation in the next lower-frequency TV channel. *Broadcasting.*

adjacent video channel—The carrier frequency containing the picture modulation in the next higher-frequency TV channel. *Broadcasting.*

adjustable resistor—Resistor with its resistance wire partly exposed so that a movable lug in contact with it can be adjusted for a desired resistance value. *Resistors.*

admittance—Lack of opposition to the flow of alternating current in a circuit. The reciprocal of impedance. Admittance is expressed in siemens (formerly mhos). Symbol is Y.

adsorption—Adhesion of the molecules of a gas, liquid, or dissolved substance to a surface.

aerial—(See antenna.)

aerodynamics—Science of forces exerted by air or other gases in motion.

a.f.—(See audio frequency.)

a.f.c.—(See automatic frequency control.)

a.g.c.—(See automatic gain control.)

aging—Allowing an electron tube, quartz crystal, permanent magnet, or other device to remain in storage for a period of time, often with voltage applied, until its characteristics become constant.

air capacitor—Capacitor using air as the dielectric between its plates. *Capacitors.*

air-core coil—A coil with nonmetallic core. *Coil Data.* (See Figure A-14.)

Figure A-14
Air-Core Coil

air-core transformer—Transformer with nonmetallic core. *Transformers.*

aircraft flutter—Fluctuations in a TV picture caused by signals reflected by a flying aircraft being received along with signals direct from the transmitter. The varying phase difference between the signals is responsible for the flutter.

air gap—The airspace between two objects which are electrically or magnetically related; especially the space between the poles of a magnet that reduces the tendency toward saturation.

airport beacon—Beacon (light or radio) located at or near an airport to indicate its location. *Navigation Aids.*

airport radar control—Surveillance portion of radar approach control engaged in pickup, holding, pattern, and similar operations. *Navigation Aids.*

airport runway beacon—Radio-range beacon that defines one or more approaches to an airport. *Navigation Aids.*

airport surface detection equipment—Short-distance radar with high resolution that shows ground plan of airport with runways, taxiways and ramps, and location of all aircraft and other traffic, moving or stationary on the airfield, for use by traffic control.

air-position indicator—Airborne computer which computes aircraft's position continuously from heading, airspeed and time, and displays it on a cockpit indicator.

air-spaced coax—Coaxial cable with air as the dielectric. The conductor is held centered in the shield by spirally wound synthetic filament, beads, or braided filament. *Transmission Lines.*

airway beacon—Beacon, other than an airport beacon, located on or near an airway to indicate the position of the airway. *Navigation Aids.*

algorithm—1. Arithmetic procedure, such as addition, subtraction and so on, done according to set rules, using pencil and paper as opposed to abacus, calculator, and the like. (Erroneously refashioned from algorism.) 2. In a computer program, a sequence of instructions to solve a specific problem in a finite number of steps.

alignment—Adjusting the tuned circuits of a receiver to obtain the desired frequency response.

alignment pin—A pin or device to ensure the correct mating of two components; e.g., a tube in its socket.

alignment tool—A nonmetallic screwdriver or socket wrench for adjusting tuned circuit trimmer capacitors or inductor cores.

alkaline cell—1. Primary cell similar to zinc-carbon cell except that the electrolyte is potassium hydroxide. 2. Edison and nickel-cadmium storage cells (q.v.). (See Figure A-15.)

Allen screw—Screw with hexagonal socket in its head, often used as a set screw. It is turned with an Allen wrench, an L-shaped hexagonal-section rod that fits in the screw's socket. (See Figure A-16.)

alligator clip—Spring-loaded metal clip used for making temporary electrical connections. (See Figure A-17.)

alloy-junction transistor—Transistor made by fusing two "dots" of p- or n-type semiconductor material to opposite sides of a wafer of n- or p-type material. The wafer is the base region, the dots the collector and emitter regions. *Transistors.*

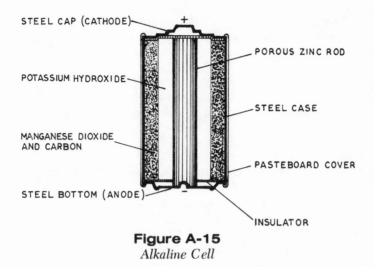

Figure A-15
Alkaline Cell

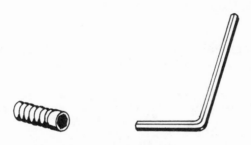

Figure A-16
Allen Screw and Wrench

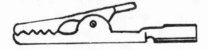

Figure A-17
Alligator Clip

all-pass network—Filter designed to introduce phase shift or delay without appreciable attenuation at any frequency. *Active Filters.*

all-wave antenna—Receiving antenna suitable for reception over a wide range of frequencies. *Antennas.*

all-wave receiver—Receiver capable of reception on all the commonly used wavebands in the broadcast and high-frequency ranges.

alnico—Alloy of aluminum, nickel, cobalt and iron, used for permanent magnets with very high flux density and magnetic retentivity, suitable for speakers, magnetrons, and so on.

alpha—Emitter-collector current gain of common-base transistor amplifier. Symbol, α; also used for angles, coefficients, attenuation constant, absorption factor. *Transistors.*

alpha cutoff frequency—Frequency at the high end of a transistor's range at which current gain of a common-base amplifier drops to three decibels below its low-frequency value.

alphanumeric—Pertaining to a character set containing both letters and numerals, and other symbols such as punctuation and mathematical symbols, used in data processing. *Computer Hardware.*

alpha particle—Particle identical to the nucleus of a helium atom emitted by radioactive materials. *Nuclear Physics.*

alpha ray—Stream of fast-moving alpha particles. *Nuclear Physics.*

alternate mode—In a dual-trace amplifier or preamplifier, a mode in which the two input signals are displayed on the oscilloscope screen on alternate sweeps.

alternating current—Current that flows first in one direction and then in the opposite direction, repeated continuously at a rate of so many cycles per second, called hertz, which is its frequency (q.v.). Abbreviated a.c.

alternating voltage—Type of voltage that reverses its polarity regularly, such as that generated by an alternator.

alternator—A generator or dynamo that converts mechanical energy to alternating current. In an automobile alternator the a.c. is then changed to d.c. by diodes so the battery can be charged.

altimeter—Instrument that measures elevation above sea level. The conventional altimeter is basically an aneroid barometer that measures the pressure of the surrounding air. As atmospheric pressure decreases with height, the difference between that at the aircraft's altitude and that at sea level can be expressed in feet or meters. Provision is made for adjusting the instrument to compensate for variation in barometric pressure. (See also absolute altimeter.)

alumina—An oxide of aluminum Al₂O₃, present in bauxite and clay, and found as different forms of corundum, including emery, sapphires, rubies, etc. It is used for insulators, substrates, and in applications where its ability to withstand high temperatures (up to 1930°C) is important.

aluminized-screen picture tube—A picture tube in which a very thin film of aluminum has been deposited on the inner side of the phosphor screen. This film is transparent to electrons but not to light, so that light is not wasted by shining into the interior of the tube. The picture is therefore brighter, with better contrast. Also used in oscilloscope cathode-ray tubes.

aluminum—Silvery, light-weight, easily worked metal, used extensively in electronic hardware. Symbol, Al; atomic weight, 26.98; atomic number, 13.

aluminum-electrolytic capacitor—Capacitor with aluminum electrodes separated by absorbent paper saturated with electrolyte. *Capacitors.*

a-m—(See amplitude modulation.)

amateur bands—Radio frequencies assigned exclusively to amateur radio operators ("hams"). *Frequency Data.*

ambient noise—Acoustic noise in a room or vicinity. *Electroacoustics.*

American National Standards Institute, Inc. (ANSI)—Organization responsible for standardization of usage in electrical and electronic drawings. Its standards are mandatory for use in the Department of Defense and adhered to closely by the electronics industry.

American Radio Relay League (ARRL)—Organization of amateur radio operators.

American wire gage (AWG)—A standard gage for designating the dimensions of solid wires. *Properties of Materials.*

am/fm receiver—Receiver that accepts either a.m. or f.m. signals.

ammeter—Instrument for measuring strength of current. Contraction of ampere meter. (See permanent magnet moving coil meter.)

ammeter shunt—Low-value resistance connected across the input of a meter movement to extend its range. (See Figure A-18.)

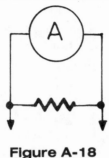

Figure A-18
Ammeter Shunt

amp—Abbreviation for ampere (q.v.).

amperage—Strength of an electric current measured in amperes.

ampere—S.I. base unit of electric current. The ampere is that constant current which, if maintained in two straight parallel conductors of infinite length and of negligible cross section, and placed one meter apart in a vacuum, would produce, between these conductors, a force equal to 2×10^{-7} newtons per meter of length. *S.I. Units.*

ampere-hour—Quantity of electricity that passes through a circuit in one hour when the current is one ampere. Used to denote amount of energy a storage battery can deliver before it needs recharging.

ampere turn—The amount of magnetomotive force produced by an electric current of one ampere flowing around one turn of a wire coil.

amplidyne—Direct-current generator used in servo systems as a power amplifier.

amplification—Increase in signal magnitude from one point to another, or the process causing this increase.

amplification factor—1. In any device, the ratio of output magnitude to input magnitude. 2. In electron tubes, the ratio of incremental plate voltage to control-electrode voltage at a fixed plate current with constant voltage on other electrodes. Symbol: μ. *Electron Tubes.*

amplifier—Device which enables an input signal to control power from a source independent of the signal and thus be capable of delivering an output which bears some relationship to, and is generally greater than, the input-signal. *Electron Tube Circuits, Transistor Circuits, Microelectronics, Operational Amplifiers.*

amplitude—The magnitude of variation of a waveform from its baseline or zero value. Amplitude may be peak, rms, maximum, etc., depending upon where it is measured. (See Figure A-19.)

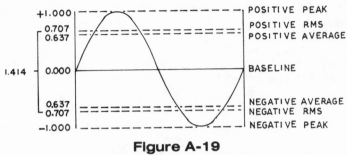

Figure A-19
Amplitude Measurements

amplitude distortion—Distortion when the output amplitude is not a linear function of the input amplitude. *Distortion.*

amplitude-frequency response—Variation of transmission gain or loss of a device with frequency.

amplitude-modulated transmitter—Transmitter for amplitude-modulated signals. *Modulation.*

amplitude modulation—Modulation in which the amplitude of a carrier is varied. *Modulation.*

a-m tuner—Device that recovers the audio modulation from a radio-frequency carrier. *Modulation.*

analog—Representation of numerical quantities by physical variables such as voltage, resistance, gear, or shaft rotation.

analog computer—Computer in which quantities are represented by physical analogues such as voltage or current, which are continuously variable, as opposed to the digital computer in which quantities are represented by digits. The analog computer measures; the digital counts. The basic circuit of the analog computer is the operational amplifier; that of the digital computer is the logic gate. *Computer Hardware.*

analog data—Data represented by analogous voltages, currents, etc., as opposed to digital data.

analog-to-digital converter—Unit for converting continuously varying voltage or current into digital data.

AND gate—Logic circuit in which both or all inputs must be true to give a true output. *Computer Hardware.*

anechoic enclosure—Enclosure that absorbs sound or radiation (such as microwave signals) instead of reflecting them.

anemometer—Gage for measuring wind velocity or pressure.

aneroid barometer—(See barometer.)

angle of lead (or lag)—Angular phase difference between two waves having the same frequency, expressed in degrees. (See Figure A-20.)

angstrom unit—Unit used for measurement of wavelength of light and other radiation, equal to 10 nanometers. The S.I. unit has superseded the angstrom unit in modern scientific usage. Symbol: Å. *Optoelectronics.*

angular velocity—Rate of change of angle, expressed in radians per second. In the case of periodic quantity, it is equal to the frequency multiplied by 2π, or $2\pi f$.

anion—Negative ion, attracted to the anode in a discharge tube or electrolytic cell. The corresponding positive ion is called a cation.

anisotropy—State of having properties, as conductivity, speed of transmission, field strength, etc., which vary according to the direction in which they are measured.

annealing—Process of heating metal, glass, etc., followed by slow cooling, to render it more ductile or less brittle.

annunciator—Visual device which indicates by pilot lights or drop indicators an existing condition or former condition of each circuit being monitored.

anode—1. Positive electrode, such as the plate of an electron tube. 2. The electrode of an electrolytic cell to which negative ions are drawn.

anode terminal—Semiconductor-diode terminal that is positive with respect to the other terminal when the diode is biased in the forward direction.

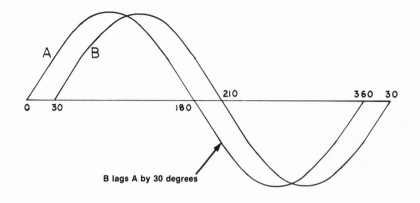

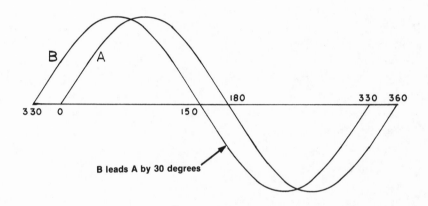

Figure A-20
Angles of Lag and Lead

anodizing—Putting a protective oxide film on a metal surface by an electrolytic process. Aluminum and magnesium are often anodized to prevent corrosion.

anomalous propagation—Freak propagation because of discontinuities in the propagation medium. *Electromagnetic-Wave Propagation.*

A-N radio range—Radio beam used by aircraft. When the aircraft is on course the pilot hears a continuous signal in his earphones. If he deviates to the right he hears the Morse code letter A (. –), to the left N (– .). The continuous on-course tone is produced by the merging of the two signals at their common boundary. Becoming obsolete in the U.S. *Navigation Aids.*

ANSI—Abbreviation for American National Standards Institute, Inc. (q.v.).

antenna—Any structure or device used to detect or radiate electromagnetic waves. *Antennas.*

ANTENNAS

Radio waves are transmitted through space without wires, but they are produced and detected by electrical circuits in which electrons flow in wires. An antenna is the means of converting radiation energy into electrical energy, and vice versa.

Half-Wave Dipole

The simplest form of antenna is the dipole. This consists of a straight wire or rod divided at the center to form a pair of terminals to which the transmitter or receiver is connected. When fed with high-frequency alternating current from a transmitter, at a given instant one terminal will be positive and the other negative. The positive terminal attracts electrons while the negative terminal repels them, so currents flow in opposite directions in the two elements. These result in an accumulation of negative charges at the outer end of the element with the negative terminal, and of positive charges (atoms that have lost electrons) at the end of the element with the positive terminal. (See Figure A-21.)

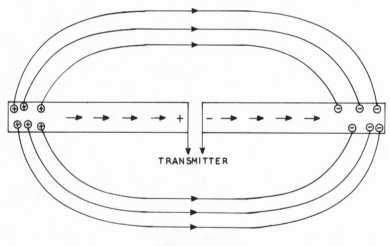

Figure A-21
How a Dipole Works

During the instant that this condition lasts, an electric field exists with curving flux lines connecting the negative and positive charges. An induced magnetic field also forms, encircling the dipole. Then the potentials on the terminals reverse and the field begins to collapse. The positive and negative charges that were at the extremities of the antenna now rush toward each other, dragging the ends of their associated flux lines with them, so that the "heads" and "tails" of the lines are brought together. At this point the charges disappear and a new field of opposite polarity builds up. The old flux lines still existing (with their tails in their mouths, so to speak) are repelled. They are now an independent electric field without associated electric charges, but carrying with them a magnetic field of equal energy, because a moving electric field always creates a magnetic field, just as a moving magnetic field creates an electric field. The electromagnetic field speeds away from the antenna at approximately 300 megameters (186,000 miles) per second. Because the flux lines originally were distributed along the dipole but not around the ends, the field is at a maximum at right angles to the long axis of the antenna.

Since the velocity of propagation is constant, the spacing between successive fields is according to the rate at which the antenna polarity changes. As this is due to the sinusoidal signal from the transmitter, the field strength as it passes a given point rises and falls in a sinusoidal manner, positive peak, negative peak, positive peak, and so on. This wavelike fluctuation is the electromagnetic wave, and its wavelength is the distance between two successive peaks of the same polarity. The relationship between wavelength (λ) in meters and frequency (f) in megahertz is given by:

$$\lambda = \frac{300}{f}$$

(For further discussion of radiation, see *Electromagnetic-Wave Propagation* and *Maxwell's equations*.)

The basic dipole has a length of approximately half the wavelength transmitted, so it is often called a half-wave dipole. The same antenna can also be used for harmonics of the fundamental frequency, so that an antenna designed for use at 3.5 megahertz will work at 7.0, 14.0, 21.0 and 28.0 megahertz as well. This explains why these bands are all allocated for amateur use, since only one antenna is required for all five.

The physical length of a dipole made of wire is calculated from the following formula, which takes into account the fact that radio waves travel slightly more slowly in the atmosphere than in a vacuum, and the presence of insulators makes a small difference as well:

$$L \text{ (feet)} = \frac{468}{f(\text{MHz})}$$

However, for frequencies over 30 megahertz, antennas more often employ tubing, so the diameter of the tubing has to be allowed for also, in the following formula:

$$L \text{ (inches)} = \frac{5905 \times K}{f(\text{MHz})}$$

where K is the ratio of half-wavelength ($\lambda/2$) to conductor diameter, from Figure A-22.

The impedance of a half-wavelength dipole is 75 ohms, which means that a transmission line of this impedance is required to connect the transmitter to the

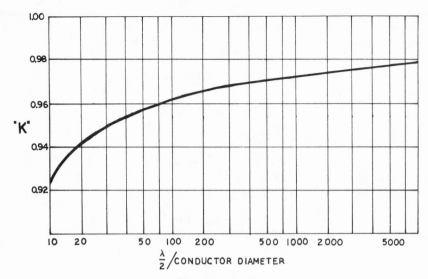

Figure A-22
Graph to Find Factor "K" for Dipoles Constructed of Metal Tubing

antenna. However, antenna coupling is treated in more detail under *Transmission Lines*.

Folded Dipole

By making the half-wave antenna into a two-wire or folded dipole, the impedance is changed to 300 ohms, which allows the use of TV ribbon twin lead as a transmission line. Indeed, the whole antenna can be made of twin lead for frequencies around 100 megahertz, but for lower frequencies the spacing must be increased: 4 inches at 14 megahertz, 8 inches at 3.5 megahertz, for instance. (See Figure A-23.)

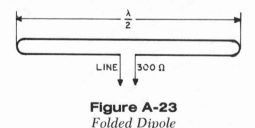

Figure A-23
Folded Dipole

Monopole Antenna

If a dipole is mounted vertically, it is possible to dispense with one element if the surface beneath is a conductor, such as the ground, or the metal body of a vehicle. The vertical portion is insulated from the "ground," of course, and the inner conductor of the coaxial transmission line is connected to it. The outer conductor is connected to the "ground" beneath the antenna. In the case of an earth ground its conductivity is usually enhanced by a system of buried conductors radiating from the center to which the outer conductor is connected. In the case of a CB base station mounted on a mast, a similar arrangement of conductors (usually three) is mounted on the mast at the foot of the monopole. In this case it is called a ground plane.

Since a monopole is only half of a dipole, its length is a quarter-wave instead of a half-wave. However, its physical length may be shortened without changing its electrical length by *loading*, using inductance or capacitance, using a coil at the base of the antenna, or a disk or "hat" mounted on the top.

Long-Wire Antennas

A long-wire antenna in some ways resembles a horizontal monopole. Any antenna more than a half-wavelength is a long-wire antenna, but its length still has to be some multiple of a half-wavelength. The formula for determining its length is:

$$L \text{ (feet)} = \frac{492(N-0.05)}{f(\text{MHz})}$$

where N is the number of half-wavelengths.

This antenna may be center-fed or end-fed. However, an end-fed long-wire antenna will show an unbalance in the feeder line for all frequencies other than the frequency band for which the antenna is designed, resulting in radiation from the feeder line.

Long-Wire Directional Antennas

Two long-wire antennas may be combined in a V, as in Figure A-24. This increases gain and directivity along a line bisecting the angle between the wires. However, a

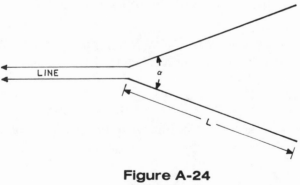

Figure A-24
"V" Antenna

better long-wire antenna, with multiband capability, is the *rhombic antenna*, which is diamond-shaped, as in Figure A-25. It is commonly used for high frequencies and has a power gain of 8 to 12 decibels compared to an isotropic antenna (hypothetical antenna with gain of 1, radiating equally in all directions). The angle Δ shown in the figure is called the vertical angle of radiation and is obtained from the geometry of a triangular path over the Earth's curvature, with the apex of the triangle placed at the

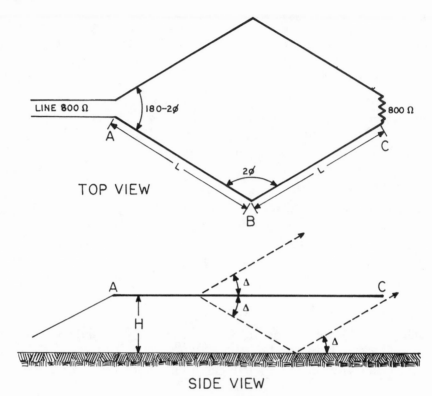

Figure A-25
Rhombic Antenna

virtual height assumed for the angle of the reflection *(see Electromagnetic-Wave Propagation).*

The gain of a rhombic antenna increases with the length (L) of each side. However, it is usually limited to under six wavelengths because longer lengths give too sharp a directivity in the vertical plane. If you know Δ and L you can design a rhombic antenna, using the graph in Figure A-26. For instance, if Δ is to be 16 degrees and L three wavelengths, you find the intersection of these values in the L=3 wavelengths curve, to give a φ of 66 degrees on the right-hand scale. The antenna height H is determined by reading the intersection of the Δ value with the H curve, which is 0.9 wavelength on the left-hand scale.

The correct value for the terminating resistance is 800 ohms, which is also the antenna impedance. The resistor used must be non-inductive and capable of safely dissipating one-half of the output power.

Discone Antennas

A discone antenna consists of a combination of a cone and a disk made of sheet metal, as shown in Figure A-27. The outer conductor of the coaxial transmission line is connected to the cone at the gap, and the inner conductor to the center of the disk. The

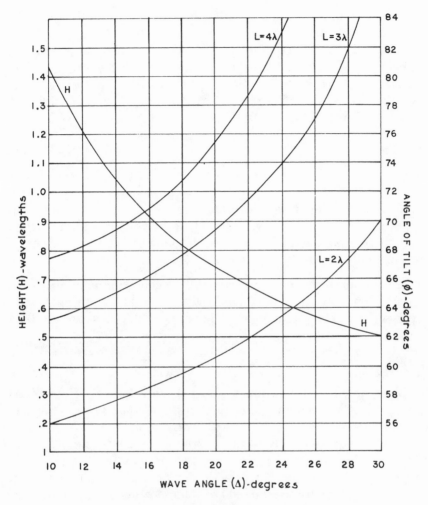

Figure A-26
Rhombic Antenna Design Chart

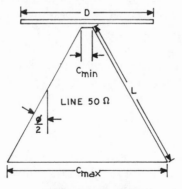

Figure A-27
Discone Antenna

smaller the cone opening C_{min} the broader the bandwidth, so the opening is usually made only a little larger than the coaxial transmission line. The gap between the cone and the disk is $0.3\,C_{min}$ and the diameter of the disk is $0.7\,C_{max}$. C_{max} depends upon the cone slant height L, which is a quarter-wavelength, and the angle ϕ. The optimum value for ϕ is 60 degrees. A discone antenna constructed with these dimensions has a standing-wave ratio (SWR) of less than 1.5 over at least a 7/1 frequency range, and an SWR of less than 2 over at least a 9/1 frequency range.

Helical Antennas

A helical antenna is a monopole antenna twisted into the shape of a corkscrew. Radiation from it will be normal (90°) as in an ordinary monopole as long as its diameter and electrical length are less than one wavelength. The main difference is that whereas the electromagnetic wave travels at the velocity of light in the vicinity of the ordinary straight antenna, the helix has a braking effect on it. This means that the axial length of the helix can be made less.

When the helix circumference is approximately a wavelength, an end-fire circularly polarized radiation pattern is produced. The more turns the helix has the more directional it will be. The pitch angle of the turns is 12.5 degrees, and the total axial length of the helix must be an integer multiple of a wavelength.

Parasytic Arrays

Since the basic antenna dimension is the half-wavelength, it is obvious that as wavelengths get smaller, antennas must do so also. But their reduced physical size can be turned to advantage, since multielement arrays which would be too massive to be practical at lower frequencies can, at higher frequencies, be mounted on a single mast or tower. For instance, the familiar television receiving antenna, consisting of a dipole and one other element, is about six feet wide. A half-wave dipole for reception of AM broadcasting would have to be from 300 to 1000 feet, according to the frequency of the station!

A multielement antenna array, whether used for transmission or reception, has higher gain and directivity, compared to the basic dipole. This is achieved by the provision of reflectors and directors, which are termed parasytic elements, behind and in front of a dipole, which is called the driven element. A variety of TV antennas have appeared in the past, including folded dipoles, conical dipoles, V-shaped dipoles, bow-tie dipoles (for UHF), all with various types of reflectors, but the present trend favors the multielement Yagi-Uda array, popularly called "Yagi," and the log-periodic dipole array, called "color antennas."

Yagi-Uda Array

Figure A-28 shows a commonly used three-element Yagi-Uda array. Using the dimensions in the figure, this antenna will have an impedance of 75 ohms. Additional directors may be added for extra gain, but no significant improvement is obtained by using more than one reflector.

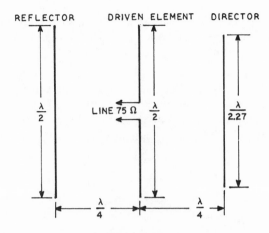

Figure A-28
3-Element Yagi-Uda Antenna

Log-Periodic Dipole Array

Figure A-29 shows a log-periodic dipole antenna, in which the lengths of the elements and the spacing between them increases logarithmically from the front to the back, while the ratio of element length to spacing remains constant. The angle α is the inverse tangent of this ratio. The log-periodic antenna is essentially independent of frequency over a broad band, and undirectional. The directivity may be further improved by using a V-pattern with the elements inclined forward at 60 degrees.

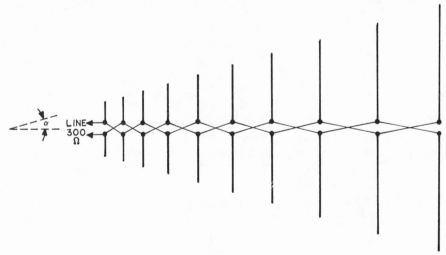

Figure A-29
Log Periodic Dipole Array

Electromagnetic Horns

At microwave frequencies the electromagnetic field is generated in the transmitter and conveyed to the outside through waveguides. By placing a suitable horn on the end of the waveguide the field is enabled to radiate into space. To design a horn as in Figure A-30, calculate L_1 from:

$$L_1 = L[1-(a/2A) - (b/2B)]$$

where a is the wide dimension of the waveguide in the H plane, and b is the narrow dimension in the E plane. Dimensions of L, A, and B in wavelengths for various gain values are obtained from the graph in Figure A-31.

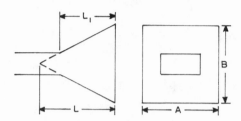

Figure A-30
Electromagnetic Horn Dimensions

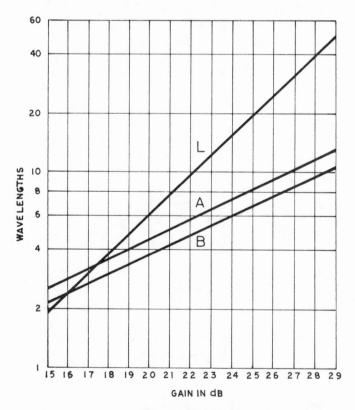

Figure A-31
Graph Used to Calculate Dimensions in Figure A-30

Parabaloid Reflectors

Horns are generally used with parabaloid reflectors, of which there are many kinds. Figure A-32 shows how to design the common dish type, which resembles, in appearance and function, the reflector in a flashlight. Other types are the horn, Cassegrain, Gregorian, cylindrical and cheese (or pillbox) reflectors, all analogous to similar optical reflectors.

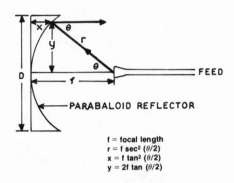

f = focal length
$r = f \sec^2 (\theta/2)$
$x = f \tan^2 (\theta/2)$
$y = 2f \tan (\theta/2)$

Figure A-32
Parabaloid Reflector

Receiving Antennas

All transmitting antennas make good receiving antennas. There are, however, a few receiving antennas which are not used for transmission. These include the ferrite or loopstick antennas used in most AM radio receivers, and other indoor or attic types. These are satisfactory for reception in strong signal areas because there is so much signal strength that even inefficient antennas gather more than enough for reception. In transmission, however, we cannot afford to waste power.

anticathode—Target of the x-ray tube, on which the stream of electrons from the cathode is focused and from which x-rays are emitted.

anticlutter gain control—Device which automatically and smoothly increases the gain of a radar receiver from a low level to the maximum, within a specified period after each transmitter pulse, so that the short-range echoes producing clutter are amplified less than long-range echoes. *Radar.*

anticollision radar—Radar system used in an aircraft or ship to warn of possible collision. *Radar.*

antihunt—Feedback signal or network in a servomechanism to prevent hunting or mechanical oscillation.

antilogarithm—The number corresponding to a logarithm (q.v.); e.g., if the logarithm of 4.5 is 0.6532125, then 4.5 is the antilogarithm of 0.6532125.

antinode—(See loop.)

antinoise carrier-operated device—(See squelch.)

antistatic agent—Means of minimizing static electricity. May be metallic grounding device, antistatic chemical spray, or chemical additive mixed with material prone to static charge. This last is part of manufacturing process.

aperiodic—Without periodic or repetitive characteristics.

aperiodic damping—Condition of a system when the amount of damping is so large that when the system is subjected to a single disturbance it comes to rest without overshoot (q.v.). Also called overdamping.

aperiodic function—Function with no repetitive characteristics.

aperture mask—(See shadow mask.)

APL—(See average picture level.)

apparent power—Power value obtained in an a.c. circuit by multiplying the effective values of voltage and current. Expressed in volt-amperes (VA).

Appleton layer—Region of highly ionized air in the ionosphere, capable of reflecting or refracting radio waves, under certain conditions, back to earth. *Electromagnetic-Wave Propagation.*

applicator electrodes—Electrodes used in dielectric heating (q.v.).

approach control radar—(See ground-controlled approach.)

aquadag—Graphite coating on inside of certain cathode-ray tubes for collecting secondary electrons emitted by the phosphor screen.

arc—Discharge of electricity through a gas, normally characterized by a voltage drop approximately equal to the ionization potential of the gas.

arc furnace—Furnace heated by arcs between two or more electrodes.

arcing—Production of an arc; e.g., at the brushes of a motor or contacts of a switch.

arc lamp—Use of an electric arc to produce a brilliant light.

arc through—In a gas tube, loss of control resulting in the flow of principal electrons in the normal direction during a scheduled nonconducting period.

A register—Accumulator for arithmetical operations in a digital computer. *Computer Hardware.*

argon—Inert gas, colorless and odorless, used in discharge tubes and some electric lamps. Symbol, Ar or A; atomic weight, 39.948; atomic number, 18.

argon glow lamp—Glow lamp containing argon gas which produces a pale blue-violet light.

arithmetic mean—Average obtained by dividing a sum by the number of its addends.

arithmetic unit—Section of a computer that performs adding, subtracting, multiplying, dividing, and comparing operations. *Computer Hardware.*

armature—Revolving part of a generator, motor, or other electromagnetic device which converts electrical energy to mechanical motion or vice versa; also the movable part of relay, bell, or buzzer.

armed sweep—(See single sweep.)

armor—Metal sheath enclosing a cable, primarily for mechanical protection.

Armstrong oscillator—Inductive feedback oscillator that consists of a tuned-grid circuit and an untuned tickler coil in the plate circuit. Control of feedback is by varying the coupling between the tickler and grid coils. (See Figure A-33.)

array—Combination of antennas with suitable spacing and with all elements excited to make the radiated fields from the individual elements add in the desired direction. *Antennas.*

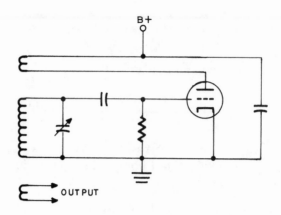

Figure A-33
Armstrong Oscillator

arrester—1. Protective device used to provide a bypass path directly to ground for lightning discharges that strike an antenna or other conductor. 2. Power-line device that is capable of reducing the voltage of a surge applied to its terminals, interrupting current if present, and restoring itself to original operating conditions.

ARRL—Abbreviation for American Radio Relay League (q.v.).

artificial antenna—(See dummy antenna.)

artwork—Pattern of microcircuit or passive-element configurations, formed of alternately clear and opaque regions in a dimensionally stable medium, made some multiple of the final mask size, usually on a mylar laminate known as Ruby Studnite. *Microelectronics.*

asbestos—Silicate of calcium and magnesium, used to insulate conductors where the temperature may be between 130 and 180 degrees Celsius.

A-scan—(See A-display.)

A-scope—(See A-display.)

aspect ratio—Television screen width-to-height ratio, which in the U.S. is 4/3.

assembler—Computer program that operates on symbolic input data to produce from such data machine instructions by carrying out functions of: translating symbolic operation codes into computer operating instructions; assigning locations in storage for successive instructions; computing absolute addresses from symbolic addresses. (See also compiler.) *Computer Software.*

assembly—Complete operating unit, such as a TV receiver, made up of subassemblies, such as the tuner, the audio section, etc.

assigned frequency band—Frequency band, the center frequency of which is the frequency assigned to the station. Band widths vary with the service. *Frequency Data.*

astable multivibrator—Free-running multivibrator, consisting of a pair of cross-connected transistors or electron tubes that continuously switch alternately from conducting to nonconducting, generating a pulse output. (See also bistable and monostable multivibrators.) *Electron Tube Circuits. Transistor Circuits.* (See Figure A-34.)

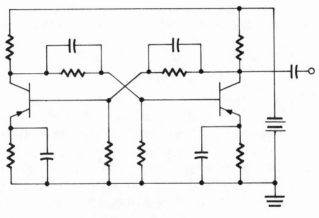

Figure A-34
Astable Multivibrator

astatic—Having no particular orientation, such as a vertical antenna.

A station—One of two transmitting stations in a Loran system. *Navigation Aids.*

astigmatism—In a cathode-ray tube, a focus defect in which electrons in different axial planes come to focus at different points.

astronautics—Science of space navigation.

asymmetrical distortion—When one half of a fluctuating wave is not a mirror image of the other half. *Distortion.*

asymptote—Straight line always approaching but never meeting a curve; tangent at infinity. (See Figure A-35.)

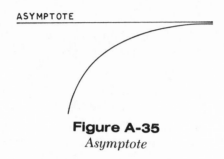

ASYMPTOTE

Figure A-35
Asymptote

asynchronous operation—Unclocked operation. *Computer Hardware.*

AT-cut crystal—Quartz-crystal slab cut at 35 degrees with respect to the Z axis. The temperature coefficient of this cut is virtually zero, and its frequency range is from 0.5 to 10 megahertz.

atmosphere—Air that envelopes the Earth. The atmosphere is broadly divided into lower and upper. The lower consists of the troposphere (0-11 miles), the stratosphere (11-30 miles), and the mesosphere (30-50 miles). The lower atmosphere is also called the homosphere, in which the proportions of molecular oxygen and nitrogen are maintained at 21 and 78 percent respectively. The upper atmosphere (heterosphere and exosphere) is characterized by the presence of atomic oxygen, helium, and hydrogen. *Electromagnetic-Wave Propagation.* (See Figure A-36.)

atmospheric duct—Stratum of the troposphere within which the variation of refractive index is such that propagation of an abnormally large proportion of any radiation of sufficiently high frequency is confined within the limits of the stratum. *Electromagnetic-Wave Propagation.*

atmospherics—(See static.)

atom—Smallest unit of a chemical element that retains the element's chemical identity. *Nuclear Physics.*

attack—In music, growth of a sound from its onset until it reaches steady-state intensity. Growth, duration, and decay comprise the envelope (q.v.) which is an important element of timbre, the distinctive quality of the sound. *Electroacoustics.*

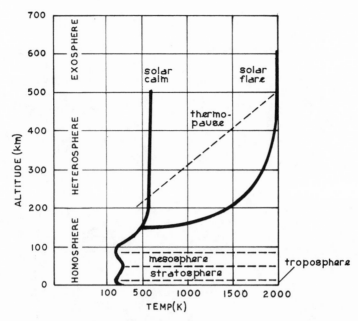

Figure A-36
Atmosphere

attenuation—Decrease in magnitude in transmission from one point to another. May be due to passage through equipment lines or space, or through an attenuator (q.v.).

attenuator—Resistive network or distributed network that reduces the amplitude of a signal without introducing appreciable distortion. *Attenuators*.

ATTENUATORS

An attenuator is a network designed to provide a known reduction in the amplitude of a signal without introducing appreciable phase or frequency distortion. To do this, it has to be constructed of resistors only, and its input and output impedances must match the source and load between which it is connected.

If an attenuator's input and output impedances are the same it is said to be symmetrical, and it can be connected either way. If they are not the same it is called a minimum-loss pad. Either type may also be balanced or unbalanced. A balanced attenuator or pad has neither side grounded; an unbalanced one has one side grounded or at zero potential.

Symmetrical attenuators, both balanced and unbalanced, are shown in Figure A-37. Underneath each is a formula for determining the resistance of each element. To perform the calculation, substitute for Z the required impedance, and for A, B, or C the value in the appropriate column in Table I corresponding to the attenuation desired.

UNBALANCED

BALANCED

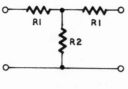

T Pad
$$R1 = ZA; \quad R2 = ZB$$

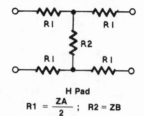

H Pad
$$R1 = \frac{ZA}{2}; \quad R2 = ZB$$

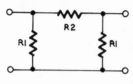

π Pad
$$R1 = \frac{Z}{A}; \quad R2 = \frac{Z}{B}$$

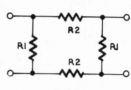

O Pad
$$R1 = \frac{Z}{A}; \quad R2 = \frac{Z}{2B}$$

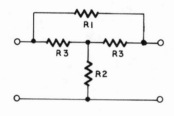

Bridged T Pad
$$R1 = \frac{Z}{C}; \quad R2 = ZC; \quad R3 = Z$$

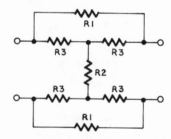

BRIDGED H PAD
$$R1 = \frac{Z}{2C}; \quad R2 = ZC; \quad R3 = \frac{Z}{2}$$

Figure A-37
Symmetrical Attenuators

UNBALANCED

BALANCED

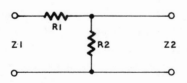

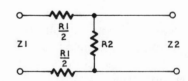

$$R1 = \sqrt{Z1\,(Z1 - Z2)}$$

$$R2 = \frac{Z1Z2}{R1}$$

Figure A-38
Minimum Loss Pads

TABLE I

ATTENUATION dB	A	B	C
−0.1	0.00576	86.9	86.4
−0.2	0.0115	43.4	42.9
−0.4	0.0230	21.7	21.2
−0.6	0.0345	14.4	14.0
−0.8	0.0460	10.8	10.4
−1.0	0.0575	8.67	8.20
−2.0	0.115	4.30	3.86
−3.0	0.171	2.84	2.42
−4.0	0.226	2.10	1.71
−5.0	0.280	1.64	1.28
−6.0	0.332	1.34	1.00
−7.0	0.382	1.12	0.807
−8.00	0.431	0.946	0.661
−9.00	0.476	0.812	0.550
−10.0	0.519	0.703	0.462
−12.0	0.598	0.536	0.335
−14.0	0.667	0.416	0.249
−16.0	0.726	0.325	0.188
−18.0	0.776	0.256	0.144
−20.0	0.818	0.202	0.111
−22.0	0.853	0.160	0.0863
−24.0	0.881	0.127	0.0673
−26.0	0.905	0.100	0.0528
−28.0	0.923	0.0797	0.0415
−30.0	0.939	0.0633	0.0327
−35.0	0.965	0.0356	0.0181
−40.0	0.980	0.0200	0.0101
−50.0	0.994	0.00632	0.00317
−60.0	0.998	0.00200	0.00100
−80.0	0.999	0.000200	0.000100
−100.0	0.999	0.0000200	0.0000100

Minimum-loss pads are shown in Figure A-38, together with formulas for calculating their resistor values. Z is always the larger impedance so this part will invariably be connected to the external unit with the higher impedance.

Ladder attenuators are cascaded sections, as shown in Figure A-39 (imagine that each shunt resister is really two resistors in parallel). All the sections attenuate equally, and the selector switch, which is marked in decibel steps, connects the input to P_o, P_1, P_2, or P_3 as required.

atto—SI prefix defined as one millionth millionth millionth of, or 10^{-18}. Symbol: a. *S.I. Units.*

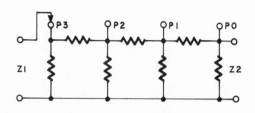

Figure A-39
Ladder Attenuator

audio—1. Frequencies corresponding to sound waves that can normally be heard by the human ear.   2. In combination with other words, relating to hearing or operation at audio frequency (q.v.). *Electroacoustics.*

audio amplifier—(See audio-frequency amplifier.)

audio frequency—Any frequency that can be detected as a sound by the human ear, or a corresponding electrical signal in the range of 15-20,000 hertz approximately. *Electroacoustics.*

audio-frequency amplifier—Amplifier for signals in the audio-frequency range (15-20,000 Hz). *Electron Tube Circuits, Transistor Circuits.*

audio-frequency choke—Choke used to impede the flow of audio-frequency currents; generally a coil wound on an iron core.

audio-frequency oscillator—Oscillator circuit for generating audio frequencies. *Electron Tube Circuits, Transistor Circuits.*

audio-frequency transformer—Transformer designed to transfer audio-frequency signals from one circuit to another. Generally an iron-core transformer. *Transformers.*

audio oscillator—(See audio-frequency oscillator.)

audio transformer—(See audio-frequency transformer.)

aural—Pertaining to the ear or to the sense of hearing. *Electroacoustics.*

aurora—Light emitted from the upper atmosphere of both the Arctic and Antarctic regions, caused by the glow of atoms in the upper atmosphere as they are hit by fast-moving streams of electrons and protons accelerated by the magnetosphere. Absorption of radio waves due to aurora activity is called auroral absorption. *Electromagnetic-Wave Propagation.*

autoalarm—Receiver tuned to either of the international distress and calling frequencies (500 and 2182 kilohertz) that automatically actuates an alarm if any signal is picked up.

autodyne reception—System of heterodyne reception using an autodyne circuit which acts simultaneously as an oscillator and heterodyne detector. (See also beat-frequency oscillator.)

automatic bass compensation—Circuit used in receivers to automatically reduce low frequencies less than medium frequencies as the volume is reduced. This compensates for the poor response of the ear to low frequencies at low volume settings.

automatic frequency control—Arrangement whereby the frequency of an oscillator is automatically maintained within specified limits.

automatic gain control—1. Type of circuit used to maintain the output of a receiver constant, regardless of variations of signal strength.   2. Radar circuit which prevents saturation of the receiver by long blocks of received signals, or by a carrier modulated at low frequency.   3. Self-acting compensating device which maintains the output of a transmission system constant within narrow limits despite wide variations in attenuation in the system.

automatic record changer—Part of a record player that drops a single record on to the turntable, sets the tone arm on the lead-in groove of the record, then at the end of the recording raises the tone arm, moves it back to and above its rest, drops another single record from the stack held on the spindle (or shelf), and continues so until the last record has been played, when it replaces the tone arm on its rest and shuts off the motor. *Recording.*

automatic scanning receiver—Receiver that automatically monitors preselected frequencies, locking on to each active channel, then resuming scan when activity ceases. Also called a scanner.

automatic shutoff—A switching arrangement that shuts off a record or tape player automatically after the last record or end of the tape (and also if the tape breaks). *Recording.*

automatic tuning—(See push-button tuning.)

automatic volume control—1. Self-acting device which maintains the output of a radio receiver substantially constant despite variations of signal strength.   2. Self-acting compensating device which maintains the output of a transmission system constant within narrow limits despite wide variations in the attenuation of the system.

automation—System or method whereby processes, controls, motions, etc., are self-operated.

automotive electronics—Electronics engineering concerned with automotive applications.

autoradio—Radio receiver installed in an automobile, using automobile power.

autotransformer—Transformer with single winding that acts as both primary and secondary. *Transformers.* (See Figure A-40.)

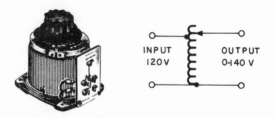

Figure A-40
Autotransformer

available power—1. Mean square of the open-circuit terminal voltage of a linear source, divided by four times the resistive component of the source impedance. 2. Maximum power that a network can deliver to a conjugately matched load.

avalanche diode—Silicon diode with a high ratio of reverse to forward resistance until avalanche (nondestructive) breakdown occurs. After breakdown, voltage drop across the diode is essentially constant and independent of current, hence the diode is used for voltage regulation. (See also zener diode.) *Semiconductors.*

a.v.c.—(See automatic volume control.)

average—(See arithmetic mean.)

average picture level—Average luminance level of the part of a television line between blanking pulses.

average voltage—Sum of the instantaneous voltages in a half-cycle of a wave, divided by the number of instantaneous voltages. In a sine wave the average voltage is equal to 0.637 times the peak voltage.

avionics—Aviation electronics.

AWG—(See American Wire Gage.)

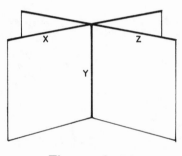

Figure A-41
Axes

axial leads—Leads that emerge from opposite ends of a tubular component and are aligned with its longitudinal axis.

axis—1. Imaginary straight line passing through a body and around which it revolves or is symmetrically arranged. 2. Any of three straight lines, termed x, y, and z, perpendicular to each other at the point of intersection, which define directions in space. (See Figure A-41.) 3. Any of three straight lines, termed pitch, roll and vertical, perpendicular to each other, which define the relations of different parts of an aircraft.

Ayrton-Perry winding—(See bifilar winding.)

Ayrton shunt—High-resistance parallel connection used to increase the range of a galvanometer without changing the damping. Also called a universal shunt.

azimuth—Angle between true north and a point on the Earth's surface (or in the horizontal plane), measured clockwise by an observer. (See Figure A-42.)

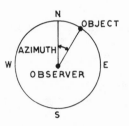

Figure A-42
Azimuth

B—1. Symbol for bel (q.v.).   2. Symbol for magnetic induction.   3. Symbol for boron.

b—1. Symbol for barn (q.v.).   2. Symbol for bit (q.v.).

B− (minus)—Negative terminal of a B battery or other source of anode voltage.

B+ (plus)—Positive terminal of a B battery or other source of anode voltage.

Ba—Symbol for barium (q.v.).

babble—Aggregate crosstalk or mutual interference in a multiple-channel system.

background noise—Total system noise independent of whether or not a signal is present. The signal is not included as part of the noise.

background radiation—Presence of any radiation, wave, or particle that gives spurious counts or signals in measuring devices.

backlash—Play between adjacent movable parts.

back porch—Portion of horizontal pedestal that follows horizontal synchronizing pulse. In a color transmission, the color burst is located on the back porch, but is absent in a monochrome transmission. *Broadcasting.*

backscatter—1. Reflected radiation from a target toward the radar transmitter. Also reflections from rain, etc., which may obscure the radar return from the desired target. 2. Propagation of extraneous signals by F- or E- region reflection in addition to desired ionospheric scatter mode. The undesired signal enters the antenna through the back lobes. *Radar, Electromagnetic-Wave Propagation, Antennas.*

back-to-front-ratio—Ratio used in connection with antennas, metal rectifiers, or any device in which signal strength or resistance in one direction is compared to that in the opposite direction.

backward diode—Type of tunnel diode which has approximately linear conduction in the reverse direction and some rectification in the forward direction. *Semiconductors.*

backward-wave oscillator—Traveling-wave tube of special design in which the traveling wave is reflected backward in phase to sustain oscillations. *Electron Tubes.*

baffle—Barrier or partition used to increase the effective length of the external transmission path between two points in an acoustic system, for example, the board on which a loudspeaker is mounted. *Electroacoustics.*

balance control—On a stereo amplifier a control that increases the volume of sound from one speaker while reducing it from the other, the total volume remaining the same.

balanced circuit—Circuit in which two sides are electrically alike and symmetrical to a common reference point, usually ground.

balanced transmission line—Transmission line having equal conductor resistances per unit length and equal impedances from each conductor to ground and to other electrical circuits. *Transmission Lines.*

balance-to-unbalance transformer—Transformer for matching a pair of lines, balanced with respect to ground, to a pair of lines not so balanced. Also called a balun or bazooka.

ballast resistor—Resistor in which resistance increases if current increases. Used to maintain constant current from powerline to receiver or amplifier despite fluctuations in powerline voltage.

ballast tube—Tube that provides low-cost automatic voltage and current regulation. As supply voltage increases, ballast tube resistance increases. Regulating action is based on high temperature coefficient of resistance element and heat-dissipation characteristics of tube.

ball bonding—Thermocompression bonding in which a small ball is formed by melting the tip of a gold wire lead, which is then bonded to a metal pad on the silicon chip by heating and compression. Also called nail-head bonding. (See Figure B-1.)

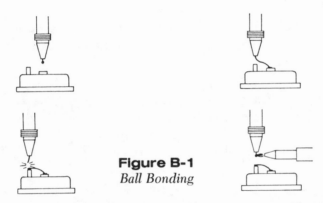

Figure B-1
Ball Bonding

ballistic galvanometer—Galvanometer with higher inertia and lower torque to make it capable of maximum transient deflection proportional of the total charge in a current pulse.

balun—(See balance-to-unbalance transformer.)

banana plug—Plug which may be fitted to the end of a test lead, or may be supplied as a molded fitting on a test lead, with an insulated plastic handle covering a brass body, and a tip consisting of four springs that are compressed when inserted in the jack so as to hold firmly. (See Figure B-2.)

Figure B-2
Banana Plug

band—Range of frequency spectrum between two limits. *Frequency Data.*

band-elimination filter—Filter that will not pass a band of frequencies. Also called band-stop, notch, or band-reject filter. *Active Filters, Passive Filters.*

bandpass—Number of hertz between the limiting frequencies at which the desired fraction, usually half-power (−3 dB) of the maximum output is obtained. *Active Filters, Passive Filters.*

bandpass amplifier—Amplifier designed to amplify uniformly a certain frequency band; e.g., chroma amplifier.

bandpass filter—Filter with a single transmission band. All other signals on either side of the band are attenuated. *Active Filters, Passive Filters.*

band-reject filter—(See band-elimination filter.)

band selector—Switch to select any one of the frequency bands in which the receiver, transmitter, or generator operates. Also called a band switch.

bandspread—1. Technique, using an auxiliary tuning control, to expand a crowded frequency band to facilitate tuning individual stations. Also called a slow-motion dial. 2. Method of double-sideband transmission in which the frequency band of the modulating wave is shifted upward in frequency so the sidebands are separated from the carrier by a greater frequency. *Modulation.*

bandstop filter—(See band-elimination filter.)

bandwidth—Range within the limits of a band. The limits are generally defined as those points where the attenuation is not more than −3 dB (0.707 times the maximum level).

bank winding—Winding in which single turns are wound successively in each of two or more layers, the entire winding proceeding from one end of the coil to the other without return. Used to reduce distributed capacitance. *Coil Data.*

bar—Unit of air pressure equal to one million dynes per square centimeter. The millibar (1/1000 bar) is used at the present time in aviation and meterology as a unit of barometric pressure (1013 millibars = 29.92 inches of mercury), but is being superseded by the pascal. 10 millibars = 1 kilopascal (kPa). *S.I. Units.*

bar generator—Generator that produces a display of stationary bars on a television screen. Used to verify vertical and horizontal linearity.

barium—Silvery-white, light-weight, alkaline-earth metal, used as a getter in electron tubes; the oxide is also used as a cathode coating because it readily emits electrons when heated. Barium titanate is used in the manufacture of capacitors (q.v.). Symbol, Ba; atomic weight, 137.34; atomic number, 56.

Barkhausen effect—Series of sudden changes in the size and orientation of ferromagnetic domains that occurs during magnetization or demagnetization.

Barkhausen oscillation—Undesired oscillation in the horizontal output tube of a television receiver, causing one or more ragged dark vertical lines to appear on the left side of the picture.

bar magnet—Permanent magnet in the form of a straight bar as opposed to a horse shoe.

barn—Unit of area used to measure the absorption cross section of atomic nuclei, equal to 10^{-24} square centimeter. *Nuclear Physics.*

barometer—Device to measure atmospheric pressure. The mercury type indicates the height in inches of a column of mercury balanced against the atmospheric pressure. The nonliquid (aneroid) type has a dial with a pointer actuated by a flexible-walled capsule, which expands or contracts with changes of pressure. It also reads in inches of mercury. Standard barometric pressure at sea level is 29.92 inches of mercury.

bar pattern—(See bar generator.)

barrel distortion—Non-linearity in cathode-ray tubes where vertical and horizontal lines near the borders of the screen bow outward.

barretter—Bolometer (q.v.) consisting of a fine wire or metal film having a positive temperature coefficient of resistivity, so that its resistance increases with temperature. Used for making power measurements in microwave devices, for which it is inserted in a barretter mount. (See Figure B-3.)

barrier layer—(See depletion layer.)

base—Region of a bipolar transistor between the emitter and the collector into which minority carriers are injected from the emitter, which controls the flow of current between emitter and collector. *Transistors.*

baseline—1. In graphical presentations, the horizontal axis (*x* axis), often representing time, zero potential, etc.   2. In radar displays, the visual line representing the track of the radar scanning beam.

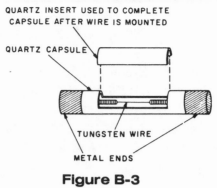

Figure B-3
Barretter

baseloaded antenna—Vertical antenna with a series impedance at the base which has the effect of increasing the electrical length of the antenna. *Antennas.*

base station—In a land mobile communications system, a fixed station that communicates with mobile stations.

basket winding—Coil winding in which adjacent turns are separated except at points of crossing.

bass—Lower audio frequencies, usually below the sound made by the middle C on the piano keyboard (261.63 Hz). *Electroacoustics.*

bass boost—Attenuation of higher audio frequencies to allow lower frequencies to predominate.

bass compensation—Emphasizing the low-frequency response of an audio amplifier at low-volume levels to compensate for reduced sensitivity of human ears to weak low-frequency sounds. *Electroacoustics.*

bass control—Tone control that attenuates the higher audio frequencies according to the setting of a potentiometer so as to give variable predominance to the lower frequencies. (See Figure B-4.)

bass-reflex enclosure—Speaker enclosure in which an opening of the proper size and position has been provided in the baffle board so that the rear wave can emerge and reinforce the bass tones from the front of the speaker.

bass response—1. Extent to which a loudspeaker or audio-frequency amplifier handles low audio frequencies.   2. Ability of any device to pick up or reproduce low audio frequencies. *Electroacoustics.*

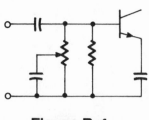

Figure B-4
Bass Tone Control

bat handle—Toggle-switch lever shaped like a miniature baseball bat.

bathtub capacitor—Capacitor having an outer metal case with rounded corners resembling those of a bathtub.

battery—Source of electrical energy produced by chemical reaction. Strictly speaking, a battery consists of two or more cells, but the term has long been applied equally to a single cell. Primary batteries are those that are discarded when discharged; secondary are rechargeable. Earlier batteries consisted of wet cells, with a liquid electrolyte; most batteries today have "dry" cells, with a jelly or paste, the principal exception being the lead-acid storage battery used in automobiles. (See Figure B-5.)

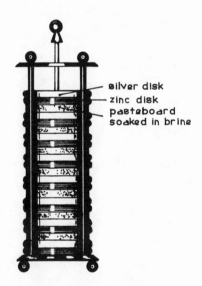

silver disk
zinc disk
pasteboard
soaked in brine

Figure B-5
The First Battery (Volta: c. 1800)

battery clip—Heavy-duty lead-plated spring clip, with screw terminal for lead, used for temporary connection to a battery or other terminal.

battery holder—Snap-in holder for battery, including contacts and terminals, to facilitate mounting one or more cells in battery-operated equipment.

battery plug—Snap-on plated brass eyelets, mounted on flat oval insulator with leads attached, for connecting nine-volt transistor battery to portable radio, etc.

bat wing—Antenna element, so called because of its shape. *Antennas.*

baud—Unit of speed of signal transmission, usually one bit per second. *Computer Hardware.*

bay—Enclosure standing about six feet high, consisting of several racks containing electronic equipment. A console may consist of one or more bays. Also called a relay rack. Standard width of panels on front of enclosure is 19 inches.

bayonet base—Base having two projecting pins on opposite sides of a smooth cylindrical base, to engage in corresponding slots in a bayonet socket (q.v.).

bayonet coupling—Non-constant impedance type connectors used for RG-55 and 58/U cables, usually called BNC connectors. Male connector has pins on opposite sides which engage in slots on the sides of the female connector, which may also be panel mounted.

bayonet socket—Socket for bayonet-base lamps or male connectors. It has slots on opposite sides and one or more contacts on the bottom, which may be spring loaded.

bazooka—(See balance-to-unbalance transformer.)

B battery—In a battery-operated vacuum-tube receiver, the battery furnishing plate and screen-grid voltages.

bcd—Abbreviation for binary-coded decimal (q.v.).

B-display—Type of radar display in which the target appears as a bright spot. Its bearing is indicated by its horizontal coordinate, its distance by its vertical coordinate. *Radar.*

beacon—Equipment emitting radio or radar beams to orient aircraft or radar.

beam—Flow of particles or waves in one direction, as in an electron beam, a radio beam.

beam lead—Silicon chip connection lead formed in place chemically and suspended over a void or space. The cantilevered end can be formed on the silicon chip and then attached to the interconnecting pattern, or it can be cantilevered from the interconnecting pattern and bonded to the chip. *Microelectronics*.

beam-power tube—Electron tube which uses directed electron beams to enhance its power-handling capability.

bearing—Horizontal angle at a given point made by lines joining the point to the object whose bearing is being measured and the reference datum. The latter may be true North, the heading of the aircraft or ship, or any other chosen reference. Bearings with respect to North are measured clockwise in degrees; those with respect to a vessel's heading are measured to left or right (port or starboard) in degrees.

beat—Superimposed waves of different frequencies "beat" together to produce other frequencies that are the sum and difference of the superimposed frequencies.

beat-frequency oscillator—Used to generate a signal that beats with that of an interrupted-carrier (Morse code) signal to produce a difference signal that can be heard by the receiving operator. Abbreviated b.f.o.

becquerel—S.I. unit of radioactive activity equal to one disintegration per second. Symbol, Bq. *Nuclear Physics*.

bel—Unit for expressing ratio of two amounts of loudness or power. (See decibel.)

bell—Bell operated by an interrupter (q.v.)

bell transformer—Small iron-core transformer mainly used to reduce powerline voltage to 10 or 20 volts for operating doorbells, thermostats, etc.

bell wire—Cotton-covered AWG 18 copper wire, formerly used for doorbell connections. Almost entirely superseded by wire with plastic insulation. *Coil Data*.

beta—Current gain of a transistor in the common-emitter configuration. Symbol β; also used for angles, coefficients, phase constant. *Transistors*.

beta cutoff frequency—Frequency at the high end of a transistor's range at which current gain of a common-emitter amplifier drops to three decibels below its low-frequency value.

beta particle—Electron or positron emitted by certain unstable atomic nuclei in the process of beta decay. *Nuclear Physics.*

betatron—Circular accelerator producing high-energy electrons or highly penetrating x-rays for use in research, industry, and medicine. *Nuclear Physics.*

bev—Abbreviation for billion electron volts; now termed gigaelectronvolt (Gev). *Nuclear Physics.*

bevatron—High-energy proton accelerator. *Nuclear Physics.*

bezel—Circular or rectangular frame for holding dial cover glass, oscilloscope graticule, etc.

b.f.o.—(See beat-frequency oscillator.)

B-H curve—Curve to show characteristics of a magnetic material, in terms of magnetizing force (H) and resulting flux density (B). *Magnetic Amplifiers.* (See Figure B-6.)

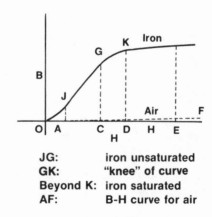

JG: iron unsaturated
GK: "knee" of curve
Beyond K: iron saturated
AF: B-H curve for air

Figure B-6
B-H Curve for Iron

bias—1. Any force (electrical, magnetic, or mechanical) applied to a device to establish a reference level.   2. D.C. potential applied to control grid of electron tube or base of transistor to establishing operating point.

bias oscillator—Oscillator used in tape recording to bias the recording head with a signal between 60 and 100 kilohertz to improve frequency response, distortion, and signal-to-noise ratio. *Recording.*

bias winding—An additional winding similar to a control winding, to apply magnetic bias to a magnetic amplifier. *Magnetic Amplifiers.*

biconical antenna—Antenna consisting of two metal cones on a vertical axis, the upper cone inverted, and the vertices meeting. Radio-frequency is fed to the point where the vertices meet. *Antennas.*

bidirectional—Operates in either direction; e.g., a field-effect transistor.

bifilar winding—Method of winding resistors, etc., with the turns running in opposite directions to make them noninductive.

bimetallic strip—Strip formed of two dissimilar metals welded or riveted together. Because of their different temperature response they expand to different lengths as temperature changes, resulting in a bending or curling proportional to the temperature. This action is used to move a pointer in a dial thermometer or actuate a switch in a thermostat.

binary-coded decimal—Decimal numbers (0 through 9) expressed in binary numbers (0 and 1). (See binary number system.)

binary number system—Positional numeral system employing only two symbols, 0 and 1, called binary digits. *Computer Hardware.*

binaural effect—An effect upon both ears.

binding energy—Energy required to separate a particle from a system of particles, especially with reference to atomic particles. *Nuclear Physics.*

binding post—Panel-mounted terminal that may accept phone tip plug, spade lug, banana plug, alligator clip, or wire strand (head screws down to clamp wire in place). (See Figure B-7.)

binomial array—Directional antenna array with maximum response in two directions. *Antennas.*

bionics—Construction of artificial systems resembling living systems.

Figure B-7
Binding Post

bipolar—Having two poles or polarities, as a transistor that uses both negative and positive charge carriers.

bistable multivibrator—Circuit having two stable states, consisting of a pair of cross-connected transistors or electron tubes that switch alternately from conducting to nonconducting when triggered, generating a high or low output. Also called a flip-flop. (See also monostable and astable multivibrators.) *Electron Tube Circuits, Transistor Circuits, Computer Hardware.* (See Figure B-8.)

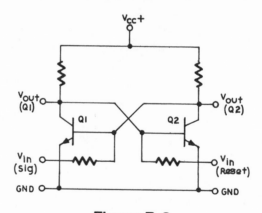

Figure B-8
Bistable Multivibrator

bit—Abbreviation for binary digit (0 or 1) in the binary number system (q.v.).

bit rate—(See baud.)

black body—Surface that absorbs all radiation falling on it. Radiation emitted from such a body when hot is black-body radiation and is an accurately known function of the temperature. *Optoelectronics.*

black box—Unit of electronic equipment where the user is only concerned with the input and output requirements, and not with the internal circuitry, which he does not need to know.

blacker-than-black level—In television, a level of greater instantaneous amplitude than the black level. *Broadcasting*.

black level—Reference black level is the amplitude level of the television signal that gives maximum black and is slightly lower than the blanking level (q.v.). *Broadcasting*.

blackout—Interruption of radio communication due to solar flares. *Electromagnetic-Wave Propagation*.

blanking—Making a channel or device noneffective for a desired interval. In television, blanking pulses cut off the electron beam between lines and frames to prevent the display of retrace lines.

blanking level—In television signals, the blanking level is at 75 percent of the peak carrier level. Also called the pedestal level. *Broadcasting*.

bleeder resistor—Power supply resistor that serves both as voltage divider stabilizer and discharge path for filter capacitors when circuit is de-energized.

blind approach—General term covering various guidance systems used by aircraft in poor visibility. *Navigation Aids*.

blip—On a cathode-ray tube screen, a spot of light or other indication of a radar target. Also called a pip. *Radar*.

block diagram—Diagram in which the units of an electronic system or of a computer program are drawn in the form of blocks or other geometric figures with connecting lines, and arrows if necessary, to show their relationship.

blocking capacitor—Capacitor used to block direct current while passing alternating current.

blocking oscillator—Relaxation oscillator consisting of an amplifier with its output fed back to its input, generally via a transformer, so that it builds up a charge on a capacitor while at the same time causing the tube or transistor to cease conduction. The charge then dissipates slowly through a resistance until the bias on the tube or transistor permits conduction, when the cycle is repeated. *Electron Tube Circuits, Transistor Circuits*.

blooming—Expansion of display on cathode-ray or picture tube, when the brightness control is advanced, is caused by low anode voltage. The potential difference between cathode and anode is too low to make the electron beam "stiff," and electrons are consequently deflected too much, both vertically and horizontally.

blue gun—In a three-gun color picture tube, the electron gun responsible for exciting the phosphor elements producing the blue primary color. (See color picture tube.)

BNC—(See bayonet coupling.)

bolometer—Device for measuring radiation intensity. It consists of a resistance bridge in which one of the arms absorbs the radiation, with a consequent rise in temperature, resulting in a change of its resistance. The adjustment required to rebalance the bridge gives the radiation intensity. (See also barretter.)

bonding—1. Attaching connections. (See ball bonding, thermal compression bonding, wedge bonding, nailhead bonding, etc.) 2. Connecting together with bonding conductors all metal parts of an airframe, automobile chassis and the like, to keep them at the same potential to prevent arcing between them from static buildup. *Semiconductors.*

Boolean algebra—Symbolic logic used by digital computers. *Computer Hardware.*

boosted B voltage—Increased direct voltage obtained by adding the rectified output from across part of the horizontal output transformer to the B power supply, to provide a total boosted B voltage of 400 or 500 volts for circuits and sections in a TV receiver that require higher voltage. The rectification is performed by the damper tube, and filtering by combined action of capacitors and resistance, as in a low-voltage power supply.

booster—When cathode emission has become low due to aging of a picture tube, emission may usually be increased to obtain some further useful life by applying greater than normal voltage to the heater. A booster for the filament voltage is a small transformer in a housing on which is a picture-tube base with pins. This is plugged into the regular socket, and another socket, connected to the booster with leads, is attached to the picture-tube base. Also called a brightener.

boot—Protective cover of rubber or similar material put on a terminal or connector to keep out moisture or other contamination.

bootstrap—Any technique or device that brings itself into a desired state by its own action ("pulls itself up by its own bootstraps"). A bootstrap circuit is a single-stage amplifier in which the output load is connected between the negative end of the plate supply and the cathode, and the input signal between the grid and the cathode. This adds the output signal potential to the input signal potential with respect to ground.

boresighting—Initial alignment of radar or other directional antenna system, using an optical procedure.

boundary—In a semiconductor, the interface between p and n type material. Also called a pn junction. *Semiconductors.*

bow-tie antenna—Dipole antenna in which the two elements are metal triangles placed point-to-point, usually with a reflector behind them. Used for u.h.f. reception. *Antennas.*

B power supply—Low-voltage power supply for electron-tube plates and screen grids. Also called B supply.

braiding—Method of interlacing three or more strands of wire or fiber, forming a narrow strip of flat or tubular flexible material. Braided wire in tubular form is used for shielding or protecting cables. In ribbon form it is used for grounding or bonding straps. Fiber braid is used as a cable covering.

brain wave—Fluctuation in the electrical activity of brain cells as evidenced on a chart called an electroencephalogram (EEG).

breadboard—Arrangement in which components are fastened temporarily to a board or chassis for experimental work.

break-before-make contact—In a switch, contacts which interrupt one circuit before establishing another.

breakdown—1. Discharge through any insulating medium. 2. Avalanche action. (See avalanche diode.)

breaker points—Contacts that make or break current flow in a distributor, as in an automobile.

break-in operation—Ability for transmitting operator to hear signals from receiving operator during the short "key-up" intervals in code transmission.

break-make contact—(See transfer contact.)

breakout—Point in a main cable where a subsidiary conductor branches off.

breezeway—In the composite television video signal, the gap on the back porch between the trailing edge of the horizontal synchronizing pulse and the start of the color burst. *Broadcasting.*

bremsstrahlung—Electromagnetic radiation produced by a sudden slowing down or deflection of charged particles (especially electrons) passing through matter in the vicinity of the strong electric fields of atomic nuclei. *Nuclear Physics.*

Brewster angle—Angle of incidence at which the reflection of parallel-polarized electromagnetic radiation at the interface between two dielectric media is zero.

bridge—Instrument for measuring electrical quantities by comparing the unknown quantity with a known quantity. When the two are equal the bridge is balanced ("nulled"). See specific type of bridge; e.g., Wheatstone bridge.

bridged-T network—Network with two series impedances of the T bridged by a fourth. *Passive Filters.*

bridge rectifier—Full-wave rectifier with four elements connected in a bridge circuit, so that direct voltage is obtained from one opposite pair of "corner" junctions when alternating voltage is applied to the other pair. (See Figure B-9.)

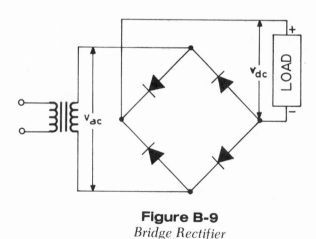

Figure B-9
Bridge Rectifier

brightness—Subjective, visual sensation related to luminance, which is the intensity of light per unit of area. In television, brightness is the Y signal. It may be varied in amplitude by using the brightness control. *Broadcasting.*

brilliance—In sound reproduction, the degree to which higher audio frequencies are reproduced. In some three-way speakers there is a brilliance control to vary the tweeter output level to achieve a proper relative volume with respect to the lower frequencies. *Electroacoustics.*

broad band—Term used to denote equipment with an essentially flat response over a wide frequency range.

broadcasting—Transmission of radio and television programs intended for the general public and distinguished from private signals directed to specific receivers.

BROADCASTING

The first broadcast program in the U.S. was on Christmas Eve, 1906, originating from Reginald Aubrey Fessenden's experimental station at Brant Rock, Massachusetts. The only people who heard it were a few ship's radio operators, they being the only listeners with the equipment to receive it. Military restrictions during World War I delayed further progress until 1920, when the first commercial station, KDKA in Pittsburgh, went on the air November 2 with the returns of the Harding-Cox presidential election. By the early 1970's there were in this country 4911 AM radio stations, 2554 FM stations, and 3354 TV stations operating a staggering total of 10,819 broadcasting stations, not counting cable TV. No reliable figures are available for the number of radio receivers, but it is estimated there were 85,000,000 television receivers in the country. No other nation came anywhere near the U.S. in the number of its broadcasting stations, but in receivers, Russia with 40,000,000 TV sets and Japan with 23,000,000 were second and third.

AM Radio Broadcasting

Standard broadcast stations in the U.S. operate in channels spaced by 10 kilohertz in the frequency band from 535 to 1605 kilohertz, and are classified as clear channel, regional channel, and local channel. Clear channel stations (Classes I and II) broadcast with powers up to 50 kilowatts over wide areas and are protected against objectionable interference. Regional stations (Class III), with powers up to 5 kilowatts, give coverage to larger cities and surrounding rural areas. Local stations (Class IV), with powers not exceeding 1 kilowatt, serve even smaller areas. Class III and IV stations are not protected from interference, which therefore may further restrict their service areas. Most stations operate on reduced power at night.

Other requirements for AM stations include:

Modulation	85-95 percent
Carrier stability	Within 20 hertz of assigned frequency
Audio response	Flat within 2 decibels from 100 to 5000 kilohertz
Audio distortion	Less than 7.5 percent for 85-95 percent modulation

FM Radio Broadcasting

FM broadcasting stations are allocated 100 channels, each 200 kilohertz wide, extending from channel 201 (88.1 megahertz) to channel 300 (107.9 megahertz). These stations are designated Class A, B, C, or D, as follows:

Class A stations have powers up to 3 kilowatts, with antennas restricted to a maximum height of 300 feet above the average terrain in their area. They use channels 221, 224, 228, 232, 237, 240, 244, 249, 252, 257, 261, 265, 269, 272, 276, 280, 285, 288, 292, and 296 only.

Class B stations operate in Zone I (northeastern U.S.) or Zone IA (Puerto Rico, Virgin Islands, and all of California south of 40°N), and are allowed up to 50 kilowatts power with an antenna height not exceeding 500 feet above the average terrain in their area.

Class C stations operate in Zone II (all of the U.S. not in Zones I or IA), and are allowed up to 100 kilowatts power at an antenna height not exceeding 2000 feet above the average terrain in their area.

Class D stations are noncommercial educational stations operating exclusively in channels 201 through 220 with a maximum power of 10 watts.

Other requirements for FM stations are:

Modulation	100 percent (frequency swing 75 kilohertz)
Center frequency stability	Within 2000 hertz of assigned frequency
Audio response	50-15000 hertz with preemphasis
Audio distortion	Less than 2.5 percent from 100-7500 hertz
	Less than 3.0 percent from 7500-15000 hertz
	Less than 3.5 percent from 50-100 hertz

Stereophonic transmission must be such that it can be received by monophonic as well as stereophonic receivers, so the left and right audio channels at the station are added to give an L + R signal, and subtracted to give an L-R, or difference signal. The L + R signal is used to modulate the main carrier so it can be used by monophonic receivers. The L − R signal is used to amplitude-modulate a 38-kilohertz carrier, which is then suppressed, so that only the sidebands are used to modulate the main carrier, as shown in Figure B-10. A pilot subcarrier of 19 kilohertz is also transmitted, which is doubled in the receiver to restore the suppressed 38-kilohertz subcarrier. The latter is recombined with its sidebands to permit recovery of the L-R signal. The L + R and L-R signals are then combined algebraically to reproduce the original L and R audio signals, as shown in Figure B-11.

As mentioned above, the total frequency swing allowed is 75 kilohertz, but only 53 kilohertz are required for stereophonic modulation. The other 22 kilohertz are used by some stations for subsidiary communications. SCA signals, as they are called (from Subsidiary Communications Authorization), are two types: (a) background music and the like for subscribing stores and offices (they need a special receiver); and (b) communications with other FM stations for relaying program material and so on.

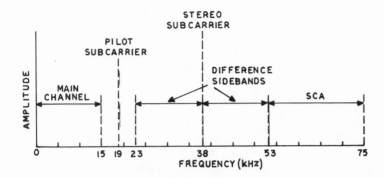

Figure B-10

FM Stereo Spectrum

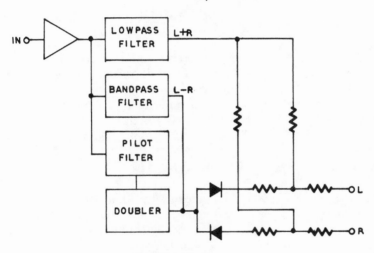

Figure B-11

Stereo Circuits in Receiver

TV Broadcasting

Television broadcasting station channel numbers and frequencies are listed in Table I. Authorized power and antenna heights are given in Table II.

Other requirements for TV stations are:

Channel width	6 megahertz
Picture carrier location	1.25 megahertz ±1000 hertz above lower boundary of channel
Aural center frequency	4.5 megahertz ±1000 hertz above visual carrier
Aural transmitter power	20 percent of peak visual power
Modulation	AM composite picture and sync signal on visual carrier and FM audio signal on aural carrier (see Figure B-12)

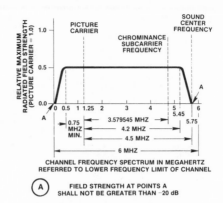

Figure B-12

Frequency Spectrum of TV Signal

Scanning lines	525 lines per frame, interlaced two to one
Scanning sequence	Horizontally from left to right, vertically from top to bottom
Horizontal scanning frequency	2/455 times chrominance subcarrier frequency (15734.264 ±0.044 hertz)
Vertical scanning frequency	2/525 times the horizontal scanning frequency (59.94)
Chrominance subcarrier frequency	3.579545 megahertz ±10 hertz
Blanking level*	75.0 ± 2.5 percent of peak carrier level
Reference black level*	92.5 ± 2.5 percent of amplitude of blanking level above reference white level
Reference white level*	Luminance signal of reference white is 12.5 ± 2.5 percent of peak carrier level
Horizontal pulse-timing tolerance	±0.5 percent of average interval
Horizontal pulse-repetition stability	±0.15 percent per second
Audio modulation (FM)	Same as FM radio
Audio distortion	Same as FM radio
Color signal	Luminance component transmitted as AM of the picture carrier, and chrominance components as AM sidebands of two suppressed subcarriers in quadrature (see Figure B-14)
I-channel bandwidth	Flat within 2 decibels to 1.3 megahertz
Q-channel bandwidth	Flat within 2 decibels to 400 kilohertz

---

*See Figure B-13.

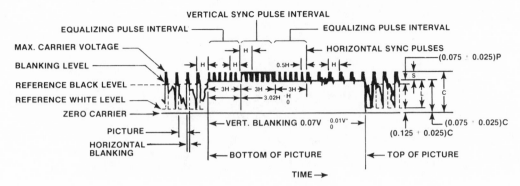

VERTICAL SYNC PULSE INTERVAL

EQUALIZING PULSE INTERVAL ——————— EQUALIZING PULSE INTERVAL

MAX. CARRIER VOLTAGE —————— HORIZONTAL SYNC PULSES

BLANKING LEVEL —————— (0.075 ± 0.025)P

REFERENCE BLACK LEVEL - - - - -

REFERENCE WHITE LEVEL ——————

ZERO CARRIER ———→

PICTURE ———→

HORIZONTAL BLANKING

3H — 3H — 3H

3.02H   H 0

VERT. BLANKING 0.07V   0.01V*

(0.125 ± 0.025)C
(0.075 ± 0.025)C

BOTTOM OF PICTURE — TOP OF PICTURE

TIME →

A FIELD 1

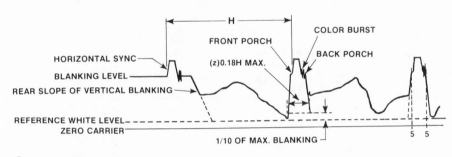

SYNC

0.5H

B FIELD 2

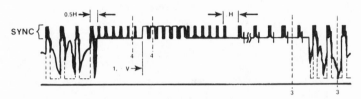

H

FRONT PORCH   COLOR BURST

BACK PORCH

HORIZONTAL SYNC ———→

BLANKING LEVEL ——————

(z)0.18H MAX.

REAR SLOPE OF VERTICAL BLANKING ——→

REFERENCE WHITE LEVEL - - - - - -
ZERO CARRIER ——————

1/10 OF MAX. BLANKING

C — DETAIL BETWEEN 3-3 IN B

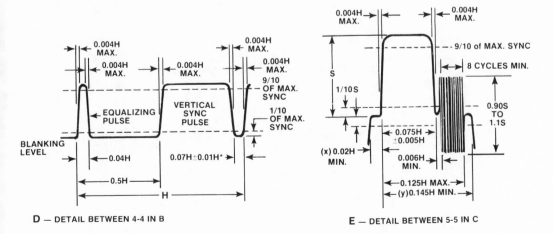

0.004H MAX.

0.004H MAX.

0.004H MAX.

0.004H MAX.

0.004H MAX.

9/10 OF MAX. SYNC

1/10 OF MAX. SYNC

EQUALIZING PULSE

VERTICAL SYNC PULSE

BLANKING LEVEL

0.04H

0.07H±0.01H*

0.5H

H

D — DETAIL BETWEEN 4-4 IN B

0.004H MAX.

0.004H MAX.

9/10 of MAX. SYNC

8 CYCLES MIN.

S

1/10 S

0.90S TO 1.1S

(x) 0.02H MIN.

0.075H ±0.005H

0.006H MIN.

0.125H MAX.
(y)0.145H MIN.

E — DETAIL BETWEEN 5-5 IN C

Figure B-13
Television Composite Signal

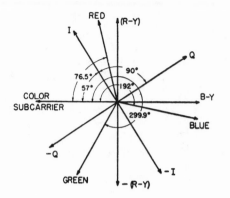

Figure B-14
Phases of Color Signal

Subsidiary signals (test, cue, and control signals) are transmitted in lines 17 through 20 of the vertical blanking interval. Coded patterns for identification of the program may be transmitted in the intervals within the first and last 10 microseconds of lines 22 through 24 and 260 through 262. The aural FM signal may also be multiplexed for communication between station control and transmitter sites as in FM radio.

TABLE I
Television Channels and Bands

Channel Number	Band (megahertz)	Channel Number	Band (megahertz)
2	54-60	36	602-608
3	60-66	37	608-614
4	66-72	38	614-620
5	76-82	39	620-626
6	82-88	40	626-632
7	174-180	41	632-638
8	180-186	42	638-644
9	186-192	43	644-650
10	182-198	44	650-656
11	198-204	45	656-662
12	204-210	46	662-668
13	210-216	47	668-674
14	470-476	48	674-680
15	476-482	49	680-686
16	482-488	50	686-692
17	488-494	51	692-698
18	494-500	52	698-704
19	500-506	53	704-710
20	506-512	54	710-716
21	512-518	55	716-722
22	518-524	56	722-728
23	524-530	57	728-734
24	530-536	58	734-740

TABLE I (continued)

Channel Number	Band (megahertz)	Channel Number	Band (megahertz)
25	536-542	59	740-746
26	542-548	60	746-752
27	548-554	61	752-758
28	554-560	62	758-764
29	560-566	63	764-770
30	566-572	64	770-776
31	572-578	65	776-782
32	578-584	66	782-788
33	584-590	67	788-794
34	590-596	68	794-800
35	596-602	69	800-806

TABLE II

Authorized Power and Antenna Height for TV Broadcasting

Channel	Maximum Power [1] (kilowatts)	Maximum Antenna Height [2] (feet)
2-6	100	2000[3]
7-13	316	2000[3]
14-69	5000[4]	2000

[1] Minimum power is 100 watts in all cases.

[2] Increased height may be authorized, but output power must then be reduced accordingly.

[3] In Zone I (same as for FM) maximum height is 1000 feet.

[4] Limited to 1000 kilowatts within 250 miles of Canadian border.

Cable Television (CATV)

Cable TV originally was used to provide television to communities having poor service or none at all, by having a receiving antenna array in a good location and amplifying and distributing the signal by coaxial cables to the subscribers' homes. Today there are CATV systems in areas of good reception that enable subscribers to dispense with antennas and also to receive a greater choice of programs. The regulations governing transmission frequencies are similar to those for TV stations, and the signal level at the subscriber's input must be not less than 2 millivolts across 300 ohms or 1 millivolt across 75 ohms.

Instructional Television

Instructional television is similar to commercial television, except that the assigned transmission frequencies are in the band from 2.5 to 2.68 gigahertz, and the power is limited to a maximum of 10 watts. Directive transmitting and receiving antenna arrays are preferred, to minimize interference.

International Broadcasting Service

U.S. international broadcasting stations serve the general public in foreign countries and are beamed to target areas. Transmitter power is at least 50 kilowatts, and several frequencies are assigned from Table III to each station to permit the best possible reception at different times of the day.

TABLE III

Band	Frequency (kHz)
A	5950-6200
B	9500-9775
C	11700-11975
D	15100-15450
E	17700-17900
F	21450-21750
G	25600-26100

broadside array—Antenna array whose direction of maximum radiation is perpendicular to the line or plane of the array.

brush—Conductor, such as metal or carbon block, which makes electrical contact with the moving part of a generator or motor.

brush discharge—Luminous electrical discharge of visible streams of charged particles through the air between the terminal of a static machine or other high-frequency high-voltage source.

brute-force filter—Untuned filter that uses large values of capacitance and inductance to smooth out pulsations in a power supply.

B & S—Abbreviation for Brown and Sharpe, alternative name for American Wire Gage (q.v.).

B-scope—Radarscope employing B display (q.v.).

B supply—(See B power supply.)

bubble memory—Memory in which information is stored in the form of magnetic "bubbles" in a substrate of garnet on which a pattern of permalloy tracks has been deposited. *Computer Hardware, Microelectronics.*

bucking voltage—Opposing voltage.

buffer—A circuit or component placed between others to isolate them from each other; e.g., a buffer stage is usually put between the master oscillator and the power amplifier in a transmitter.

bug—1. Semiautomatic telegraph sending key. 2. Circuit malfunction in development stage. 3. Surveillance device used in espionage.

bulk eraser—A device for erasing an entire reel of magnetic tape, using a 60-hertz erasing field which is slowly reduced as the reel rotates.

bump contacts—Contacts on a chip that protrude sufficiently to register with terminal pads on a board when the chip is inverted and bonded to them in flip-chip bonding (q.v.).

buncher—Input resonant cavity in a klystron or velocity modulated tube. Sometimes called a buncher resonator. *Electron Tubes.*

buried layer—Highly conductive diffused layer under the collector area of an integrated-circuit transistor, used to lower collector resistance. *Microelectronics.*

burn-in—Operation of newly fabricated device to stabilize components and expose defects, if any.

burst—Sudden increase in the strength of a signal. (See also color burst.) *Electromagnetic-Wave Propagation.*

burst transmission—Method of radio transmission in which signals are stored for a given time, then transmitted at from ten to one hundred times the normal speed. The received signals are recorded, then slowed down to their normal rate for the user.

bus—1. Solid metal or wire conductor, uninsulated, sometimes called a busbar. 2. In a computer, principal channels connecting major elements; e.g., address bus, control bus, data bus. *Computer Hardware.*

Butterworth filter—Filter designed for flattest possible response in the passband. *Passive Filters.*

button—In a carbon microphone, the metal capsule containing the carbon granules. *Telephone.*

button silver-mica capacitor—Capacitor consisting of a stack of silvered mica discs enclosed in button-shaped metal capsule. *Capacitors.*

buzzer—Signaling device consisting of an interrupter (q.v.).

BX cable—Flexible metal-sheathed cable with two inner insulated conductors, mainly used for electrical wiring.

bypass—Shunt path around some other circuit element; e.g., a bypass capacitor shunting a resistor effectively shorts out the resistor for a.c., but not for d.c.

B-Y signal—In television, the color difference signal which, when combined with the Y signal, excites the blue gun to display the color blue on the picture-tube screen. *Broadcasting.*

byte—A group of eight bits. *Computer Hardware.*

C—1. Symbol for coulomb (q.v.). 2. Symbol for capacitor or capacitance (q.v.). 3. Symbol for Celsius (q.v.), written °C.

c—Velocity of light (2.9979250×10^8 m/s). *Constants.*

C– (minus)—Negative terminal of a C battery or other source of grid-bias voltage.

C+ (plus)—Positive terminal of a C battery or other source of grid-bias voltage.

CAB—Abbreviation for Civil Aeronautics Board.

cabinet—Enclosure for electronic equipment.

cable—Bundle of insulated wires.

cable clamp—Supporting device for a cable.

CAD—Abbreviation for computer-aided design.

cadmium—Silver-white soft metal, used to electroplate other metals to protect them from corrosion, and also in the nickel-cadmium rechargeable cell (q.v.) and standard cell (q.v.). Symbol, Cd; atomic weight, 112.40; atomic number, 48.

cadmium sulfide (CdS)—Substance whose resistance decreases in proportion to the intensity of light falling on it. Used in photoconductive cells (q.v.).

calculator—Machine for performing arithmetical operations. Formerly mechanical, now generally electronic. (See Figure C-1.)

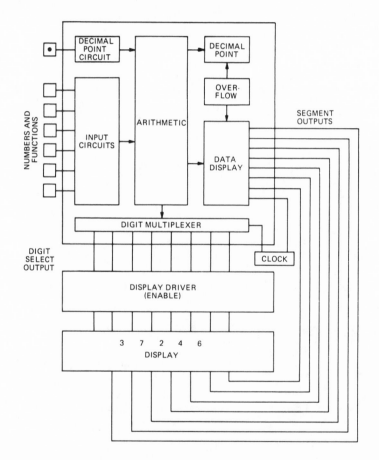

Figure C-1
Calculator

calibration—1. Comparing a measurement instrument with a standard instrument of greater precision to determine its percentage of error, usually termed its accuracy. 2. Adjusting an instrument to the maximum accuracy that it is capable of.

call—Transmission to establish communication with another station.

call sign—The identifying letters (and figures, if any) of a transmitter. International call-sign prefixes have been established for each country. The U.S. uses AAA through ALZ, KAA through KZZ, NAA through NZZ and WAA through WZZ.

canal ray—Positive ions emanating from the anode in a vacuum tube and flowing to the cathode.

candela—S.I. base unit of luminous intensity. The candela is the luminous intensity, in a perpendicular direction, of a surface of 1/600 000 of a square meter of a blackbody at the temperature of freezing platinum (2045 K, or 3223° F) under a pressure of 101 325 pascals (29.92 inches of mercury). Symbol, cd. *S.I. Units.*

capacitance—That property of a capacitor which determines how much charge can be stored in it for a given potential difference across its terminals. *Capacitors.*

capacitive coupling—(See resistance-capacitive coupling.)

capacitive reactance—Impedance offered by a capacitor to alternating current. Symbol, X_c. $X_c = \dfrac{1}{2\pi fC}$ ohms.

capacitor—Device that stores electrical energy when a voltage is applied. It consists of two conductors separated by an insulating material called a dielectric. *Capacitors.*

CAPACITORS

Operation of a Capacitor

Capacitor Behavior in a D.C. Circuit—A capacitor is a component used in electronic circuits to store electrical energy. It consists essentially of two plates of conducting material separated by an insulating material called a dielectric. When the plates of a capacitor are connected to a battery, electrons flow from the negative terminal of the battery on to the plate connected to it. The surplus electrons on this plate repel electrons from the other plate back to the positive terminal of the battery, and at the same time the positive terminal of the battery attracts electrons from the plate connected to it, leaving a shortage of electrons on this plate, so it has a positive charge. This positive charge attracts more electrons from the negative terminal of the battery to the plate with the negative charge. Thus, there will be a surplus of electrons on one plate and a shortage on the other.

The charges on the plates act on the electrons in the dielectric material between them. The charge on the negative plate repels them toward the positive plate. The charge on the positive plate attracts them with equal force. As the dielectric is an insulator the electrons cannot actually leave their atoms, but their orbits become distorted toward the positive plate. Since the electrons are negative, this has the effect of increasing the negative pressure on the positive plate, repelling yet more electrons from this plate which in consequence becomes more positive and attracts more electrons from the battery to the negative plate. Thus, the dielectric increases the amount of charge the capacitor plates can hold. The effect varies with different dielectric materials, according to how much their atomic electron orbits can be bent out of shape.

The amount of electricty that can be stored depends not only on the dielectric, but also upon the distance between the plates and their overlapping area. (The latter can be increased by using more plates.) The "electrical size" of the capacitor is called its capacitance, measured in farads (q.v.), and is given by:

$$C = 0.0885 \frac{KS(N\text{-}1)}{d} \tag{1}$$

where
    C = capacitance, in picofarads,
    K = dielectric constant (from Table I),
    S = area of one plate, in cm^2,
    N = number of plates,
    d = thickness of the dielectric, in cm.

Electrons will continue to flow until each capacitor plate is at the same potential as the battery terminal connected to it. However, as the capacitor charges, the charging current decreases. This is due to the fact that the actual voltage producing the current is equal to the source voltage less the capacitor voltage, so as the latter builds up the difference between them gets less, and the actual voltage (which is the voltage forcing electrons through the circuit) goes down. If the battery is suddenly disconnected from the capacitor, the charges will remain on the capacitor plates, and in this way electricity is stored in the capacitor. It may be discharged subsequently by connecting a wire between the two plates, when the surplus electrons on the negative plate will flow to neutralize the shortage on the positive plate.

Capacitor Behavior in an A.C. Circuit—The basic operation of a capacitor is exactly the same when it is connected to an alternating-voltage source. At the start of each cycle there will be a certain number of electrons on both plates, and the electrons in each atom of the dielectric will be revolving around the nucleus in a normal manner. However, when the electrons begin to move out of one plate and on to the other, the orbits of the dielectric electrons will be distorted toward the positive plate, as explained above. Thus, although the electrons flowing on to the negative plate do not cross to the other plate they force the dielectric electrons toward it; and, although the dielectric electrons do not reach it either, their increased negative pressure drives more electrons out of the positive plate.

As the first half-cycle of the alternating voltage ends, the charges on the plates disappear and the dielectric electrons resume their normal orbits. During the following half-cycle the capacitor plates charge again, but with the opposite polarity, and the dielectric electron orbits are distorted in the opposite direction, pushing electrons out of the former negative plate. So, although the dielectric electrons themselves only move back and forth with each alternation of charging current, like the heads of spectators at a tennis match, they force the free electrons to flow out of each plate in turn as it becomes positive. Since this is what happens anyway in a wire carrying alternating current the effect is as if the electrons flowed back and forth through the capacitor.

But there are two other effects which make a considerable difference. Since work must be done to move the dielectric electrons back and forth, an opposing force must be overcome. This force is called capacitive reactance, and it is expressed in ohms, like resistance. However, capacitive reactance and resistance are by no means the same. Reactance varies with frequency, and is given by:

$$X_c = \frac{1}{2\pi fc} \tag{2}$$

where X_c = capacitive reactance, in ohms,
π = 3.1416,
f = frequency of alternating current, in hertz,
C = capacitance of capacitor, in farads.

The amount of the electric charge on a capacitor is given by:

$$Q = E\,C \tag{3}$$

where   Q = amount of charge, in coulombs,
E = voltage of capacitor when charged,
C = capacitance of capacitor, in farads.

The amount of the charge is also given by:

$$Q = I_{ave} \times t \tag{4}$$

where   Q   = amount of charge, in coulombs,
I_{ave} = average current flow during charging, in amperes,
t   = duration of the charging, in seconds.

Combining (3) and (4) gives:

$$I_{ave} \times t = EC \tag{5}$$

or

$$I_{ave} = \frac{EC}{t} \tag{6}$$

This last equation (6) means that if E and C are held constant, I_{ave} varies inversely with t, or the average current increases as the charging time decreases. But the charging time gets less as the frequency rises, so I_{ave} increases with frequency. This is the same thing as saying X_c varies inversely with the frequency, as expressed in formula (2) above.

The other effect arises from the fact that the charging current is at its maximum at the beginning, when the capacitor voltage is zero, and tapers off as the capacitor voltage rises, becoming zero when the capacitor voltage equals the source voltage. The

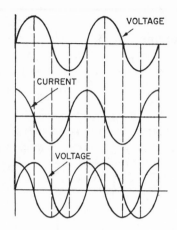

Figure C-2

Phase Relationship Between Voltage and Current in a
Capacitive Circuit

current alternations are, therefore, out of step with the voltage alternations, leading them by 90 degrees, as shown in Figure C-2. This angle is called the phase angle, and is 90 degrees only in a purely reactive circuit. Since all circuits contain some resistance, however small, the value of the phase angle is given by:

$$\theta = \text{arc tan } \frac{X_c}{R} \tag{7}$$

where θ = the phase angle, in degrees,
            X_c = the capacitive reactance from formula (2), in
               ohms,
            R   = non-reactive resistance, in ohms.

The cosine of the phase angle is the power factor of the capacitor, or the measure of its energy loss due to leakage (conductance between the plates through the dielectric), and series resistance (due to skin effect of plates and leads, as well as dielectric losses other than leakage). In most capacitors the phase angle θ is so close to 90 degrees that the angle $90 - \theta$ degrees, called the phase difference ϕ, is used instead, in which case power factor is given by:

$$PF = \tan \phi \tag{8}$$

As long as ϕ is less than 3 degrees, tan ϕ is essentially equal to cos θ.

Types of Capacitor

Capacitors can be classified as low, medium or high-loss, as in Table I. The low-loss capacitor is preferred for critical applications where stability is important, as in resonant circuits and filters, and for high-frequency operation, since losses increase with frequency. The medium-loss capacitor is used for most other purposes. In power supply filters, where large values of capacitance are generally required, electrolytic capacitors are employed.

Table I—Capacitor Categories

Category	Dielectric Material	K*
Low loss	Air Polystyrene Mica Glass Low-K ceramic	1 2.56 6-8 8 5-168
Medium loss	Plastic film (Mylar, etc.) Paper High-K ceramic	3.16 1.5-3.3 $\leqslant$1500
Electrolytic	Aluminum oxide Tantalum pentoxide	6-8

*Based on 1 as the dielectric constant of a perfect vacuum.

Variable Air-Dielectric Tuning Capacitors are of the general form shown in Figure C-3. The stator is insulated from the frame, so the rotor is generally at ground potential. The rotor plates are either symmetrical, as in (a) of Figure C-4, or offset, as in (b). The type in (a) is called a straight-line capacity plate, since the capacitance varies linearly with the overlapping area. The type in (b) is called a straight-line frequency plate, because it was designed to provide an even distribution of frequencies on a tuning dial. The only difference between receiving and transmitting capacitors is that the latter must have greater spacing between the plates to prevent arcing. Since variable capacitors are used at high frequencies they must be carefully constructed to keep losses low.

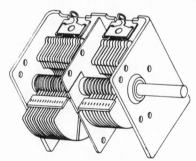

Figure C-3
Air-Dielectric Tuning Capacitor

SHADED AREAS INDICATE ACTIVE SURFACES

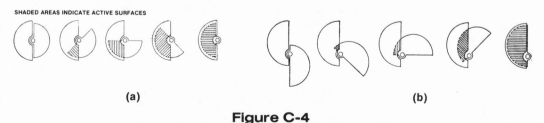

(a)

(b)

Figure C-4
(a) Straight-Line Capacitu Plate
(b) Straight-Line Frequency Plate

Fixed Capacitors are most commonly made of alternate layers of conducting plates separated by a dielectric layer. By using ribbon-like strips of these materials and rolling them up, a larger capacitance can be obtained without increasing the volume. This method of construction is shown in Figure C-5, and is generally used for paper and plastic-film capacitors, the plates being made of aluminum foil. The leads emerge straight out from the ends (axial leads), or at right-angles to the ends (radial leads), and are attached to the foil by conducting disks, as in the figure.

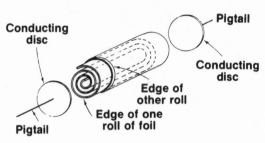

Figure C-5

Construction of Paper or Plastic-Film Fixed Capacitor with Axial Leads

When the dielectric material is mica or ceramic, it cannot be rolled up, so these capacitors are made of very thin plates of the dielectric on which the metal has been deposited as a thin film.

All fixed capacitors (except fixed air-dielectric types used in high-power transmitters) are encapsulated to protect them from moisture and other environmental hazards. The covering varies from wax-impregnated paper to extremely hard plastic.

Aluminum Electrolytic Capacitors are made of two strips of etched aluminum alloy, separated by gauze soaked in an electrolyte such as a solution of boric acid, glycerine and ammonia, in paste form, rolled up, and placed in a container that is then sealed to prevent evaporation. Before assembly, one of the foil strips is treated chemically to produce an oxide coating on its surface. This coating is only molecules thick, but nevertheless is the dielectric layer between the plate it coats (the anode) and the electrolyte, which is the other capacitor plate. (The second strip of foil is not a plate; it is provided to make contact with the electrolyte.)

Tantalum Electrolytic Capacitors may consist of (1) a sintered porous anode of tantalum powder housed in a silver or silver-plated container, with an electrolyte of sulfuric acid; (2) the same, but with a neutral electrolyte; or (3) etched foil plates and gauze soaked in electrolyte, constructed in the same way as aluminum electrolytic capacitors. In these capacitors the dielectric is an oxide coating on the tantalum powder grains or foil.

Polarization—Unlike other capacitors, electrolytic capacitors are polarized. They must be used only in circuits having d.c., pulsating d.c., or a.c. riding on d.c., and must be installed so that the anode, or plate with the oxide coating (marked +, or having a red lead), is connected to the positive side of the voltage source.

NOTE

Charts for determining reactance values for various frequencies and capacitances are given under *Coil Data*.

capacitor-input filter—Power-supply filter in which a capacitor is connected across the rectifier output.

capacitor microphone—(See *Recording*.)

capacitor-start—Motor in which a capacitor is connected in series with an auxiliary winding for starting. When the motor reaches a predetermined speed the auxiliary circuit is opened automatically.

capacity—Maximum output capability of a battery, expressed in ampere-hours (q.v.).

capstan—Spindle or shaft in a tape recorder that presses the tape against the pressure roller (capstan idler) and is rotated by the motor to draw the tape through the machine at a constant speed. *Recording*.

capture effect—Phenomenon where two FM signals with the same frequency are present, but the receiver selects only the stronger and completely rejects the weaker.

carbon—Nonmetallic element existing in crystalline form (diamond and graphite), or amorphous form (charcoal). In compounds with hydrogen, oxygen, nitrogen, and a few other elements, it comprises about 18 percent of all the matter in living things. Symbol, C; atomic weight, 12.011; atomic number, 6.

carbon-dioxide laser—Powerful gas laser used in cutting, drilling, welding, etc. *Lasers*.

carbon microphone—Microphone in which varying sound pressure on carbon granules affects their resistance, so as to produce fluctuations in an electric current passing through them that correspond to the sound waves causing them. *Telephone*.

carbon resistor—(See *Resistors*.)

carbon/silicon carbide transducer—Silicon carbide, or carborundum, thermocouples (q.v.) are used for temperature measurement in steel furnaces and ladles. The carbide element is enclosed in a carbon sheath to protect it from corrosion.

carcinotron—Backward-wave oscillator capable of continuous-wave output of hundreds of watts through X band (5.2 - 10.9 GHz). *Electron Tubes.*

cardiac monitor—Oscilloscope for monitoring a patient's heart action.

cardiac pacemaker—Device that stimulates the heart muscle with a regular progression of electrical impulses.

cardioid—Heart shaped. Said of some antenna radiation patterns or microphone response patterns.

card punch—Device that punches cards used in data processing. *Computer Hardware.*

card reader—Device used to read punched-card data into a computer. *Computer Hardware.*

carrier—Radio-frequency signal that can be modulated in various ways to transmit information. *Broadcasting.* (See also charge carriers.)

cartridge—Device inserted in a phonograph pickup head, that consists of a stylus that tracks in the record groove and vibrates in accordance with the recording; and a small coil attached to it that vibrates between the poles of a magnet, generating corresponding electrical signals that are amplified and applied to a speaker system. *Recording.*

cartridge fuse—(See fuse.)

cascade—Arrangement of two or more similar circuits (such as amplifiers), in which the output of one stage provides the input for the next.

cascode amplfier—Amplifier using a neutralized grounded-cathode or emitter input stage followed by a grounded-grid or base output stage. A cascode amplifier is often used in tuners, because it has high gain, low noise, and high input impedance. (See Figure C-6.)

Cassegrainian antenna—Special type of paraboloidal reflector antenna developed for communications with satellites. It reduces the amount of microwave energy that spills over past the edge of the reflector. *Antennas.* (See Figure C-7.)

cassette—Small plastic magazine for holding a reel of magnetic recording tape,

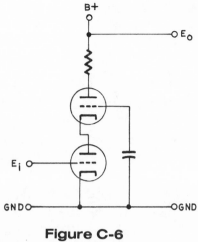

Figure C-6
Cascode Amplifier

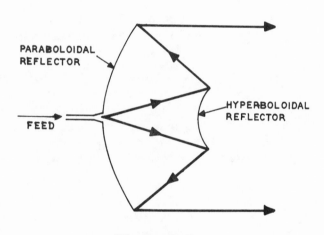

Figure C-7
Cassegrainian Antenna

designed to be inserted into a tape player, played, and rewound without removing the reel or the tape. Blank tape cassettes are also available for recording. Video cassettes are used for recording and replaying TV programs. *Recording.*

Cassiopeia noise—Brightest radio supernova remnant, known as Cassiopeia A, which was a star that exploded violently in AD 1702, and now emits intense radiation at 30 cm and 3 $W/m^2/Hz \times 10^{29}$.

cathode—Negative terminal or electrode through which electrons enter a d.c. load, such as an electrolytic cell, or an electron tube; and the positive terminal of a battery, or other source of electrical energy, through which they return. *Electron Tubes.*

cathode bias—Biasing an electron tube by making the cathode more positive with respect to ground than the grid, usually by placing the biasing resistor in the cathode return circuit. *Electron-Tube Circuits.*

cathode coupling—Using a resistor or other element in the cathode circuit of an electron tube to couple energy into the next stage. *Electron-Tube Circuits.*

cathode emission—Process whereby electrons are emitted from the cathode in an electron tube. *Electron Tubes.*

cathode follower—Electron-tube circuit in which the input is applied between the control grid and ground, and the output is taken from across a resistor between cathode and ground. The circuit does not invert the signal, has high input impedance, low output impedance and negative feedback. The gain is less than unity. Also called a grounded-plate amplifier. *Electron-Tube Circuits.*

cathode protection—Method of preventing corrosion of a metal by connecting it electrically to a more easily corroded metal, such as magnesium, which sustains the attack, thus leaving the less active metal unharmed.

cathode-ray oscilloscope—Device employing a narrow beam of electrons focused on a fluorescent screen. The beam is deflected horizontally and vertically to produce a luminous image on the screen that is analogous to the electrical voltages applied to the deflection plates. Because almost any physical phenomenon can be converted into a corresponding voltage, the oscilloscope can be used in nearly all forms of physical investigation. (See oscilloscope.)

cathode rays—Stream of electrons leaving the cathode in an electron tube. When focused on a hard target, cathode rays produce X rays; when incident on a phosphor, they cause it to luminesce. *Electron Tubes.*

cathode-ray tube (CRT)—Electron tube used in oscilloscope, television receiver, or radar equipment to display data in visible form. *Electron Tubes.*

cathode sputtering—(See sputtering.)

cation—Positive ion that is attracted by the cathode.

CATV—Abbreviation for Community Antenna Television, or cable TV. *Broadcasting.*

catwhisker—Small wire used to make contact with a sensitive point on the crystal in an early form of radio receiver known as a crystal set.

cavity resonator—Any region bounded by conducting walls within which resonant electromagnetic fields may be excited. *Electron Tubes, Waveguides and Resonators.*

C band—Microwave frequency band from 3.9 to 6.2 GHz. *Frequency Data.*

CCD—(See charge-coupled device.)

C display—Radar indication consisting of a bright spot on a rectangular graticule. The horizontal coordinate gives the azimuth, the vertical elevation. *Radar.*

cell—(See particular type of cell; e.g., solar cell, standard cell, battery, etc.)

Celsius temperature scale—Scale based on 0° for the freezing point of water, and 100° for the boiling point of water, at standard sea-level atmospheric pressure, hence it is sometimes called the centigrade scale. Since the scale has the same-sized intervals as the Kelvin scale, temperature in degrees Celsius can be converted to temperature in kelvins by the addition of 273.15. *Conversion Factors, S.I. Units.*

center tap—Connection at the electrical center of an inductor or resistor.

centigrade temperature scale—(See Celsius temperature scale.)

centimetric waves—Microwaves with wavelengths of 1 to 10 centimeters (3 to 30 GHz).

central processing unit (CPU)—Control section and arithmetic logic unit of a computer. *Computer Hardware.*

ceramics—1. Clay consisting principally of alumina, fired and glazed, used for insulators.   2. Barium titanate or lead titanate zirconate, are piezoelectric (q.v.) and used as transducers.   3. Semiconductors such as gallium arsenide, cadmium sulfide, silicon carbide, etc., used for rectifiers, solar cells, photocells, thermistors, and so on.   4. Ferrites (iron oxide compounds) used for magnetic devices. *Capacitors, Semiconductors.*

Cerenkov counter—Radiation counter that makes use of visible light emitted by relativistic charged particles moving in a medium of index of refraction n with a velocity >c/n. *Nuclear Physics.*

cermet—Mixture of ceramic insulating materials and metals; e.g., silicon monoxide and chromium. Used for fabrication of thin-film resistive elements.

cesium—Silvery white, most reactive, softest of all metals, liquid in a warm room. Used in photoelectric cells, television cameras, plasma propulsion engines for deep-space exploration, atomic clocks, and as a getter (q.v.) to remove traces of gas from vacuum tubes after sealing. Symbol, Cs; atomic weight, 132.905; atomic number, 55.

chad—Confetti produced by punching cards or tape.

chaff—Tinsel scattered by aircraft to jam enemy radar.

chain radar beacon—Radar beacon with fast recovery time to permit simultaneous use by a number of radars. (See also transponder.) *Navigation Aids.*

chain radar system—Number of radar stations on missile range, linked by data and communication lines, for tracking and data collection.

chain reaction—Series of nuclear fissions, each initiated by a neutron produced in a preceding fission. *Nuclear Physics.*

channel—1. General term for the vehicle or route used to transmit any type of information electronically.   2. Electrically conductive path between two separated regions of a semiconductor material. *Transistors.*

channel selector—Tuning switch on television receiver.

character—Letter (alpha), figure (numeric), or other symbol printed or displayed to present information. *Computer Hardware.*

characteristic—1. In a common logarithm (q.v.), the number on the left of the decimal point.   2. Distinguishing feature or quality of something.

characteristic curve—Graph to show the inherent property of a device. For example, the relationship between grid voltage and plate current in an electron tube.

characteristic impedance—Ratio of voltage to current at every point along a transmission line on which there are no standing waves. *Transmission Lines.*

character reader—Input device that reads characters directly. *Computer Hardware.*

Charactron—Cathode-ray tube for displaying data output from a computer in alphanumeric form. *Computer Hardware.*

charge—Quantity of electricity that flows in electric currents or that accumulates on the surfaces of dissimilar nonmetallic substances that are rubbed together vigorously. Electric charges are positive or negative, are a basic property of matter, occur in discrete natural units, and are conserved. Like charges repel each other, unlike charges attract each other. The unit of electric charge is the coulomb (q.v.). (See also capacitance.)

charge carrier—Particle having electric charge, the motion of which constitutes an electric current. In metals, the charge carriers are free electrons (negative charge carriers). In a semiconductor there are also positive charge carriers called holes. (A hole is the absence of an electron in a bond between atoms.) In gases and liquids, the charge carriers are positive and negative ions (q.v.). *Semiconductors.*

charge-coupled device (CCD)—Junctionless semiconductor device consisting of a silicon substrate covered by a layer of silicon dioxide, which in turn is capped with a row of closely spaced aluminum electrodes. Application of a potential to these electrodes results in surface depletion regions ("potential wells") beneath each electrode. The applied potential should be negative for an n-type substrate, positive for a p-type. Charge carriers of opposite polarity can then be stored in the depletion regions. If a more negative (or positive, as the case may be) pulse is applied to an electrode adjacent to one under which a charge is stored, the charge will spill over into the deeper potential well. In this way a charge can be shifted along the chain of depletion regions. Ones and zeros are indicated by presence or absence of charge, which can be injected at the input of the chain by radiation-generated electron-hole pairs, and read out by the reverse process at the output. CCD's can provide memory, shift register, delay line, and imaging functions. *Computer Hardware.* (See Figure C-8.)

charger—Device that changes alternating powerline voltage into pulsating direct current to charge storage batteries.

charge-storage tube—(See electrical readout tube.)

chassis—In a radio or television receiver, all working parts and their mounting, but excluding the cabinet or enclosure.

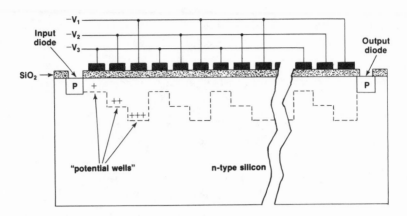

Figure C-8
Charge-Coupled Device

chassis ground—Conducting connection to chassis or frame, or equivalent chassis connection of a printed-circuit board. Chassis ground may be at substantial potential with respect to the earth or structure in which the chassis or frame is mounted. (See also ground.)

Chebishev filter—Bandpass or band-reject filter with extremely steep-sided attenuation curve, characterized by ripple in the pass band. *Passive Filters.*

check bit—Binary check digit. (See error-correcting code.)

checksum—Sum of all binary digits in a computer program, used to verify program.

cheese antenna—Antenna with parabolic reflector resembling a semicircular slice of cheese, that produces a fan beam. Also called a pillbox antenna. *Antennas.* (See Figure C-9.)

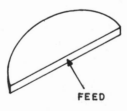

FEED

Figure C-9
Cheese Antenna

chemofacture—Fabrication by chemical means; e.g., printed wiring is chemofactured. (See also etching.)

chip—Single substrate on which all the active and passive elements of an electronic circuit have been fabricated using the semiconductor technologies of diffusion, passivation, masking, photoresist, and epitaxial growth. Also called a die.

chirp—Objectionable shift in frequency with keying, in using a transmitter with unregulated power supplies for code. Variations in the voltage of the oscillator as it is keyed cause changes in the effective input capacitance and so cause the frequency to shift.

choke—Inductor used to prevent the flow of alternating current in a particular circuit. Iron-core chokes are common in power supply filters. (See also swinging choke, filter reactor.) *Rectifiers and Filters.* (See Figure C-10.)

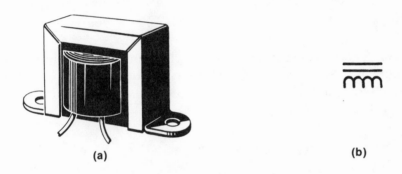

(a)   (b)

Figure C-10
(a) Typical Iron-Core Choke
(b) Iron-Core Choke Schematic Symbol

choke-input filter—Power supply filter in which a choke is the first element in series with the d.c. output of the rectifier. *Rectifiers and Filters.*

chopped mode—In an oscilloscope with two vertical input channels, switching the two amplifiers on and off alternately by means of an astable multivibrator running at (for example) 100 kilohertz, so that the signals in each channel are displayed apparently simultaneously.

chopper—Device for interrupting d.c. or light beam at a fixed rate so that the resultant pulses can be applied to a transformer, a.c. amplifier, or used as square waves.

Christmas-tree effect—TV picture symptom due to severe horizontal-frequency error.

chroma—Purity of a color, determined by its degree of freedom from white or gray;

also used for the color signal in television receiver as distinct from the Y or brightness signal.

chroma amplifier—Amplifier that separates the color signal from the rest of the video signal in a color TV receiver, and amplifies it. Also called the bandpass amplifier.

chromatic aberration—Property of lenses that causes various frequencies in a beam of light to be focused at different points, thus causing a spectrum to appear.

chromaticity diagram—(See color diagram.)

chrominance signal—Chrominance subcarrier sidebands transmitted along with the luminance (brightness or monochrome) signal in compatible color transmission. *Broadcasting.*

chromium—Hard steel-gray metal that takes a high polish and is used in alloys to increase strength and corrosion resistance. Also used for corrosion control of metal surfaces (chromium plating and anodizing). Symbol, Cr; atomic weight, 51.996; atomic number, 24.

circuit—Path for transmitting electric current.

circuit breaker—Automatic switch that opens an electric circuit if abnormal current conditions (such as an overload) occur.

circulator—Device used where a microelectronic radar transmitter and receiver share the same antenna, and signals picked up by the antenna go only to the receiver. (See also duplexer.)

citizen's band radio—Communications system involving the use of radio for communication or remote-control purposes within a range of 10 to 15 miles. *Frequency Data.*

clamp circuit—Circuit which holds the direct voltage at some desired level. See also d.c. restorer.

class-A amplifier—Amplifier operated so that plate or collector current flows continuously throughout each electrical cycle. *Electron-Tube Circuits, Transistor Circuits.*

class-AB amplifier—Amplifier operated so that plate or collector current flows appreciably more than half but less than all of each electrical cycle. *Electron-Tube Circuits.*

class-A insulating material—Materials such as cotton, silk, and paper suitably impregnated to withstand temperatures not exceeding 105°C.

class-A station—FM broadcasting station limited to a maximum of 3 kilowatts effective radiated power. *Broadcasting.*

class-A0 emission—Transmission with no modulation.

class-A1 emission—Telegraphy with no modulation (on-off keying of carrier).

class-A2 emission—Telegraphy using audio-modulated carrier.

class-A3 emission—Telephony, double sideband, full carrier.

class-A3A emission—Telephony, single sideband, reduced carrier.

class-A3B emission—Telephony, two independent sidebands, reduced carrier.

class-A3J emission—Telephony, single sideband, suppressed carrier.

class-A4 emission—Facsimile transmission.

class-A4A emission—Facsimile transmission with single sideband, reduced carrier.

class-A5C emission—Television transmission with vestigial sideband.

class-A7A emission—Multichannel voice-frequency telegraphy, with single sideband, reduced carrier.

class-B amplifier—Amplifier operated close to cutoff, so that plate or collector current flows only during approximately half of each electrical cycle. *Electron-Tube Circuits, Transistor Circuits.*

class-B insulating material—Materials such as mica, glass fiber, asbestos, etc., to withstand temperatures not exceeding 130°C.

class-B station—FM broadcasting station limited to a maximum of 50 kilowatts of effective radiated power, operated in Zones I or IA. *Broadcasting.*

class-C amplifier—Amplifier operated so that plate or collector current flows for appreciably less than half of each electrical cycle. *Electron-Tube Circuits.*

class-C insulating material—Materials consisting only of mica, porcelain, glass, quartz, and similar inorganic substances, able to withstand temperatures of 220°C and above.

class-C station—FM broadcasting station limited to a maximum of 100 kilowatts of effective radiated power, operated in Zone II. *Broadcasting.*

class-D station—FM educational broadcasting station limited to a maximum transmitter output power of 10 watts. *Broadcasting.*

class-F emission—FM transmission: F0, F1, F2 indicate telegraphy: F3 indicates telephony; F4 indicates facsimile; F5 indicates television; F6 indicates four-frequency diplex telegraphy.

class-F insulating material—Materials such as mica, glass fiber, asbestos, etc., to withstand temperatures not exceeding 155°C.

class-H insulating material—Materials such as silicone elastomer, mica, glass fiber, asbestos, etc., to withstand temperatures not exceeding 180°C.

class-O insulating material—Materials such as cotton, silk and paper without inpregnation for temperatures not exceeding 90°C.

class-P emission—PM transmission: P0, P1D, P2D, P2E, P2F, indicate telegraphy; P3D, P3E, P3F, P3G, indicate telephony.

clear—(See reset.)

clear channel—AM broadcasting stations in classes I and II. *Broadcasting.*

click filter—(See key-click filter.)

clipper—Circuit to limit the amplitude of a signal to a desired value. Also called a slicer. *Rectifiers and Filters.*

clipping—Form of distortion in which the peaks of a signal are flattened, as when an amplifier is overdriven.

clock—Timing pulse generator. *Computer Hardware.*

closed-circuit television (CCTV)—Cable TV. *Broadcasting.*

closed-loop feedback—Means of regulating the operation of a device using a sample of the output fed back to the input. *Feedback.*

clutter—Echoes due to rain, waves, trees, etc., that interfere with the observation of desired signals on a radar display. *Radar.*

CMOS—Abbreviation for complementary metal-oxide-semiconductor device (q.v.).

CMRR—Abbreviation for common-mode rejection ratio (q.v.).

coaxial—Having a common axis. Used of coaxial cable, coaxial connector, coaxial line, etc.

coaxial cable—High-frequency electrical conducting line used to transmit telegraph, telephone, television, and other signals. It may be a single line made up of a pencil-sized copper tube with a single wire at its center, held in position by insulating spacers; a braided wire sheath enclosing a single wire in a dielectric covering; or a cluster of such lines combined in a larger cable. *Transmission Lines.*

coaxial speaker—Speaker system in which tweeter and midrange speakers are mounted on the axis of and inside the cone of the woofer. (See Figure C-11.)

cobalt—Silver-white metal (when polished), with a bluish tinge, used mainly in magnetic alloys such as Alnico (aluminum, nickel and cobalt, plus iron), high-temperature steel, and low-expansion alloys. The man-made isotope cobalt-60 is radioactive (5.3-year half-life), but natural cobalt-59 is stable. Symbol, Co; atomic weight, 58.9332; atomic number, 27.

Cockroft-Walton generator—Rectified-voltage accelerator used to develop high bombarding energy for nuclear research. *Nuclear Physics.*

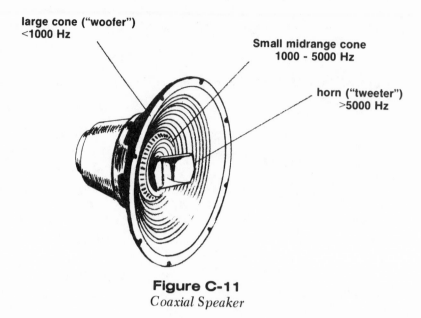

large cone ("woofer")
<1000 Hz

Small midrange cone
1000 - 5000 Hz

horn ("tweeter")
>5000 Hz

Figure C-11
Coaxial Speaker

code—1. Morse code, a system of dots, dashes and spaces, representing letters, numerals, etc., in telegraphy. *International Morse Code.* 2. Binary code, where each decimal digit or letter of the alphabet is represented by binary digits. *Computer Hardware.*

coefficient—1. Number, constant for a given substance, used as a multiplier in measuring the change in some property of the substance under given conditions: as, the coefficient of expansion. 2. The number prefixed as a multiplier to a variable or unknown quantity: as, 6 is the coefficient in 6x.

coercive force—Opposing magnetic intensity required to reduce the residual magnetic induction of a substance to zero.

coherence—Light in a laser beam may be said to have temporal coherence (monochromaticity) and spatial coherence (directionality), but these definitions are very general and should really be termed quasi-coherence. *Lasers.*

coherent detection—In digital data modulation, where the receiver is phase-locked with the transmitter. *Modulation.*

coherent pulse-operation—Method of pulse operation in which the phase of the radio-frequency wave is maintained through successive pulses. *Electron Tubes.*

coherent radar—Means of distinguishing between fixed and moving targets in pulsed doppler radar. *Radar.*

coil—One or more loops of wire wound spirally, often around a cylindrical cardboard or iron core, and exhibiting the property of inductance (q.v.). Also called an inductor. *Coil Data.*

COIL DATA

Single-Layer Wound Coil—The formulas for winding a radio-frequency, air-core coil with one layer of closely wound enamel wire are:

$$L = \frac{(rN)^2}{9r + 10l} \tag{1}$$

$$N = \frac{\sqrt{L(9r + 10l)}}{r} \tag{2}$$

where L = self-inductance of coil, in microhenries,
N = total number of turns,
r = mean radius of coil, in inches, or centimeters,
l = length of coil in inches, or centimeters.

Multi-Layer Wound Coil—The formulas for winding a radio-frequency, air-core coil with more than one layer of closely wound enamel wire are (3) below and (2) above:

$$L = \frac{0.8\,(rN)^2}{6r+9l+10b} \tag{3}$$

where L = self-inductance of coil, in microhenries,
N = total number of turns,
l = length of coil in inches, or centimeters,
b = (outside radius)–(inside radius), in inches, or centimeters,
r = mean radius of coil, in inches, or centimeters.

Data for these calculations are given in the following table for copper wire sizes from AWG#10 to AWG#40.

Inductance, Capacitance, and Reactance Charts—The charts in Figures C-12, C-13, and C-14 may be used to determine unknown values of frequency, inductance, capacitance, and reactance. Any one unknown value may be found if two of the others are known. Using the correct chart for the frequency, lay a straightedge across it so it passes through the two known values, and find the required value where it intersects the appropriate scale.

ENAMELED COPPER WIRE TABLE

Gage (AWG/BS)	Nominal Diameter (in)	Nominal Diameter (in)	Turns/inch	Turns/cm	Ohms/1000'	Ohms/100 m
10	0.1019	2.588	9.6	3.8	0.9989	0.3277
11	0.09074	2.305	10.7	4.2	1.260	0.4134
12	0.08081	2.053	12.0	4.7	1.588	0.5210
13	0.07196	1.828	13.5	5.3	2.003	0.6572
14	0.06408	1.628	15.0	5.9	2.525	0.8284
15	0.05707	1.450	16.8	6.6	3.184	1.045
16	0.05082	1.291	18.9	7.4	4.016	1.318
17	0.04526	1.150	21.2	8.3	5.064	1.661
18	0.04030	1.024	23.6	9.3	6.385	2.095
19	0.03589	0.9116	26.4	10.4	8.051	2.641
20	0.03196	0.8118	29.4	11.8	10.15	3.330
21	0.02846	0.7229	33.1	13.03	12.80	4.200
22	0.02535	0.6439	37.0	14.6	16.14	5.300
23	0.02257	0.5733	41.3	16.3	20.36	6.680
24	0.02010	0.5105	46.3	18.2	25.67	8.422
25	0.01790	0.4547	51.7	20.4	32.37	10.62
26	0.01594	0.4049	58.0	22.8	40.81	13.39
27	0.01420	0.3607	64.9	25.6	51.47	16.89
28	0.01264	0.3211	72.7	28.7	64.90	21.29
29	0.01126	0.2860	81.6	32.1	81.83	26.88
30	0.01003	0.2548	90.5	35.6	103.2	33.86
31	0.008928	0.2268	101	39.8	130.1	42.68
32	0.007950	0.2019	113	44.5	164.1	53.84
33	0.007080	0.1798	127	50.0	206.9	67.88
34	0.006305	0.1601	143	56.3	260.9	85.60
35	0.005615	0.1426	158	62.2	329.0	107.9
36	0.005000	0.1270	175	68.9	414.8	136.1
37	0.004453	0.1131	198	78.0	523.1	171.6
38	0.003965	0.1007	224	88.2	659.6	216.4
39	0.003531	0.08969	248	97.6	831.8	272.9
40	0.003145	0.07988	282	111.0	1049.0	344.2

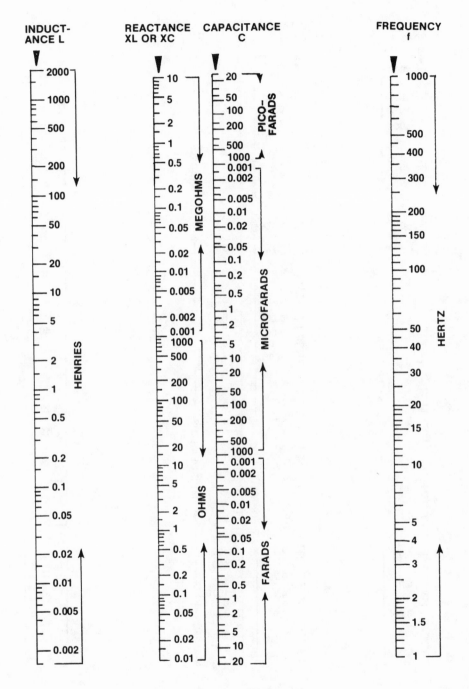

Figure C-12

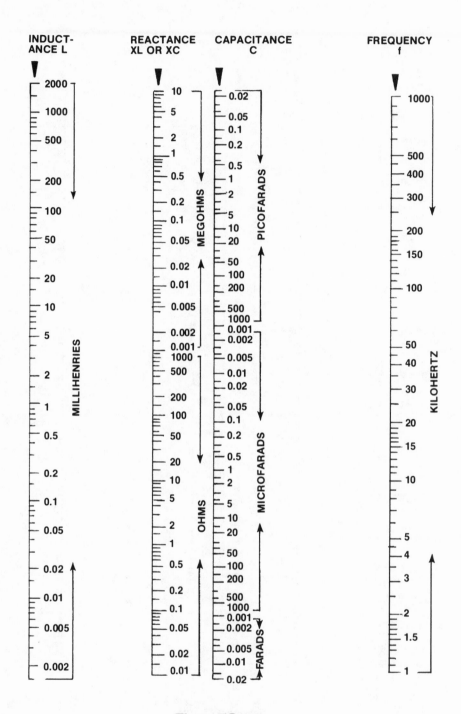

Figure C-13

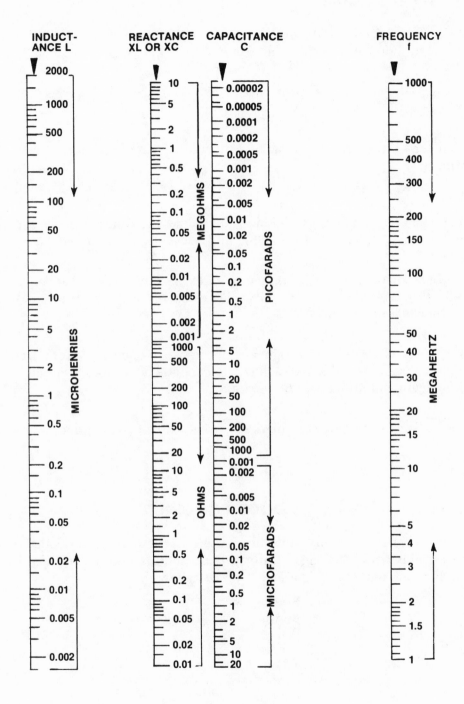

Figure C-14

cold-cathode gas tube—Glow-discharge or arc-discharge tube in which the cathode is not heated to induce electron emission. *Electron Tubes.*

collector—Region of transistor that is reverse biased with respect to the base. *Transistors.*

collimator—Device for changing light diverging from a point source into a parallel beam.

color—Electromagnetic radiation with a wavelength between 750 and 370 nanometers is perceived by the human eye and elicits in the mind of a standard observer a color image corresponding to the wavelength. *Optoelectronics.*

color-bar generator—Test instrument used to produce colored bars on a television screen to enable the operator to evaluate the receiver's response and to make adjustments if necessary.

color burst—Color reference signal of not less than eight cycles at a frequency of 3.579545 megahertz following each horizontal pulse of the composite signal during a color-television transmission. *Broadcasting.*

color coding—Using colors to mark electronic parts. *Color Coding.*

COLOR CODING

Fixed non-metallic resistors and some other components are color-coded to indicate their values instead of having the information printed or stamped on them. The colors used and their numerical significance are as follows:

Black	0
Brown	1
Red	2
Orange	3
Yellow	4
Green	5
Blue	6
Violet	7
Gray	8
White	9

On a *composition resistor* there are three or four colored bands around the resistor body, adjacent to one end. Starting with the one nearest the end, the first two colors are the significant figures of the resistor value in ohms, and the third is the power of ten of the multiplier. For example, blue (6), red (2), and yellow (4) should be read as 62×10^4, or 620 kilohms. If the third band is gray (or silver), or white (or gold), the multiplier is 10^{-2}, or 10^{-1}, respectively.

If there is no fourth band the tolerance is 20%. Otherwise, silver denotes 10%, gold 5%. Other tolerances may be indicated by using the applicable colors, but these are seldom encountered as most precision resistors have their values stamped or printed on them.

Some *film-type* resistors also may be color-coded, in which case an additional color may be added in front, to provide for three significant figures, followed by a multiplier and tolerance.

Capacitors, for the most part, are not color-coded. Exceptions are some *ceramic* and *mica* types. Ceramic *tubular* capacitors may be coded in the same way as composition resistors, except that an additional color is added in front of the others to indicate the *temperature characteristic.* The colored bands may also be replaced by colored dots. Ceramic *disks* have these arranged around the edge, to be read clockwise. Some very small tubular capacitors have only three dots, which give two significant figures and the multiplier only.

Mica capacitors are now mostly not color-coded. However, the flat rectangular molded silvered-mica type may employ an arrangement of six colored dots arranged three on each side of an arrow, or some other mark to indicate which way to read them. With the arrow pointing to the right the upper three dots (reading from left to right) indicate: EIA standard (this is a white dot, which may be omitted) and two significant figures. The bottom row (reading from right to left) gives the multiplier, tolerance, and type.

The significant figures always give the value in picofarads. The temperature characteristic (ceramic capacitors), or type (mica capacitors), is denoted as follows:

COLOR	TEMPERATURE CHARACTERISTIC	TYPE (MFR'S SPEC.)
Black	NPO	A
Brown	N033	B
Red	N075	C
Orange	N150	D
Yellow	N220	E
Green	N330	
Blue	N470	
Violet	N750	

Diodes used in signal circuits are often color-coded because of their small size. Diode designations are always given, for example, in the form 1N914A. The 1N– is common to all and does not have to be indicated; therefore, the color bands denote the figures and letter following. Suffix letters are coded as follows:

Black	(no suffix)
Brown	A

Red	B
Orange	C
Yellow	D
Green	E
Blue	F
Violet	G
Gray	H
White	J

The colored bands are grouped at the cathode end of the diode, and should be read from that end.

Transformer leads are color-coded as follows:

POWER TRANSFORMERS

Primary (tapped)   *Secondary*

Black (common)

Black/yellow (centertap)

Black/red

High voltage
{
Red
Red/yellow (centertap)
Red
}

Primary (untapped)

Rectifier filament
{
Yellow
Yellow/blue (centertap)
Yellow
}

Two blacks leads

Amplifier filament #1
{
Green
Green/yellow (centertap)
Green
}

Amplifier filament #2
{
Brown
Brown/yellow (centertap)
Brown
}

Amplifier filament #3
{
Slate
Slate/yellow (centertap)
Slate
}

IF TRANSFORMERS

Primary   *Secondary*

Blue (plate)

Red (B+)

Green (grid or diode)

Violet (full-wave diode)

White (grid or diode return, AVC, or ground)

AUDIO & OUTPUT TRANSFORMERS

Primary	*Secondary*
Blue (plate)	Green (grid or voice coil)
Red (B+)	Black (return or voice coil)
*Blue or brown (plate)	*Green or yellow (grid)

color diagram—Diagram to show the relationship between hue, saturation and brightness of colors, in which colors are fully saturated around the border, but contain more and more white toward the center, which is all white. (See Figure C-15.)

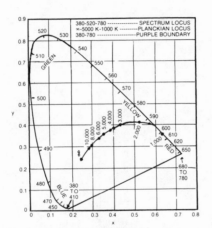

Figure C-15
Color Diagram

color-difference signal—Primary-color signal less the luminance portion, designated R-Y, B-Y, or G-Y as the case may be.

color killer—Circuit in television receiver to disable the color section during reception of monochrome programs so that no colored interference appears on the screen.

color picture tube—Cathode-ray tube for displaying colored television programs that has a screen with three phosphors arranged in a pattern of dots or stripes, which produce the colors red, blue, and green to the extent they are excited by their corresponding electron guns. Electrons from each gun are prevented from striking the wrong phosphor by a perforated or slotted shadow mask. (See Figure C-16.)

color signal—In compatible color television, the chrominance portion of the transmission. *Broadcasting.*

---

*Pushpull only.

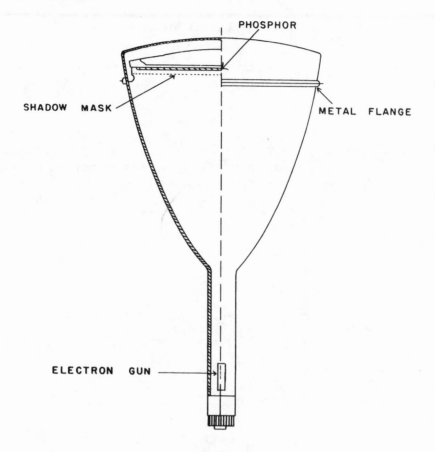

Figure C-16
Color Picture Tube

color temperature—Temperature of a universal radiator (blackbody) when radiating light of some color, expressed in kelvins, is the color temperature for that wavelength. *Optoelectronics.*

Colpitts oscillator—Oscillator employing capacitive feedback. *Transistor Circuits.*

common-base amplifier—Transistor amplifier circuit in which the base is in the return path for both input and output circuits. Also called grounded-base amplifier. *Transistor Circuits.*

common-collector amplifier—Transistor amplifier circuit in which the collector is in the return path for both input and output circuits. Also called an emitter follower or grounded collector amplifier. *Transistor Circuits.*

common-emitter amplifier—Transistor amplifier circuit in which the emitter is in the return path for both input and output circuits. Also called a grounded-emitter amplifier. *Transistor Circuits.*

common-mode rejection ratio—Ratio of common-mode input voltage to common-mode output voltage, expressed in decibels. A figure of merit for a differential amplifier (q.v.).

common-mode signal—Signal appearing simultaneously and in phase at both input terminals of a differential amplifier (q.v.).

communications receiver—Receiver used in communications between radio stations.

community antenna television (CATV)—System comprising an advantageously located receiving antenna array and distribution network of amplifiers and coaxial cables to subscribers' homes. Also called cable television. *Broadcasting.*

commutator—Device used on electric motors or generators to maintain a unidirectional current. (See Figure C-17.)

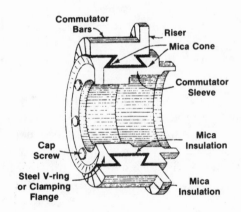

Figure C-17
Commutator (Cutaway View)

comparator—Circuit for comparing two voltages and indicating their difference or agreement. Used in feedback circuits, digital voltmeters, counters, etc. *Rectifiers and Filters, Feedback.*

compiler—Program to translate a source-language program a human easily understands into a machine-language program a computer understands. *Computer Software.*

complementary MOS (CMOS)—Logic family that uses PMOS and NMOS technology in a complementary arrangement on a common substrate. Almost immune to noise, operates from a wide range of supply voltages, consumes less than a microwatt while switching at moderately fast speed. *Microelectronics.*

complex number—Pair of numbers (x,y), united symbolically, for purposes of technical convenience, in the form x - yj, where j is the imaginary number $\sqrt{(-1)}$. Multiplication of a complex number by j turns the corresponding displacement through a right angle.

compliance—1. Degree to which a phono-cartridge stylus can be deflected, so a measure of its low-frequency tracking.   2. Flexibility of a speaker-cone suspension, so a measure of its ability to reproduce low-frequency sounds.

component—Any functional part of an assembly.

composite picture signal—Complete television video signal. *Broadcasting.*

composition resistor—(See *Resistors.*)

compression—Process in which low-amplitude signals are amplified more than high-amplitude signals, so that quiet sounds are raised and loud sounds lowered, or brightness adjusted in a television picture.

computer—Automatic electronic machine that performs calculations. There are two types. The analog computer operates on the principle of similarity in proportional relations (analogues), and manipulates potential differences. The digital computer operates in terms of two-state logic, using switching and storage devices that can be in only one of the two states, usually designated "1" or "0". More than 90 percent of computers are digital. *Computer Hardware, Computer Software.*

computer hardware—The physical parts of a computer system: electronic circuits, input and output devices, power supplies, and so on. *Computer Hardware.*

COMPUTER HARDWARE

Types of Computer

There are three types of computer: *digital, analog,* and *hybrid.* Digital computers count discretely, never varying or responding in degree, but only to the two binary digits, 0 and 1. Analog computers measure the values of continuous variables and convert these to voltages that are proportional to those values. The principal building block of the analog computer is the operational amplifier (q.v.). Hybrid computers are a combination of digital and analog, in which the two techniques play equal roles. Microminiaturization of computer circuitry has led to the digitization of many operations formerly performed by analog computers, so that the latter tend now to be used only for custom-tailored single purposes. The term computer today almost invariably refers to a digital computer.

Digital Computer Hardware

In some cities at Christmas it is customary to leave the lights on in certain of the windows of City Hall, so that at night they form a gigantic illuminated cross. From the spectator's point of view this is very symbolic. From the point of view of the City Hall electrician inside, it means only that specified switches are in the on position and others are in the off position, in accordance with the Mayor's instructions.

Figure C-18 shows how the basic operation of a digital computer closely parallels this Christmas story. The *user* (Mayor) gives instructions via an *input/output device* (telephone) to the *controller* (electrician), who in turn manipulates the switches until the *data* (cross) are stored in *memory* (City Hall). In this sequence, the I/O device, controller, switches, and memory are all actual physical objects and are therefore called the *computer hardware*. The data however exist only in the form of a pattern which is termed *computer software*.

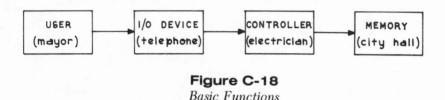

Figure C-18
Basic Functions

Of course the hardware has no intelligence of its own. It is a mechanical system, operated by outside intelligence by means of a *program*, which is a sequence of instructions to the controller (which is *not* an intelligent electrician, but a *central processing unit*). When a computer prints out some sequence of letters, such as "WRONG, TRY AGAIN," it is meaningful to the user, but the meaning does not arise mysteriously from within the computer; it comes from the way the programmer assigned letters to processes going on in the computer. It is really a dialogue between the user and the programmer, in which the programmer has stored all the responses required in the computer. The computer's instructions then make it select the appropriate response for any input. If computers are ever made that program themselves without human intervention it may be a different story!

The most basic computer process is storage in memory. Primary storage is in *registers* in the controller itself; secondary storage is in auxiliary memory devices, such as magnetic tape, punched cards, and so on. A register consists of a set of *flip-flops* (q.v.).

A flip-flop is a bistable multivibrator, whose output is in either a *zero* or a *one* state, like a switch that is off or on. Electrically speaking, this means that the output of the

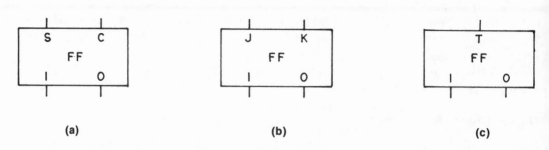

(a)   (b)   (c)

Figure C-19

Three Most Commonly Used Flip-Flops:

(a) *S - R flip-flop: sets and resets (clears) according to which input is 1.*

(b) *J - K flip-flop: same as (a), except that if both inputs are 1, circuit will complement (reverse outputs).*

(c) *T flip-flop: complements (reverses outputs) whenever triggered by a 1 at T.*

flip-flop is zero volts or five volts (this is the most usual value). Figure C-19 gives the symbols for commonly used flip-flops. The number of flip-flops in a register is generally eight, so that it can store eight *bits* of information. Eight bits of information are called a *byte*. (Some special registers store sixteen bits, however.)

The term bit means *binary digit*. In binary arithmetic there are only two digits, 0 and 1, which can easily be represented by the zero or one state of a flip-flop. In decimal arithmetic ("ordinary arithmetic") there are ten digits, so these have to be converted into binary form for the computer to use. In a decimal number the digit furthest to the right is what it says it is; the next to the left is *ten times* as great as it says it is, the next to the left again is *ten times ten*, and so on. This is because the decimal system has a *base* of ten.

In the binary system the base is *two*. The digit furthest to the right is what it says it is (0 or 1). The next to its left is *two times* what it says it is, the next to the left again *two times two*, and so on. In the binary number 111 the value of the right-hand digit is 1, the value of the next to the left is $1 \times 2 (=2)$, and the next again is $1 \times 2 \times 2 (=4)$. The value of the binary number 111 in decimal notation is 4+2+1, or 7. So the computer stores 111, the binary equivalent of 7, by setting the first three flip-flops in a register to their one state, while leaving the other five in their zero state. The register then "reads" 00000111.

Suppose the user instructs the controller to add to this number another number, say 00001010 (decimal equivalent is 10). If this were done with pencil or paper, the procedure would be to place the second number under the first and add each column, starting at the right, as in ordinary addition:

$$
\begin{array}{r}
00000111 \\
00001010 \\
\hline
00010001
\end{array}
$$

In the first column, $1 + 0 = 1$. In the second column, $1 + 1 = 0$, carry 1. In the third column, $1 + 0$, plus the carry from column 2, or $1 + 0 + 1 = 0$, carry 1. In the fourth column, $0 + 1$, plus the carry from column 3, or $0 + 1 + 1 = 0$, carry 1. In the fifth column, $0 + 0$, plus the carry from column 4, or $0 + 0 + 1 = 1$.

To perform this operation in the computer a circuit is required for each column which gives a 0 output when two 0's are added, and also when two 1's are added, but gives a 1 output when a 1 and a 0 are added. This circuit must also be able to transfer any carry to the next column. Such a circuit is made by combining *logic gates*, as shown in Figure C-20. This circuit consists of three AND gates, two inverters, and one OR gate. Each of these gates is a switch with two inputs and one output. If a 1 bit (5 volts) is present at *both* inputs of an AND gate it will switch and its output will be a 1 bit, but it will not switch for any other combination of inputs. If a 1 bit is present at *either* (or both) inputs of an OR gate, it will switch, and its output will be a 1 bit, but it will not switch if both its inputs are 0. The principle involved in each of these gates is illustrated in Figure C-21, where the AND gate is represented by two switches in series, and the OR gate by two switches in parallel. In reality these switches are transistors which are turned on or off by control signals that apply the correct biases to their bases so they saturate or cut off. An inverter is a simple common-emitter amplifier that reverses the polarity of a bit applied to its base, so that a 1 changes to a 0, and vice versa.

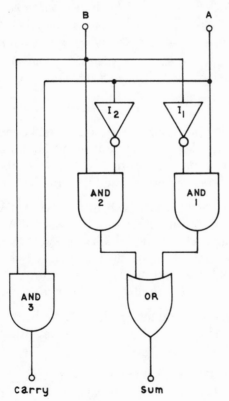

Figure C-20
Circuit for Adding ("half adder")

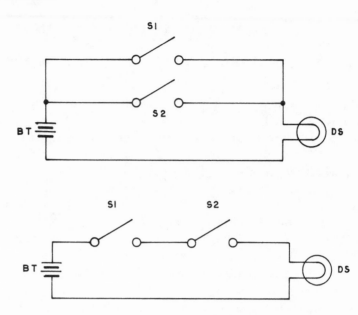

Figure C-21

Illustrating the Difference Between an AND Gate and an OR Gate by Switches Connected in Series and in Parallel

- *An AND gate is like two switches in series; both must be on before current from battery BT can reach lamp DS.*

- *An OR gate is like two switches in parallel; either can complete the path for current from battery BT to reach lamp DS.*

Returning to Figure C-20, if the circuit is required to add 1 and 0, these two bits appear at A and B. The 1 at A is applied to one input of AND_1, and the 0 bit on B goes to I_1, where it is changed to a 1, so both inputs of AND_1 have 1 bits on them. AND_1's output is therefore 1, which is then applied to one input of the OR gate.

The 0 bit at B appears on one input of AND_2, and the 1 bit at A, after being inverted by I_2, appears as another 0 bit on the other input of AND_2. AND_2, therefore, does not switch and continues to give a 0 output, which applied to the second OR input. But as there is already a 1 on the first OR input, the OR gate switches to give a 1 output. This appears at the terminal marked SUM, from which it is transferred to the corresponding location in the accumulator register (described later).

The terminals A and B are also connected directly to the two inputs of AND_3, which generates the carry if both inputs are 1. In this case the output remains 0, so there is no carry.

This is not by any means the whole story of arithmetical operations in the *arithmetic and logic unit* (ALU) in the controller, but it serves to illustrate how gates are used. Gates may have more than two inputs, and there are other versions that are used for convenience, but all are based on the same principles: AND gates must have 1 bits on

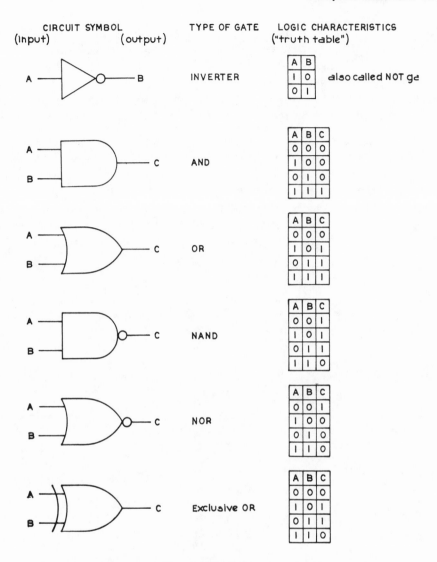

| CIRCUIT SYMBOL
(Input) (output) | TYPE OF GATE | LOGIC CHARACTERISTICS
("truth table") |

Figure C-22

Principle Types of Logic Gates (AND, OR, NAND, and NOR gates may have more than two inputs)

all inputs before they switch, OR gates need only 1 bit on any input. The types of gates that are generally used are given in Figure C-22.

The computer controller or central processing unit (CPU) employs a very large number of logic gates and memory arrays. In earlier models these were fabricated out of discrete components, resulting in huge machines fifty feet long and eight feet high, weighing tons. They also consumed power by the kilowatt. Today, far more advanced circuitry is provided on silicon chips a mere 3/16-inch square. These are called microprocessors, and they consume only milliwatts of power.

A microprocessor duplicates the CPU of a "conventional" computer, but because of its low cost, compact size and limited power consumption, is now used extensively in all types of "intelligent" systems that include data acquisition and control, data communication, human interface (terminals, point-of-sale, etc.), computation, and the like. So popular have they become that selecting the right one for a particular application out of the large number available has become a major problem.

As already mentioned, the CPU or microprocessor is responsible for manipulating the data fed into the computer in accordance with its instructions. These instructions are provided in a list or sequence called a program which is stored in the computer memory. Each instruction consists of a group of bits, usually 16, stored in a register called a *memory address.* The addresses of the instructions to be used are stored in proper sequence in the *program counter* (see Figure C-23). The controller gets the address of the next instruction from the program counter, obtains the data stored in

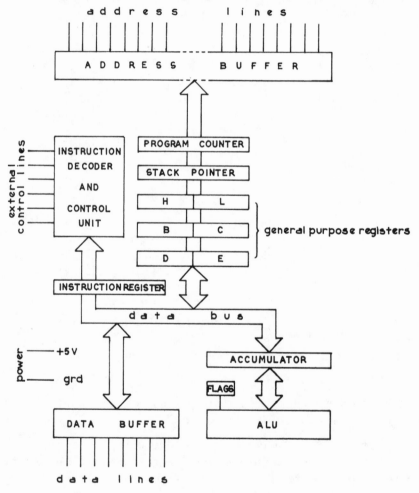

Figure C-23
Central Processing Unit

that memory address, and transfers it to the *instruction register*. This is done by means of three communication channels called the *address bus*, the *control bus*, and the *data bus*. The instruction address in the program counter is placed on the address bus and readies that storage location to yield the instruction data. A signal on the control bus then enables the data to be transferred to the data bus. Another control signal fills it into the instruction register. Here it is held while the controller decodes it and issues further control signals to perform the instruction.

The instruction is usually to do something with the data stored in the *data buffer*, which may have come from an input device or from memory. The instruction may be to perform an arithmetic operation such as the one already described, the result being stored temporarily in the *accumulator*. This is a temporary storage location, used as a kind of "running total" until the operation is completed, after which the result goes back into the memory address specified. Associated with the ALU and accumulator is a set of *condition codes*, also called *status flags*. Each of these is a one-bit register that is set high or low (1 or 0) to indicate something about the result in the accumulator: its sign (+ or −), if it is all zeros, if there is a carry, an overflow, and so on. Five or six flags are the usual number.

Sometimes there is a frequently used subroutine in the program that requires several instructions, always in the same sequence. For added speed, these are stored in adjacent memory addresses and called a *stack*. Instead of each having to be accessed separately, they are addressed as if they were one memory location, and this address is stored in the *stack pointer*. The controller has to use only the single address to call for the entire stack.

The other registers shown in Figure C-23 are called *general purposes registers*, to be used as required. There is an odd way of designating the registers in a controller. The A register is the accumulator, usually called accumulator rather than A register. Then there are the B, C, D, E, etc., registers, plus special-purposes ones such as an H register (high-order byte) and L register (low-order byte), or whatever takes the designer's fancy.

Figure C-23 also indicates a number of external connections. These include a clock, power supply, data input/output, and so on. The clock is a generator for timing pulses with a repetition rate of one or more megahertz for synchronizing operations in the controller so it works properly. The power supply needs no explanation. The other lines into the control unit are all concerned with stopping or interrupting the program for the entry of external data, and restarting it afterwards. Figure C-24 shows how the CPU transfers data between the various locations in the computer system.

The secondary-memory devices that can be connected to the CPU include read-only memories (ROM's), random-access memories (RAM's), charge-coupled devices (CCD's), magnetic tapes, disks, punched cards, and so on. ROM's, RAM's, and CCD's are IC's that can be mounted on a printed-circuit board along with the microprocessor.

A *ROM* is a memory whose contents are permanently fixed beforehand and cannot be

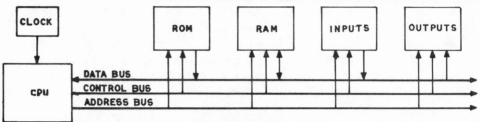

Figure C-24

How the CPU Transfers Data:

(1) *The CPU places the address bits of the memory location input/output device required on the address bus. This opens the communication channel between the CPU and that address.*

(2) *The CPU places the appropriate command voltage on the control bus to read from or write into memory, display or accept input data, as the case may be.*

(3) *The data bits now appear on the data bus, whence they are transferred by the next instruction to one of the CPU registers for manipulation.*

All these operations are kept in step by synchronizing pulses from the clock.

changed by the CPU. It is used for the storage of reference data, microinstructions, various codes and so on. It is nonvolatile, or static, which means it doesn't lose its data when power is removed.

A *RAM* is a memory whose contents can be written in or read out at will by the CPU. It is volatile, or dynamic, so its contents will be lost if power is terminated. (See charge-coupled device for description and illustration.) Magnetic tapes, disks, punched cards, and so on all require mechanisms to operate them that are similar to tape recorders, phonographs, and the like.

Full-sized magnetic tapes are usually 0.5 or 1.0 inches wide and 2400 feet long, with 7 or 9 channels, wound on 10.5-inch diameter reels, and operated by means of a tape transport. Owing to the necessity of running the tape forward and backward to access information, it is a rather slow device taking several minutes to go from one end of the tape to the other. However, it is a cheap way to store permanently large quantities of data (several hundred million bits per reel). *Magnetic tape cassettes* are used also for small-scale operations.

The *bubble memory* to some extent is analogous to the magnetic-tape memory, but on a vastly smaller scale. Instead of having mechanical moving parts, it consists of a garnet chip on which a microscopic "railroad" of Permalloy has been deposited. Permalloy is an alloy of nickel and iron that is easily magnetized and demagnetized. When it is subject to a rotating magnetic field the "rails" and "ties" of the track are magnetized with alternating polarities in such a way that if there were a permanently magnetized "wagon" on the track it would be drawn along it.

To store a data bit a magnetic pulse is applied to the memory input point. This pulse creates in the garnet a tiny permanent magnet or "bubble" perpendicular to the track. This bubble is really a knot or whorl in the internal magnetic field of the garnet, but is just as permanent as an air bubble in the liquid in a glass tube, and is just as free to move. Being weightless, it flies off down the track at high speed, with other bubbles formed by succeeding pulses racing after it.

By clocking the generating pulses the train of bubbles and the gaps between them is spaced to reproduce the sequence of data bits to be stored. This is similar to the sequential storage of data on a moving magnetic tape, except that the tape has been replaced by a moving magnetic field.

Bubble memory technology has some tremendous advantages compared to other systems. These include storage density up to a billion bits per square inch, access time an order of magnitude faster than any other sequential system, and high reliability due to the absence of moving parts. (For further details and diagrams, see *Microelectronics.*)

Magnetic disk memories, unlike phonograph records, do not use a continuous spiral track, but a very large number (200 per inch) of concentric magnetic tracks with the capability of storing 4000 bits per inch. Recording and reading are the same in principle as for magnetic tape, but the pickup is by scanning across the disk, a very much faster method. A new type of disk made of flexible plastic is called a *floppy disk*. It resembles a 45-rpm record, and has a storage capacity of two million bits with a random-access time of less than 500 milliseconds. It is also more economical and easier to ship or store than the older metal disks.

Punched cards are a very widely used input/output medium for digital computers and are the only type of memory (except punched paper tape) that can be altered by hand, or punched or duplicated by machine without using the computer. The standard IBM card has positions for 960 holes (12 rows in 80 columns), and these are punched by a machine at a rate of 100 to 240 cards per minute. This machine can be operated under computer or human control. A card reader (usually the same machine) reads all the holes simultaneously at from 100 to 2000 cards per minute.

The most commonly used *punched tapes* are made of paper or Mylar, and have 5 to 8 hole sites per row. Each row runs across the tape and represents a character. At a hole site, a hole is a 1 bit, no hole is a 0 bit. Reading is done by a wire "brush" making contact through the holes, or photoelectrically which is faster. Speeds of up to 2000 characters per second are obtained in reading, but punching is about ten times as slow.

Computers also supply their information in written form. One output device is an automatic typewriter with a type ball. Another, called a *line printer*, prints an entire line at a time, and can do this at the rate of 1000 lines a minute. It is often used to "dump" the entire contents of a computer memory. Each type of printer uses "endless" concertina-folded paper to produce a printout. The sheets can be separated and bound in bookform and are so used by accountants, planners, and so on. The corresponding input device is a typewriter keyboard (modified for use with a computer), which sends coded electrical impulses to the CPU when a key is depressed. Instead of printing on paper the data may be displayed on a cathode-ray tube or plasma display. Many home computers use a television receiver.

Everyone is familiar with the odd-shaped characters printed on checks to enable them to be read by the bank's computer. These characters are printed with magnetic ink. However, other systems use ordinary printer's ink and the printed characters are read optically. There are also those that read ordinary print by scanning.

Inputs to computers often have to be made from remote terminals, using the telephone line. Digital signals, however, cannot be sent reliably this way because of noises on the line. So a *modem* (modulator/demodulator) is connected at each end of the phone line to convert the computer data into a signal that can be transmitted reliably, and then back to the original digital data system.

computer program—Ordered sequence of instructions to enable a computer to perform a task. *Computer Software.*

computer security—Means of providing against unauthorized access to computer memory, program, etc., by special codes, voice recognition, and so on.

computer software—Programs, data, and stored values that exist as intangible patterns, as opposed to physical hardware. *Computer Software.*

COMPUTER SOFTWARE

As explained under *Computer Hardware,* computer software refers to intangible *patterns* in programs, memories and data, as opposed to physical devices that can be handled. It was also stated that a program was a sequence of instructions to the controller. Some *language*, however, has to be used to write these instructions.

Machine language is the only language computers understand, regardless of their country of origin. This language, of course, consists of the binary-digit patterns that directly control the operation of the functional elements of the controller. A program could be written in machine language and fed directly into the computer, but this could be extremely tedious, time consuming, and prone to errors hard to detect and difficult to correct. Programs, therefore, are generally written in another language and have to be translated into machine language when they are stored in the computer. There are two levels of programming language: assembler language and higher-level language.

Assembler language is intermediate between higher-level and machine language, a sort of "pidgin" language. In assembler language the programmer generally writes one program instruction or constant for each memory address occupied by the program. Each instruction is divided into a few easily understood *fields* (groups). For instance, the command "load" might be written as the mnemonic LDA (depending on which assembly language is being used), which is much simpler than its machine-language equivalent, 00111010. Assembler language is therefore a kind of shorthand, which ignores much of the detail of machine language. When the program in assembler language is ready to be entered into the computer, the computer itself translates the instructions into machine language using a previously loaded assembler program.

Higher-level language is also called *compiler language*. There are many different types, and the compiler (the user) will select the one that best suits his purpose. This language allows him to use ordinary mathematical format, or plain English (or French, German, Russian, etc.), in accordance with fairly simple rules. He does not have to concern himself with the internal affairs of the controller at all. When the higher-level language program is ready, it is run on the computer, which has already been loaded with a *compiler* program, which translates it into assembler language. The assembler-language program is then translated into machine language, as already explained.

There are five classes of higher-level languages:

    (1) Fundamental algorithmic and procedural languages;

    (2) Variants of (1) used for time-sharing and remote-control systems;

    (3) String and list-processing languages;

    (4) Simulation languages;

    (5) Process-control languages.

Fundamental algorithmic and procedural languages are two main types, mathematical and nontechnical business-oriented. Fortran (Formula Translation), Algol (Algorithmic Language), and Jovial (Jules' Own Version of the International Algorithmic Language) are of the first type. Cobol (Common-Business-Oriented Language) is of the second type. It uses precise, easily learned natural words and phrases, and can be used by anyone with minimal training. PL/1 (Program Language No. 1) incorporates the advantages of both Fortran and Cobol.

Time-sharing languages are used wherever there are several terminals using the same computer. The most well-known is BASIC (Beginner's All-purpose Symbolic Instruction Code), commonly used for business and commercial operations. It is less powerful than Fortran and Algol, but more than adequate for most non-scientific requirements. Quicktran is more like Fortran, but with time-sharing characteristics that include storage of lengthy jobs until the computer can handle them. (CAL (Conversational Algebraic Language) is a time-sharing, mathematical-problem-solving language.

List-processing languages are designed mainly for nonnumerical purposes, such as simulation of human-problem-solving, information retrieval, teaching programs, and so on. Snobol (string-oriented Symbolic Language) is mainly used for text editing and linguistics. COMIT was designed for translation of human languages. Lisp (List-Processing) is the basic language in this class. IPL-V is similar to Lisp, but more machine oriented. Slip (Symmetrical List Processing) is an extension of Fortran and is not really a separate language. Another Fortran extension is Formac (Formula Manipulation Compiler), designed strictly for algebraic operations.

Simulation languages are used for programs that formerly would have been done by analog computers. Simscript is based on the principle that if the entities involved are incorporated in the program together with their attributes, then, when the user enters into the computer the scenario for an event, the program will develop the logical sequence of results arising from it. GPSS (General-Purpose System Simulator) is of the same type, developing simulation modes through the use of, and in the terms of, block diagrams. Dynamo is an information feedback program that, from information relating to the state of a system at a given time, can determine its future state.

Process-control languages enable computers to control industrial processes, such as the operation of machine tools. APT (Automatically Programmed Tools) can be used for such purposes as to having a numerical control drafting machine produce detail blueprints while the robot tool is making the part. Similar applications include computer typesetting in conjunction with electronic photocomposition.

Flow Charts

The computer, as yet, cannot exercise judgment or common sense, so it must be meticulously instructed as to how to handle every contingency and every possible exception to anticipated results that could occur. In preparing a program, therefore, the programmer needs a road map that shows every turn and twist and detour that might be encountered. This road map is the flow chart.

A full-scale set of flow charts for an advanced program is a mammoth work, with a large team of programmers working on it for months. The principles involved, however, can be illustrated by a very simple example, as shown in Figure C-25, which gives a flow chart for an algorithm (arithmetical procedure) to sum all the numbers from 1 to 10. The oval symbols at top and bottom are always used to indicate the beginning and end of the program. Rectangles indicate manipulation of data, diamonds are used for tests. Flowlines with arrows show where the program goes next. Arrows inside the rectangles show data manipulation.

condenser—(See capacitor.)

conductance—Ability of a substance to conduct electricty, measured by the ratio of the current to the applied electromotive force, or A/V. Expressed in siemens (S), it is the reciprocal of resistance, or 1/R.

conduction—Transfer of electricity through a conductor (q.v.).

conduction band—Range of energy that freely moving electrons have within the structure of a solid. *Semiconductors.*

FLOW CHART   MEANING OF STEPS

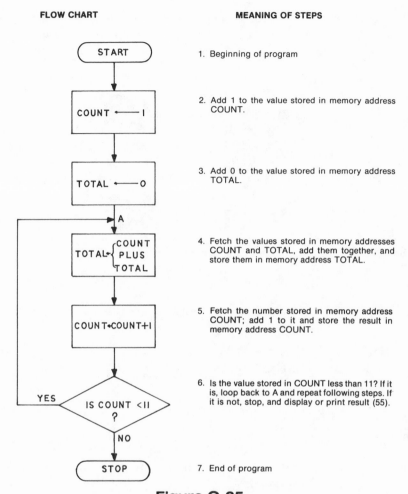

1. Beginning of program

2. Add 1 to the value stored in memory address COUNT.

3. Add 0 to the value stored in memory address TOTAL.

4. Fetch the values stored in memory addresses COUNT and TOTAL, add them together, and store them in memory address TOTAL.

5. Fetch the number stored in memory address COUNT; add 1 to it and store the result in memory address COUNT.

6. Is the value stored in COUNT less than 11? If it is, loop back to A and repeat following steps. If it is not, stop, and display or print result (55).

7. End of program

Figure C-25

Flow Chart Illustrating Program Instructing Computer to Add Together All the Whole Numbers from 1 to 10

conduction electrons—Freely moving electrons in the conduction band of a solid.

conductivity—Reciprocal of resistivity, characterizes materials on how well an electric current flows in them. (See also conductance.)

conductor—Material that will allow electricity to pass readily through it. Most metals are good conductors, silver being the best. (See also wire.)

cone—Speaker diaphragm that changes electrical currents to sound waves. *Electroacoustics.*

connector—Coupling devices for joining two cables, or connecting a cable to a piece of equipment. Connectors are male or female according to whether they plug into or receive the mating connector. (See Figure C-26.)

Figure C-26
Connectors

console—Cabinet for electronic equipment that usually stands on the floor.

constant-current chart—Chart used to illustrate constant-current characteristics of RF amplifiers. *Electron-Tube Circuits.*

constant-current source—Circuit that acts to resist changes in the current despite variations of load.

constant-K filter—Selective network designed to pass signals within a certain frequency range. *Passive Filters.*

constants—Numbers, values, or objects with fixed magnitudes. *Constants.*

CONSTANTS

Quantity	Symbol	Value		SI Units
Velocity of light	c	2.9979250	$\times 10^{8}$	m/s
Electron charge	e	1.6021917	$\times 10^{-19}$	C
Planck's constant	h	6.626196	$\times 10^{-34}$	J s
Avogadro's number	N	6.022169	$\times 10^{26}$	1/kmol
Atomic mass unit	amu	1.660531	$\times 10^{-27}$	kg
Electron rest mass	m_e	9.109558	$\times 10^{-31}$	kg
Proton rest mass	M_p	1.672614	$\times 10^{-27}$	kg
Neutron rest mass	M_n	1.674920	$\times 10^{-27}$	kg
Magnetic flux quantum	Φ_o	2.0678538	$\times 10^{-15}$	T m^2
Faraday's constant, Ne	F	9.648670	$\times 10^{7}$	C/kmol
Rydberg constant	R_∞	1.09737312	$\times 10^{7}$	1/m

CONSTANTS (continued)

Quantity	Symbol	Value		SI Units
Bohr radius	α_0	5.2917715	$\times 10^{-11}$	m
Classical electron radius	r_0	2.817939	$\times 10^{-15}$	m
Bohr magneton	μ_B	9.274096	$\times 10^{-24}$	J/T
Electron magnetic moment	μ_e	9.284851	$\times 10^{-24}$	J/T
Proton magnetic moment	μ_p	1.4106203	$\times 10^{-26}$	J/T
Nuclear magneton	μ_n	5.050951	$\times 10^{-27}$	J/T
Compton wavelength of electron	λ_C	2.4263096	$\times 10^{-12}$	m
Compton wavelength of proton	$\lambda_{C,p}$	1.3214409	$\times 10^{-15}$	m
Compton wavelength of neutron	$\lambda_{C,n}$	1.3196217	$\times 10^{-15}$	m
Boltzman's constant	k	1.380622	$\times 10^{-23}$	J/K
First radiation constant	c_1	4.992579	$\times 10^{-24}$	J m
Second radiation constant	c_2	1.438833	$\times 10^{-2}$	m K
Gravitational constant	G	6.6732	$\times 10^{-11}$	N m^2/kg^2

constant time-delay filter—Active filter to give approximately constant time delay over the range $0 < \omega < \omega_0$. *Active Filters.*

contacts—Switch elements that open or close to interrupt or permit the flow of electric current. Made of brass or copper, often plated with tin, silver or gold, depending upon the usage and reliability desired. When given a coating of mercury, contacts are said to be wetted. Where high voltage is present, contacts are immersed in oil to reduce arcing.

continuous wave (CW)—Unmodulated carrier. In CW transmission the unmodulated carrier is transmitted whenever the operator depresses the key. It is rendered audible at the receiver by beating it with a beat-frequency oscillator (BFO) to obtain an audio-frequency signal. (See also Class A1 emission.)

continuous-wave Doppler radar—Type of radar that can distinguish between moving and fixed targets. *Radar.*

continuous-wave laser—Laser that emits a continuous beam of coherent light, as opposed to pulsed light. *Lasers.*

contrast—Sharp gradations of tone, as between light and dark areas on a television screen.

control—Means of regulating the operation of any equipment; e.g., volume control, tone control, brightness control. Controls may be manual or automatic.

control grid—Electron-tube electrode that regulates the plate current. *Electron Tubes.*

controller—One of the major units in a computer, that which interprets and carries out the instructions in a program. *Computer Hardware.*

convergence—Rays of light or beams of electrons coming together to a point. Used with reference to the intersection of the electron beams in a color picture tube as they pass through the shadow mask. See also color picture tube.

conversion factor—Number used by multiplication to change a unit of measure to other units. *Conversion Factors.*

CONVERSION FACTORS

To convert	to	multiply by		conversely, multiply by	
abampere	ampere	1.00	$\times 10^1$	1.00	$\times 10^{-1}$
abcoulomb	coulomb	1.00	$\times 10^1$	1.00	$\times 10^{-1}$
abfarad	farad	1.00	$\times 10^9$	1.00	$\times 10^{-9}$
abhenry	henry	1.00	$\times 10^{-9}$	1.00	$\times 10^9$
abmho	siemens	1.00	$\times 10^9$	1.00	$\times 10^{-9}$
abohm	ohm	1.00	$\times 10^{-9}$	1.00	$\times 10^9$
abvolt	volt	1.00	$\times 10^{-8}$	1.00	$\times 10^8$
ampere (1948)	ampere	9.998 35	$\times 10^{-1}$	1.000 17	$\times 10^0$
angstrom	meter	1.00	$\times 10^{-10}$	1.00	$\times 10^{10}$
atmosphere	pascal	1.013 25	$\times 10^5$	9.869 23	$\times 10^{-6}$
bar	pascal	1.00	$\times 10^5$	1.00	$\times 10^{-5}$
barn	meter^2	1.00	$\times 10^{-28}$	1.00	$\times 10^{28}$
British Thermal Unit (mean)	joule	1.055 87	$\times 10^3$	9.470 86	$\times 10^{-4}$
calorie (thermochemical)	joule	4.184	$\times 10^0$	2.390	$\times 10^{-1}$
calorie (kilo-gram, thermo-chemical)	joule	4.184	$\times 10^3$	2.390	$\times 10^{-4}$
circular mil	meter^2	5.067 074 8	$\times 10^{-10}$	1.973 525 2	$\times 10^9$
coulomb (1948)	coulomb	9.998 35	$\times 10^{-1}$	1.000 17	$\times 10^0$
cubic foot	meter^3	2.831 69	$\times 10^{-2}$	3.531 46	$\times 10^1$
cubic inch	meter^3	1.638 71	$\times 10^{-5}$	6.102 36	$\times 10^4$
cubic yard	meter^3	7.645 55	$\times 10^{-1}$	1.307 95	$\times 10^0$
curie	becquerel	3.70	$\times 10^{10}$	2.70	$\times 10^{-11}$
degree	radian	1.745 329 2	$\times 10^{-2}$	5.729 577 9	$\times 10^1$
dyne	newton	1.00	$\times 10^{-5}$	1.00	$\times 10^5$
farad (1948)	farad	9.995 05	$\times 10^{-1}$	1.000 50	$\times 10^0$

CONVERSION FACTORS (continued)

To convert	to	multiply by		conversely, multiply by	
faraday (carbon 12)	coulomb	9.648 70	$\times 10^4$	1.036 41	$\times 10^{-5}$
fathom	meter	1.828 8	$\times 10^0$	5.468 07	$\times 10^{-1}$
fermi	meter	1.00	$\times 10^{-15}$	1.00	$\times 10^{15}$
fluid ounce (U.S.)	meter^3	2.957 353	$\times 10^{-5}$	3.381 402	$\times 10^4$
foot	meter	3.048	$\times 10^{-1}$	3.281	$\times 10^0$
foot-candle	lumen/meter^2	1.076 391 0	$\times 10^1$	9.290 304 3	$\times 10^{-2}$
foot-lambert	candela/meter^2	3.426 259	$\times 10^0$	2.918 635	$\times 10^{-1}$
gal	meter/second^2	1.00	$\times 10^{-2}$	1.00	$\times 10^2$
gallon (U.S. liquid)	meter^3	3.785 411 8	$\times 10^{-3}$	2.641 172 1	$\times 10^2$
gamma	tesla	1.00	$\times 10^{-9}$	1.00	$\times 10^9$
gauss	tesla	1.00	$\times 10^{-4}$	1.00	$\times 10^4$
gilbert	ampere turn	7.957 747 2	$\times 10^{-1}$	1.256 637 0	$\times 10^0$
grain	kilogram	6.479 891	$\times 10^{-5}$	1.543 236	$\times 10^4$
henry (1948)	henry	1.000 495	$\times 10^0$	9.995 052	$\times 10^{-1}$
horsepower (electric)	watt	7.46	$\times 10^2$	1.34	$\times 10^{-3}$
inch	meter	2.54	$\times 10^{-2}$	3.94	$\times 10^1$
inch of mercury (32° F)	pascal	3.386 389	$\times 10^3$	2.952 998	$\times 10^{-4}$
joule (1948)	joule	1.000 165	$\times 10^0$	9.998 350	$\times 10^{-1}$
knot	meter/second	5.144 444 4	$\times 10^{-1}$	1.943 844 5	$\times 10^0$
lambert	candela/meter^2	3.183 098 8	$\times 10^3$	3.141 592 7	$\times 10^{-4}$
liter	meter^3	1.00	$\times 10^{-3}$	1.00	$\times 10^3$
lux	lumen/meter^2	1.00	$\times 10^0$	1.00	$\times 10^0$
maxwell	weber	1.00	$\times 10^{-8}$	1.00	$\times 10^8$
mho	siemens	1.00	$\times 10^0$	1.00	$\times 10^0$
micron	meter	1.00	$\times 10^{-6}$	1.00	$\times 10^6$
mil	meter	2.540	$\times 10^{-5}$	3.937	$\times 10^4$
mile (U.S. statute)	meter	1.609 344	$\times 10^3$	6.213 712	$\times 10^{-4}$
mile (nautical)	meter	1.852	$\times 10^3$	5.400	$\times 10^{-4}$
millibar	pascal	1.00	$\times 10^2$	1.00	$\times 10^{-2}$
millimeter of mercury (0° C)	pascal	1.333 224	$\times 10^2$	7.500 615	$\times 10^{-3}$
neper	decibel	8.686	$\times 10^0$	1.151	$\times 10^{-1}$
newton/meter^2	pascal	1.00	$\times 10^0$	1.00	$\times 10^0$
oersted	ampere/meter	7.957 747 2	$\times 10^1$	1.256 637 0	$\times 10^{-2}$
ohm (1948)	ohm	1.000 495	$\times 10^0$	9.995 052	$\times 10^{-1}$
ounce force (avoirdupois)	newton	2.780 138 5	$\times 10^{-1}$	3.596 943 1	$\times 10^0$
ounce mass (avoirdupois)	kilogram	2.834 952 3	$\times 10^{-2}$	3.215 074 6	$\times 10^1$

CONVERSION FACTORS (continued)

To convert	to	multiply by		conversely, multiply by	
pieze	pascal	1.00	$\times 10^3$	1.00	$\times 10^{-3}$
pint (U.S. liquid)	meter^3	4.731 764 7	$\times 10^{-4}$	2.113 376 4	$\times 10^3$
poise	newton second /meter^2	1.00	$\times 10^{-1}$	1.00	$\times 10^1$
pound force (avoirdupois)	newton	4.448 221 6	$\times 10^0$	2.248 089 4	$\times 10^{-1}$
pound mass (avoirdupois)	kilogram	4.535 923 7	$\times 10^{-1}$	2.204 622 6	$\times 10^0$
poundal	newton	1.382 549 5	$\times 10^{-1}$	7.233 014 0	$\times 10^0$
quart (U.S. liquid)	liter	9.463 529 5	$\times 10^{-1}$	1.056 688 2	$\times 10^0$
quart (U.S. liquid)	meter^3	9.463 529 5	$\times 10^{-4}$	1.056 688 2	$\times 10^3$
rad	joule/kilogram	1.00	$\times 10^{-2}$	1.00	$\times 10^2$
rayleigh	1/second meter^2	1.00	$\times 10^{10}$	1.00	$\times 10^{-10}$
roentgen	coulomb/ kilogram	2.579 76	$\times 10^{-4}$	3.876 33	$\times 10^3$
rutherford	becquerel	1.00	$\times 10^6$	1.00	$\times 10^{-6}$
slug	kilogram	1.459 390 3	$\times 10^1$	6.852 176 5	$\times 10^{-2}$
square inch	meter^2	6.452	$\times 10^{-4}$	1.550	$\times 10^3$
square foot	meter^2	9.290	$\times 10^{-2}$	1.076	$\times 10^1$
square yard	meter^2	8.361	$\times 10^{-1}$	1.196	$\times 10^0$
statampere	ampere	3.335 640	$\times 10^{-10}$	2.997 925	$\times 10^9$
statcoulomb	coulomb	3.335 640	$\times 10^{-10}$	2.997 925	$\times 10^9$
statfarad	farad	1.112 650	$\times 10^{-12}$	8.987 554	$\times 10^{11}$
stathenry	henry	8.987 554	$\times 10^{11}$	1.112 650	$\times 10^{-12}$
statmho	siemens	1.112 650	$\times 10^{-12}$	8.987 554	$\times 10^{11}$
statohm	ohm	8.987 554	$\times 10^{11}$	1.112 650	$\times 10^{-12}$
statvolt	volt	2.997 925	$\times 10^2$	3.335 640	$\times 10^{-3}$
stilb	candela/meter^2	1.00	$\times 10^4$	1.00	$\times 10^{-4}$
torr (O°C)	pascal	1.333 22	$\times 10^2$	7.500 64	$\times 10^{-3}$
unit pole	weber	1.256 637	$\times 10^{-7}$	7.957 748	$\times 10^6$
volt (1948)	volt	1.000 330	$\times 10^0$	9.996 701	$\times 10^{-1}$
watt (1948)	watt	1.00 165	$\times 10^0$	9.998 350	$\times 10^{-1}$
yard	meter	9.144	$\times 10^{-1}$	1.094	$\times 10^0$

Temperature conversion formulas

Fahrenheit to Celsius:

$$t_c = \left(\frac{5}{9}\right)(t_f - 32)$$

Celsius to Fahrenheit:

$$t_f = \frac{9t_c}{5} + 32$$

Celsius to Kelvin: $\qquad t_k = t_c + 273.15$

Kelvin to Celsius: $\qquad t_c = t_k - 273.15$

converter—Circuit that converts radio-frequency signals to intermediate-frequency signals, by simultaneously generating a local-oscillator signal and mixing it with the RF signal to obtain heterodyne (sum and difference) frequencies, one of which is selected to be the IF signal. (See Figure C-27.)

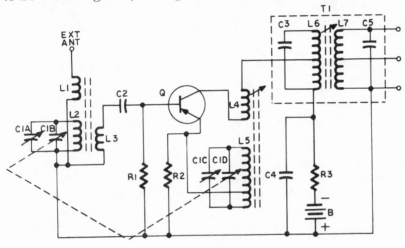

Figure C-27
Converter

From *Complete Guide to Reading Schematic Diagrams,* by John Douglas-Young, First Edition, ©1972, page 123. Reprinted with permission of Parker Publishing Company, Inc.

copper—Reddish metal, extremely ductile, unusually good conductor of electricity, highly resistive to action of the atmosphere and seawater. Alloyed with zinc (brass), tin (bronze), zinc and nickel (nickel "silver"). Symbol, Cu; atomic weight, 63.546; atomic number, 29.

copper oxide—Cuprous oxide (Cu_2O) is used as a semiconductor rectifier and as a catalyst for combustion of automobile exhaust gases.

cord—Flexible, insulated conductor.

cordwood module—Precursor of integrated circuits, in which miniature discrete components are mounted vertically between an upper and a lower board, and encapsulated to form a cube with all leads projecting from its base for mounting on a printed-circuit board.

core—Magnetic material placed within and around a coil to provide a path of lower reluctance for magnetic flux.

core loss—Sum of hysteresis and eddy-current loss (q.q.v.) in a magnetic core.

core memory—Banks of small ferrite rings, wired into a matrix, that can be switched between the two states of retentivity, 1 and 0. No power is needed to maintain the magnetic field, so information is not lost when power is off. Superseded by RAM's and ROM's (q.v.). *Computer Hardware.*

corner reflector—Reflector consisting of two square grids mounted at an angle to each other behind a bowtie antenna to increase antenna gain. *Antennas.*

corona—Faint glow appearing around an electrical conductor at high voltage, due to ionization of the air.

cosmic noise—Galactic noise (q.v.). (See also Cassiopeia noise.)

cosmic rays—High energy particles reaching the Earth from all directions from outer space. They consist of atomic nuclei mostly of hydrogen (protons), the majority of which are believed to originate from remnants of supernova explosions in the Galaxy.

COS/MOS—(See complementary MOS.)

coulomb—Unit of electric charge that is transferred each second by an electric current of one ampere, and approximately equivalent to 6.24×10^{18} electrons. *S. I. Units.*

counter—1. Electronic instrument that accumulates input pulses in binary form and then converts them to decimal form for a digital readout. The heart of a counter is a set of decade counting units (q.v.). 2. Radiation counter used to count ionizing particles. (See also Cerenkov, Geiger, and scintillation counters.) *Nuclear Physics.*

coupling capacitor—Capacitor used to couple two stages or circuits.

coupling coefficient—Degree of coupling between two circuits. In coils the coefficient of coupling (K) is given by:

$$K = M/(L_1 L_2)^{\frac{1}{2}}$$

where M is the mutual inductance value (q.v.), and L_1 and L_2 the self-inductances of the coils.

covalent bond—Type of linkage between two atoms arising from the electrostatic attraction of their nuclei for the same electrons. In common usage the term refers to the electron-pair bond in which each of the bonded atoms contributes an electron. *Semiconductors.*

cps—Cycles per second. (See hertz.)

CPU—Central processing unit (q.v.). *Computer Hardware.*

critical frequency—Highest frequency at which a vertically incident wave will be reflected by an ionospheric layer. *Electromagnetic-Wave Propagation.*

crossover distortion—Distortion in a push-pull amplifier, where the output signal goes through the zero reference level with a jog. The two half-cycles do not coincide exactly at that point.

crossover network—Speaker filter circuit that routes low frequencies to the woofer and high frequencies to the tweeter.

cross talk—Coupling of the signal in one channel into another.

crowbar—Fast-acting protection of circuit components against fault voltages and currents.

CRT—Abbreviation for cathode-ray tube (q.v.).

cryogenics—Deals with the production of low temperatures and the utilization of low-temperature phenomena. The cryogenic temperature range is from –150°C down to –273°C (absolute zero). At very low temperatures some materials become superconductive. The temperature at which their superconductivity begins is called the critical or transition temperature.

crystal—Solids are generally classified as crystalline or amorphous. In the crystalline solid its molecules are arranged in a geometrical, regular, three-dimensional lattice. This lattice is reflected by the number and orientation of external surfaces, or crystal faces, and by the crystal symmetry. Some crystals exhibit piezoelectricity (q.v.), which makes them useful as transducers, oscillators, and filters.

crystal diode—Semiconductor diode. *Semiconductors.*

crystal oscillator—Oscillator in which the frequency is largely determined by a crystal. (See Figure C-28.)

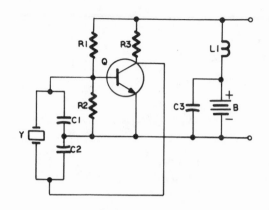

Figure C-28
Crystal Oscillator (Colpitts Type)

From *Complete Guide to Reading Schematic Diagrams*, First Edition, page
86. ©1972. Reprinted with permission of Parker Publishing Company, Inc.

crystal oven—Container for a crystal, with a heating element and thermostat, that keeps the crystal at a constant temperature to maintain its oscillation frequency stability.

crystal set—Simple receiver with no amplifying stages that detects (demodulates) AM broadcast signals with a crystal diode.

Curie point—Temperature at which certain magnetic materials undergo a sharp change in their magnetic properties, so that only a weak magnetic behavior (paramagnetism) remains. The Curie point for magnetite is 570°C; iron, 750°C; cobalt, 1120°C.

current—Any movement of electric charge carriers (electrons, protons, ions, or holes). Electric current in a wire (by electrons) is from negative to positive, although in some contexts the "conventional" direction (positive to negative) is used. Direct current flows in one direction only; alternating current periodically reverses direction. Electric current always generates an accompanying magnetic field. (See also ampere.)

current probe—Clip-on probe that converts the magnetic field around a conductor to an alternating voltage proportional to the current in the conductor, thus avoiding cutting the wire to insert an ammeter. Current probes are made for both d.c. and a.c.

curve tracer—Type of oscilloscope capable of displaying one or more characteristic curves of two-or-three-terminal devices and IC's.

cutoff—1. Minimum bias required to stop the flow of current through an electron tube or other device.   2. Frequency above or below which a low- or high-pass filter fails to respond; denoted f_c, that point on the response curve where the amplitude response has decreased to 1/2 of its maximum value.

CW—Abbreviation for continuous wave (q.v.).

cybernetics—Study of control and communication in machines and physiological systems.

cycle—One complete period of the reversal of an alternating current from positive to negative and back again.

cyclotron—Device for producing high-energy particles of various kinds, including electrons, protons, alpha particles, and heavy ions, with energies on the order of 20-25 megaelectronvolts. *Nuclear Physics*.

Czochralski process—Method of growing a single crystal of semiconductor material by slowly withdrawing a seed crystal from a melt, while the melt is held slightly above the melting point of the material. *Microelectronics*.

D—1. Symbol for deuterium (q.v.).   2. Class designation letter for crystal diode or breakdown diode.

d—Symbol for day.

d-a converter—Device for converting digital data to analog data.

damped wave—Wave in which each successive cycle decreases in amplitude with respect to its predecessor.

damper—Diode used in horizontal output circuit of television receiver to suppress oscillations (after the first sharp positive alternation that is required to reshape the deflection field), which otherwise would ride on the deflection-current sawtooth and cause vertical bars to appear in the picture.

damping—1. In resonant circuits, the decay of oscillations due to the resistance in the circuit.   2. In a meter, the manner in which the pointer comes to rest after a change in the value of the measured quantity. Periodic damping is where the pointer oscillates about the final position before coming to rest. Aperiodic, or overdamping, is where there is no oscillation or overshoot of the pointer.

damping factor—Ratio of any one amplitude to that next succeeding it in the same sense of direction.

Daniell cell—Primary cell with a constant electromotive force of about 1.1 volts, having as its electrodes copper in a copper sulfate solution and zinc in dilute sulfuric acid or zinc sulfate, the two solutions being separated by a porous partition. (See Figure D-1.)

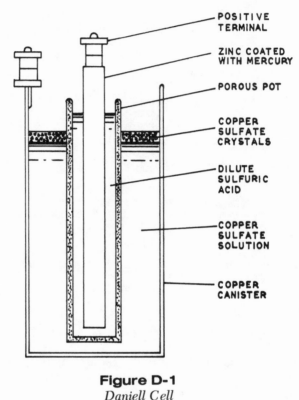

POSITIVE
TERMINAL

ZINC COATED
WITH MERCURY

POROUS POT

COPPER
SULFATE
CRYSTALS

DILUTE
SULFURIC
ACID

COPPER
SULFATE
SOLUTION

COPPER
CANISTER

Figure D-1
Daniell Cell

dark conduction—Residual electrical conduction in a photo-sensitive substance when not illuminated.

darlington amplifier—Pair of transistors in which the collectors are tied together and the emitter of the first transistor is directly coupled to the base of the second. Overall current gain is equal to the product of the separate gains of the two transistors and can be several thousand times. (See Figure D-2.)

D'Arsonval movement—(See permanent-magnet moving-coil movement.)

data—1. Things known or assumed. Singular form is datum. 2. Basic elements of information stored in memory or being processed in a computer.

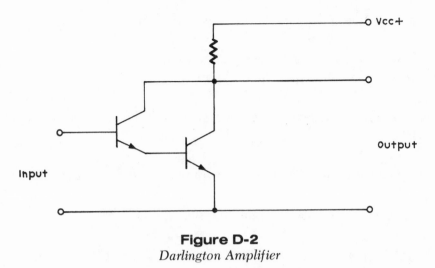

Figure D-2
Darlington Amplifier

data display—Visual presentation of processed data by cathode-ray tube or plasma display.

data link—Communication channel for digital data.

data processing—Operations involving the handling of data, usually in accordance with logical rules, in a series of discrete steps.

dataset—(See modem.)

dB—Abbreviation for decibel (q.v.).

dBa—Abbreviation for decibels adjusted, with reference to −90 dBm.

dBm—Abbreviation for decibels above or below one milliwatt. 0 dBm=1 m W across a 50-ohm circuit.

dB meter—Meter with scale marked in decibels, used in audio-frequency measurements.

d.c.—Abbreviation for direct current. Also dc, DC.

dc amplifier—(See direct-coupled amplifier.)

d.c. beta—D.c. current gain of transistor.

d.c.c.—Abbreviation for double cotton-covered (wire).

d.c. component—If pulsating d.c. is considered as consisting of two separate voltages mixed together—a.c. and d.c.—the d.c. part is the d.c. component.

d.c. restoration—In a video amplifier, reinsertion of a d.c. brightness level after it has been lost.

DCTL—Abbreviation for direct-coupled transistor logic. *Microelectronics.*

DCU—Abbreviation for decade counting unit (q.v.).

D-display—Radar display in which vertical position of blip gives target's elevation, and horizontal position its azimuth. Coarse range information is also provided by vertical position of blip in broad azimuthal trace. *Radar.*

deadbeat—Overdamped.

dead reckoning—Determining the position of an aircraft or ship by deduction from the record of courses flown or sailed, the distance made, and the known or estimated drift.

dead space—When a receiver is tuned through the frequency at which a code signal is being transmitted, that point at which the beat-frequency oscillator is silent (zero beat).

debugging—Procedure for detecting errors in a computer program.

decade box—Test equipment accessory consisting of a box with several rotary switches, each having ten positions, and a set of precision resistors, capacitors, or inductors, as the case may be, that are selectable by the switch in combinations to give from 0 to 10 times the value at position 1. Each set of components is ten times the value of the set on the next switch below it, and vice versa, so any value can be selected, by appropriate switch settings, from the total range available.

decade counting unit (DCU)—Counting device consisting of cascaded flip-flops that count from 0 through 9, and then recycle. Each figure displayed in the counter's digital readout reflects the state of its associated DCU.

decametric waves—High-frequency waves with wavelengths of 10 to 100 meters (3 to 30 MHz).

decay constant—The radioactive decay constant is the fraction of nuclei of a radioactive material disintegrating in unit time. *Nuclear Physics.*

Decca—Aviation navigation system that operates on low frequency with a range of 200-300 miles. *Navigation Aids.*

decibel (dB)—The decibel is used to express the ratio between two amounts of power, P_1 and P_2, existing at two points, according to the formula:

$$dB = 10 \log_{10} (P_1/P_2)$$

It is also used to express current and voltage ratios: $dB = 20 \log_{10} (V_1/V_2) = 20 \log_{10} (I_1/I_2)$. Also used to express the relative loudness of sounds. *Electroacoustics.*

decimal number system—Positional number system with ten numerals (0, 1, 2, 3, 4, 5, 6, 7, 8, 9) and a dot (decimal point). The numerals are positioned according to powers of ten; e.g., the number 543.21 can be shown as:

$$
\begin{array}{ll}
5 \times 10^2 & (= 500) \\
4 \times 10^1 & (= 40) \\
3 \times 10^0 & (= 3) \\
2 \times 10^{-1} & (= .2) \\
1 \times 10^{-2} & (= .01) \\
\hline
& 543.21
\end{array}
$$

decimetric waves—Ultra-high frequency (UHF) waves with wave-lengths of 1 to 10 decimeters (0.3 to 3 GHz).

decimillimetric waves—Microwaves with wavelengths of 0.1 to 1 millimeter (0.3 to 3 THz).

decoder—1. Arrangement of gates used to convert data from one code to another. Also called coder, translator, or matrix. *Computer Hardware.* 2. Device for decoding coded signals. 3. Device for converting physical signals into symbols suitable for use by the recipient; e.g., speaker, teleprinter, CRT. Also termed signal-to-recipient decoder.

decoupling—Preventing transfer or feedback of energy from one circuit to another.

Dectra—Adaptation of Decca (q.v.).

dedicated—Set apart for some special use. A curve tracer is an oscilloscope dedicated to displaying the characteristic curves of transistors, with special circuitry to facilitate this, and therefore not designed for other uses.

dee—Hollow metal electrode in a cyclotron (q.v.).

de-emphasis—Signal-to-noise improvement in FM systems of the high-frequency end of the passband, accomplished by passing the modulating signal at the transmitter through a pre-emphasis network, and then passing the output of the discriminator or ratio detector in the receiver through a de-emphasis network to restore the original signal-power distribution. (See also Dolby system.) (See Figure D-3.)

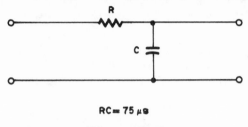

Figure D-3
De-Emphasis Circuit

definite integral—The definite integral of f(x) is the difference between two values of the integral of f(x) for two distinct values of the variable x, usually represented by the notation

$$\int_{x_0}^{x_1} f(x)dx$$

deflection—Movement of the scanning beams across and down the fluorescent screen of a picture tube, or angular displacement of the electron beam in a cathode-ray tube.

deflection factor—In an oscilloscope, the amplitude of input signal required to produce an amplitude of one graticule division on the screen when the volts/div control is set to its lowest value (minimal attenuation).

deflection plates—Two pairs of plates in a cathode-ray tube, to which potentials are applied that deflect the electron beam vertically and horizontally. This method is called electrostatic deflection.

deflection sensitivity—In an oscilloscope, the number of millimeters of screen deflection per volt of deflection potential.

deflection yoke—Assembly of two pairs of coils placed externally around the neck of a picture tube to deflect the electron beams horizontally and vertically. This method is called electromagnetic deflection.

degeneration—(See feedback.)

delayed sweep—In an oscilloscope, a sweep that has been delayed either by a predetermined period or by a period determined by an additional independent variable.

delay line—Type of transmission line used to delay a signal for a fixed time (e.g., the Y signal in a color TV receiver must be delayed about $0.9\mu s$ to arrive at the matrix simultaneously with the chrominance signals). Delay lines are of two types: *lumped parameter* (see Figure D-4), consisting of cascaded L C networks, as used in color TV receivers; and *distributed* (see Figure D-5), a special type of coaxial cable, used in oscilloscopes.

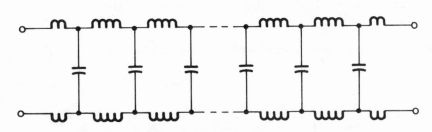

Figure D-4
Lumped Parameter Delay Line

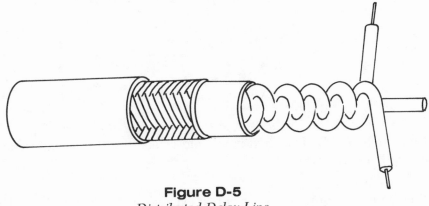

Figure D-5
Distributed Delay Line

delta—Greek letter Δ (cap) or δ (small), used to denote increment or decrement (cap or small), determinant (cap), permittivity (cap), density, angles.

delta-connected system—Three-phase system used in power distribution lines and low-voltage transmission lines. (See Figure D-6.)

Deltamax—Magnetic core material composed of 45-50 percent nickel, the remainder iron.

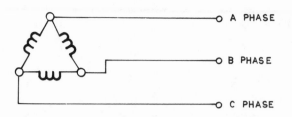

Figure D-6
Delta-Connected System

demodulation—Process of separating modulating signal from modulated carrier.

demodulator probe—Oscilloscope probe for displaying the modulating signal on a high-frequency carrier. (See Figure D-7.)

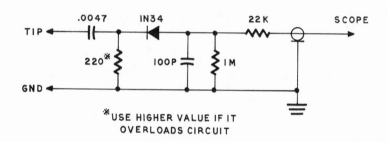

Figure D-7
Demodulator Probe

depletion mode FET—Insulated-gate field-effect transistor that is normally *on* (conducting). *Transistors.*

depletion region—Dipole charge layer that exists near the boundary between p and n regions in a semiconductor, and produces the potential barrier to keep the mobile charges in place. *Semiconductors.*

depolarizer—Chemical used in some primary cells to prevent formation of hydrogen bubbles on the positive electrode. The layer of bubbles would interfere with the flow of current in the cell.

deposition—Process of applying a material to a base (substrate) by means of vacuum, electrical, chemical, screening, or vapor methods. *Microelectronics.*

derivative—Rate of change, or instantaneous velocity, of a function with respect to a variable.

destructive readout—Memory where stored information must be written back in immediately after it is used, or it will be lost.

detection—(See demodulation.)

detector probe—(See demodulator probe.)

deuterium—Isotope of hydrogen. Symbol, D; atomic weight, approximately 2. Heavy water is deuterium oxide (D_2O).

diac—Two-terminal transistor-like semiconductor device that exhibits bistable switching for either polarity of a suitably high applied voltage. Used mostly in conjunction with a triac (q.v.) for a.c. phase-control circuits (light dimmers, motor-speed control, etc.). *Semiconductors.*

dial pulse—Signal consisting of series of from 1 to 10 pulses produced when dialing a telephone. *Telephone.*

dial tone—Continuous signal (350 or 440 Hz) generated at local telephone exchange to tell caller to dial desired number.

diamagnetism—The property of any substance subjected to an external magnetic field, whereby the electrons orbiting in its atoms are speeded up or slowed down in such a way as to oppose the external field, in accordance with Lenz's law (q.v.). In some substances this effect is masked by a weak magnetic attraction (paramagnetism), or a very strong attraction (ferromagnetism).

diathermy—Use of radio-frequency heating to relieve muscle soreness and pain, and to destroy or localize bacterial infection of tissues. Frequencies range from ultrasonic to microwave and may produce horizontal herringbone interference patterns on a TV picture.

dichroic mirror—Used in color television to reflect one color while allowing others to pass through.

die (plural, dice)—Single component or device separated from a silicon wafer. Also called a chip.

die bonding—Method of attaching a die or chip to its package.

dielectric—Substance used between the plates of a capacitor to increase its capacitance. *Capacitors.*

difference channel—FM stereophonic subchannel. *Broadcasting.*

differential—The differential of a function equals the derivative of the function multiplied by the differential of the independent variable.

differential amplifier—Amplifier consisting of two identical transistors or electron tubes connected to respond to the difference between two input signals while blocking identical input signals. *Microelectronics, Electron Tube Circuits, Transistor Circuits.*

differentiator—RC circuit with short time constant that develops a sharp positive and negative pulse from a single rectangular pulse. (See Figure D-8.)

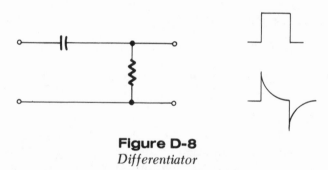

Figure D-8
Differentiator

From *Complete Guide to Reading Schematic Diagrams,* First Edition, page 189. ©1972. Reprinted with permission of Parker Publishing Company, Inc.

diffraction—Spreading of waves around obstacles, most pronounced when the wavelength is comparable to the linear dimensions of the obstacle.

diffusion—1. Thermal process by which impurities are deliberately introduced into a material, such as phosphorus into silicon. *Microelectronics.* 2. Process of spreading out of charge carriers from regions of excess density in a semiconductor. *Semiconductors.*

digit—1. Any of the ten Hindu-Arabic figures from 0 to 9. 2. Either of the numerals 1 or 0 used in binary arithmetic (binary digit, or bit).

digital circuit—Transistor switch that goes from off to on (cut off to saturated), without remaining for any appreciable time at any point between. The two states are usually denoted by the binary digits 0 and 1.

digital computer—Calculating and data processing machine designed to operate on numbers in their binary form, using digital circuits. *Computer Hardware.*

digital multimeter (DMM)—Multimeter (q.v.) with digital readout.

digital readout—Means of displaying a value by decimal figures, as opposed to using a dial and pointer.

digital-to-analog converter—Unit that changes a digital signal to an analog signal; i.e., a binary input to a corresponding voltage or current output.

diode—Two-terminal device that conducts more readily in one direction than the other. *Semiconductors, Electron Tubes.*

diode detector—Circuit with diode used to demodulate a modulated radio-frequency signal. (See Figure D-9.)

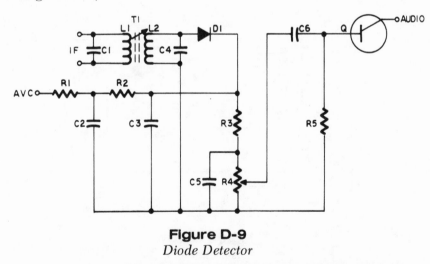

Figure D-9
Diode Detector

dipole—1. Antenna consisting of two aligned conductors, slightly separated at the center for connecting a transmission line, and approximately half a wavelength long overall. *Antennas.* 2. Pair of equal and opposite electric charges, the center of which are not coincident (electric dipole). 3. Tiny magnet of microscopic to subatomic dimensions, equivalent to a current of electric charge flowing in the form of a loop (magnetic dipole).

direct-coupled amplifier—Amplifier in which the output of one stage is connected to the input of the next stage without using intervening coupling components. *Microelectronics.*

direct current—Flow of electric charges that does not change direction. Abbreviated d.c.

direction finder—Rotatable-loop-antenna system on aircraft that measures angle between aircraft axis and direction of ground station. Also called radio compass or airborne direction finder (ADF). *Navigation Aids.*

director—Parasitic antenna element. *Antennas.*

discone antenna—Antenna consisting of a cone beneath a horizontal disk. *Antennas.*

discrete component—Separate and individual component, such as a resistor or capacitor, not part of an integrated circuit.

discriminator—Type of FM demodulator circuit. (See Figure D-10.)

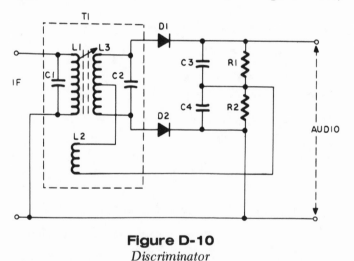

Figure D-10
Discriminator

From *Complete Guide to Reading Schematic Diagrams*, First Edition, page
139. ©1972. Reprinted with permission of Parker Publishing Company, Inc.

dish antenna—Paraboloid reflector antenna. *Antennas.*

diskette—(See floppy disk.)

display—Visual presentation of data, such as a digital readout, cathode-ray tube screen, plasma display, and so on.

dissipation factor (D)—Reciprocal of storage factor (Q), given by R/X, or tan δ, where R is the resistive element and X the reactive element of the impedance Z.

distortion—Departure of waveform from ideal shape. (See harmonic distortion, intermodulation distortion, phase distortion, or transient distortion.) *Distortion.*

DISTORTION

Distortion is any change in a signal that alters its basic waveform, or the relationship between various frequency components. It usually degrades the quality of the signal, making music "tinny" or "boomy," and speech less intelligible, as if the speaker "had his mouth full of mush." Distortion in a video signal results in smeared pictures, loss of fine detail, ghosts, or wrong colors. There are four main types of distortion: *frequency, phase, amplitude*, and *intermodulation*.

Frequency distortion results if a receiver or amplifier does not pass all signal frequencies equally. In the audio range this produces sounds lacking proper balance between bass and treble. In a TV receiver the picture loses low frequencies (which also include sync signals) or fine detail.

Phase distortion occurs mostly in equipment using coils and capacitors. There is always some phase shift in the transmission of different frequencies through reactive circuits, so that various frequencies applied simultaneously at the input may be advanced or delayed in time, and appear with changed phase relationships at the output. This is of little importance at audio frequencies but can cause wrong colors to appear in the TV picture. It can be a very serious problem in pulse signals.

Amplitude distortion is where the positive or negative peaks (or both) of a signal are flattened or clipped because of some nonlinearity in amplification. Whenever the shape of a wave is changed in this manner the harmonic content of the original signal is changed, so this form of distortion is also called *harmonic distortion*. It is the type of distortion most noticeable to the human ear, and therefore the leading complaint in the audio range. In TV receivers this problem frequently causes poor synchronization or linearity.

Harmonic distortion is measured with a distortion analyzer, which applies a distortionless signal to the amplifier input. The analyzer first measures at the amplifier output the amplitude of the fundamental frequency, all the harmonic components, and the noise present. It then employs a rejection filter to remove the fundamental, and measures the residual output. The ratio of the two measurements gives the total harmonic distortion when substituted in the following formula:

$$\text{THD} = \frac{\sqrt{\Sigma[(\text{harmonics})^2 + (\text{noise})^2]}}{\sqrt{\Sigma[(\text{fundamental})^2 + (\text{harmonics})^2 + (\text{noise})^2]}}$$

Intermodulation distortion arises when a heterodyne or audio beating effect produces appreciable sum and difference frequency components within a signal. This beating produces appreciable sum and difference frequency components which are not harmonically related to the original frequencies, so it is not amenable to harmonic analysis as in the case of amplitude distortion, but it has similar objectionable effects on sound reproduction.

Localizing distortion is performed by using the same general procedures as for other defects. The complaint must be verified by a performance test, and a visual inspection made for surface defects, after which standard trouble-shooting methods should be employed to isolate the problem to a section, stage, circuit, and part.

Sometimes the owner of the equipment confuses interference caused by external causes with distortion. It is important to make sure that this is not the case before starting to look for a problem that is not there. Modern receivers are very selective, but satisfactory reception of a favored station may be spoilt by the transmission of another station close by or more powerful. This can usually be dealt with by a change in orientation of the set, if its internal antenna is directive, or the installation of a directive external antenna.

An inspection for surface defects is usually not very revealing, because most distortion arises from causes that are more subtle than gross defects of the kind that cause a complete stoppage. However, if the unit employs electron tubes, these should be tested and replaced where necessary. The speaker should also be examined to see if the cone has been torn or crushed. Another defect that is sometimes found is a worn volume control. The owner of the equipment may consider its erratic operation to be distortion, although it would really cause noisy, intermittent reception.

If the foregoing "eyeballing" does not reveal the cause of the problem, it will be necessary to resort to section and stage isolation procedures. For convenience, any receiver, whether television or radio, can be considered to consist of two main sections: the audio (or video) section, which includes everything between the output of the second detector and the loudspeaker (or picture tube); and the r.f. section, which includes everything between the antenna and the first audio (or video) stage. There are two ways of isolating the defective stage: the first is to clear one of them of any fault, and the second is to find the one that is at fault.

The best way to do this with a stereo system is to play a record or tape. If there is no distortion, the amplifier is not at fault. If there is distortion when playing the AM or FM radio, the problem is in either the AM or FM r.f. section. On the other hand, if the distortion is the same when playing a record or tape, then the amplifier is to blame. Play each channel separately to see which one is responsible for the distortion.

In a simple radio receiver, the output of the second detector should be connected to the input of a signal tracer. If one is not available, another receiver's audio section can be used, or the amplifier of a stereo system. The radio being tested is tuned to a broadcast station and the output sound compared by turning the volume control down while that on the other unit is turned up, and vice versa. An exactly similar procedure can be used with two television receivers. If the signal is clear and free from distortion at the output of the second detector, the r.f. section must be good, and the audio section must be at fault. In a TV receiver, the same applies if the picture is now satisfactory: the r.f. section is good, so the video section must be at fault.

To locate the audio stage responsible for the distortion, first disconnect the speaker, and in its place connect a resistor with the same impedance and power-handling capability. In a stereo system, turn the balance control all the way toward the channel being tested. Then, if using a signal tracer or other audio amplifier, transfer its probe or input lead to the output of the first audio stage in the section being tested. If distortion is now heard, it must be originating in this stage. If the sound is still undistorted, transfer the connection to the input of the following stage. If distortion is heard here, it indicates a leaky coupling capacitor. However, if no distortion is heard, transfer the connection to the output of the stage, as before.

If no distortion is located in any stage (including across the speaker terminals), the resistor should be disconnected and the speaker reconnected. If the distortion returns, the trouble must be in the speaker itself, including its cross-over network.

A more sophisticated method requires the use of an audio signal generator and an oscilloscope. The type of signal generator to use depends upon the type of distortion suspected. A sine-wave generator will reveal amplitude distortion, a square-wave generator frequency distortion. A dual-channel oscilloscope is preferable, but not essential. The test equipment is connected as shown in Figure D-11.

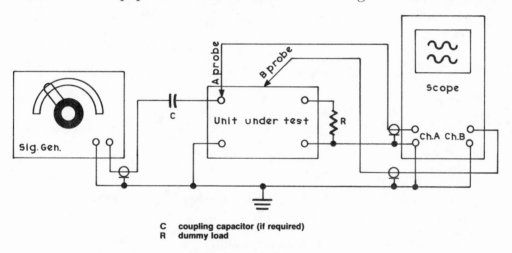

C   coupling capacitor (if required)
R   dummy load

Figure D-11
Connections for Sine or Square-Wave Tests

When using the sine-wave generator it should be set to a frequency of 1 kHz. The oscilloscope sweep should be adjusted to show two or three cycles of the waveform. With one channel connected to the input of the amplifier, the other should be connected in turn to the same points as were used in the signal-tracer procedure. The vertical attenuators of the oscilloscope should be readjusted each time to maintain waveforms of equal amplitude, while the signal generator voltage should be kept as low as is consistent with obtaining a useable signal so as to avoid overloading the amplifier. As soon as distortion is observed, the defective stage has been located. Patterns of distortion and their probable causes are shown in Figure D-12.

	PROBABLE CAUSES
 Normal waveform	Circuit operation normal
 Negative peaks clipped	*With PNP transistor:* low bias; battery voltage low; weak transistor; high-value load resistor. *With NPN transistor:* high bias; weak or high resistance battery; leaky transistor; leaky coupling capacitor; low-value load resistor.
 Positive peaks clipped	*With PNP transistor:* high bias; weak or high resistance battery; leaky transistor; leaky coupling capacitor; low-value load resistor. *With NPN transistor:* low bias; battery voltage low; weak transistor; high-value load resistor.
 Both peaks clipped	Overdrive; weak battery; weak transistor.

Figure D-12

Various Types of Sine-Wave Distortion

The square-wave generator is used in a similar manner. Patterns of distortion and their probable causes are shown in Figure D-13.

If the preliminary tests showed that the trouble was in the r.f. section, the same procedures used in the audio section should be applied in this section also. If an r.f. signal tracer is available, the receiver (if a radio) should be tuned to a station that causes the loudspeaker output to sound distorted. The volume should then be turned down, and the signal tracer probe connected to the input of the first stage, and the tracer tuned to the signal. The signal is then sampled at the input and output of each stage, working toward the second detector. When going from the mixer input to the mixer output the signal tracer must be retuned to the i.f. The defective stage reveals itself when the distortion is heard. If no distortion is heard in any r.-f. or i.-f. stage, a check should be made at the second detector output, with the audio section of the tracer, to see if detection has introduced distortion.

The oscilloscope and signal generator procedure may also be employed, using an r.f. signal generator. If the oscilloscope frequency is not high enough, an r.f. probe must

WAVEFORM	INPUT			PROBABLE CAUSES
	LF	MF	HF	
⊓‾⊔	X	X	X	Circuit operation normal
	X	X		Poor low-frequency response
			X	Poor medium-frequency response
	X	X	X	Partially open coupling or emitter bypass capacitor
		X	X	Phase shift
	X	X		Extremely poor low-frequency response
		X	X	Extremely poor medium-frequency response
	X	X		Open coupling capacitor
			X	Open interstage transformer
			X	Open transistor
	X			Accentuated low-frequency response
		X		Accentuated medium-frequency response
			X	Accentuated high-frequency response, causing phase shift
	X	X		Partially-open bypass capacitor
	X			Defective decoupling filter
		X		Defective filter capacitor
		X		Defect in feedback network
			X	Defect in frequency-compensation network
			X	Changed value of loading resistor (video amplifier)
	X	X	X	Accentuated response at fundamental frequency of input
	X			Changed value of coupling or bypass capacitor
	X	X		Defective interstage transformer
	X	X	X	Defective feedback network
			X	Poor lead dress
		X		Partially-open bypass capacitor
			X	Open bypass or decoupling capacitor
			X	Defective shielding
	X			Poor medium-frequency response
		X	X	Poor high-frequency response
	X	X	X	Excessive distributed capacitance
	X	X		Defective interstage transformer
			X	Increased value of load resistor
			X	Open peaking coil (video amplifier)
	X			Peak in medium-frequency range
		X		Peak in high-frequency range
			X	Peak in frequency response beyond normal range
	X	X		Changed value of bypass capacitor
	X	X		Defective interstage transformer
	X	X	X	Poor shielding
	X	X	X	Poor lead dress
	X	X		Defect in feedback network
	X	X	X	Defective bypass or decoupling capacitor
		X	X	Open load resistor (video amplifier)
			X	Wrong value or misadjusted peaking coil
			X	Improperly loaded interstage transformer

Figure D-13

Square Wave Distortion Reference Chart
for Various Input Frequencies

be used with a modulated r.f. signal. The signal is sampled as just described, and the defective stage is identified by the point at which distortion is observed.

Phase distortion may be examined with an oscilloscope and signal generator connected as in Figure D-11, using both channels of a dual-channel instrument. The phase difference will appear on the screen as shown in Figure D-14. By using different frequencies it can be ascertained if the phase shift is the same over the amplifier bandwidth, or if it varies with frequency. In the latter case, phase distortion is occurring. If the oscilloscope is a single-channel instrument, phase shift can be displayed by connecting the Y input to the input of the amplifier and the X input

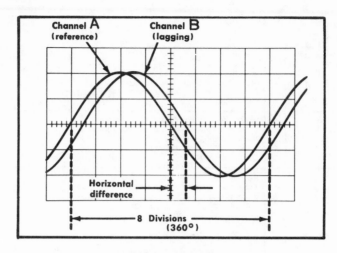

Figure D-14
Measuring Phase Shift with a Dual-Channel Scope

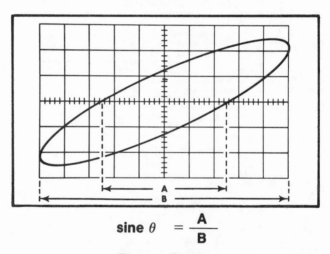

$$\text{sine } \theta \;=\; \frac{A}{B}$$

Figure D-15
Measuring Phase Shift with a Single-Channel Scope

(external horizontal input) to the test point. A Lissajous figure similar to that in Figure D-15 will result.

Intermodulation distortion can be determined by applying two sine waves to the amplifier at the same time, for example 400 and 1500 Hz. By connecting a suitable filter at the output to suppress all frequencies except that of the difference signal (1100 Hz), the amplitude of the induced intermodulation can be measured. If this is less than one percent of the total output it is generally considered satisfactory.

distress frequency—Frequency reserved for distress calls. Telegraph calling frequency is 500 kHz and 156.8 MHz; telephone calling frequencies are 2182 kHz and 156.8 MHz. Survival craft and equipment use 243.0 MHz.

distributed capacitance—In any circuit, capacitance that exists independently of capacitors, within or between components, conductors, etc., e.g., capacitance between turns in an inductor, or between conductive pattern and substrate in a microelectronic circuit.

distributed delay line—(See delay lines.)

disturbance—Noise or interference from any cause that degrades signal transmission. *Electromagnetic-Wave Propagation.*

diversity—Method of using alternate channels at different frequencies, or with different antennas, to overcome fading problems in high-frequency signaling. *Electromagnetic-Wave Propagation.*

D layer—Lowest layer of the ionosphere (50-90 km), existing only during daylight, attenuating high-frequency waves, absorbing medium-frequency waves, and reflecting low-frequency waves. *Electromagnetic-Wave Propagation.*

Dolby system—Noise reduction technique for tape recordings, in which low sound levels are boosted prior to recording and then reduced on playback, thus practically eliminating tape hiss. Also used in FM broadcasting, when it may be called Dolby FM. (See also de-emphasis.) Dolby A is for professional use (it has four separate frequency channels); Dolby B is the type most often provided in home entertainment equipment (one channel only). It offers 10 dB of noise reduction above 5 kHz. "Dolby" is the trademark of Dolby Laboratories, Inc. *Recording.*

domain—(See ferromagnetic domain.)

donor—Impurity atom such as phosphorus from Group V of the periodic table which, when added to a semiconductor, donates a conduction electron, leaving behind a positive charge. The material is then said to have been doped, and as the majority charge carriers are electrons with a negative charge, the semiconductor is called *n* type. *Semiconductors.*

doping—Adding donor or acceptor impurity elements to intrinsic semiconductors to produce extrinsic material with majority charge carriers. *Semiconductors.*

Doppler effect—Change in the observed frequency of a wave because of the relative motion of the observer and the wave source. Where they are approaching each other the observed frequency increases in proportion to the closing speed; where they are receding from each other it decreases in proportion to the opening speed. *Radar, Navigation Aids.*

dose—Quantity of radiation energy absorbed by human tissue. One joule of energy absorbed by one kilogram is equal to one gray (Gy). *Nuclear Physics.*

dosimeter—Instrument for measurement of dose. For α particles, thin-window ion chambers and scintillation counters are used. For β and γ particles, Geiger-Muller counters, ion chambers and scintillation counters are used. Personal dosimeters are small charged ion chambers that indicate amount of charge lost because of radiation. Photographic film worn on the body also gives an estimation of the degree of exposure by the amount of blackening. *Nuclear Physics.*

dot generator—Color television receiver servicing instrument that tests beam convergence by producing a pattern of dots on the screen. White dots indicate proper convergence, colored dots misconvergence.

dot matrix—Type of display using dots to form characters.

double conversion—Receiver in which two frequency conversions are performed. Also termed double superheterodyne.

double-heterojunction laser—Type of semiconductor laser. *Lasers.*

double-pole, double-throw switch (DPDT switch)—Six-terminal switch that can connect one pair of terminals to either of the two other pairs. (See Figure D-16.)

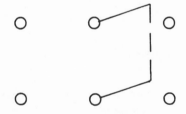

Figure D-16
Double-Pole, Double-Throw Switch

double-pole, single-throw switch (DPST switch)—Four-terminal switch that can connect one pair of terminals to another pair. (See Figure D-17.)

double sideband (DSB)—Type of modulation in which the carrier is suppressed, but both sidebands are transmitted. *Modulation.*

doubly balanced modulator—Circuit for obtaining the 90-degree phase difference and subcarrier suppression of the I and Q signals used in color television broadcasting.

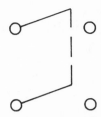

Figure D-17
Double-Pole, Single-Throw Switch

drain—Section of field-effect transistor in which the flow of majority carriers terminates, corresponding to the collector of a bipolar transistor.

drift—Movement of charge carriers in a semiconductor in response to an applied voltage. *Semiconductors.*

drift mobility—Average drift velocity per unit electric field in a homogeneous semiconductor. *Semiconductors.*

driven element—Element in antenna array to which the transmission line is connected. *Antennas.*

driver—Stage immediately preceding a power amplifier.

dropping resistor—Resistor used to decrease a voltage to a desired value.

dry cell—Battery cell with paste or jelly electrolyte.

DTL—Abbreviation for diode transistor logic. *Microelectronics.*

dual-beam oscilloscope—Oscilloscope with cathode-ray tube that produces two electron beams so that two functions can be displayed simultaneously.

dual in-line package (DIP)—Integrated-circuit package with plastic rectangular housing and a row of pins down each side. *Microelectronics.* (See Figure D-18.)

dual-trace—Multi-trace operation in which a single beam in a cathode-ray tube is shared by two signal channels, by switching rapidly between them so both displays appear to be on the screen simultaneously.

dummy antenna—Device which simulates a real antenna, used for checking power

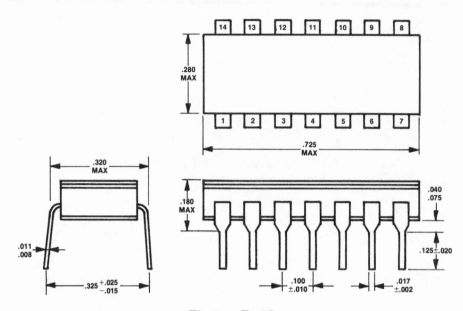

Figure D-18
Typical Dual In-Line Package (DIP)

From *Technician's Guide to Microelectronics,* by John Douglas-Young, page 46.
©1978. Reprinted with permission of Parker Publishing Company, Inc.

output and making adjustments under load without applying power to the actual antenna.

duplex—Type of operation that permits simultaneous operation in both directions (e.g., a telephone).

duplexer—Device, such as a TR tube, that permits a radar transmitter and receiver to use the same antenna while protecting the receiver from the destructively high level of power of the transmitter output. *Radar.*

duty cycle—Ratio of *on* time to *off* time.

dye laser—Liquid laser using an organic dye. *Lasers.*

dynamic—Relating to energy or physical force in action, as opposed to static. A dynamic test is made with power applied, a static with power disconnected.

dynamic microphone—(See moving-coil microphone.)

dynamic pickup—(See moving-coil pickup.)

dynamic range—Difference between overload and minimum acceptable signal levels, expressed in decibels.

dynamic speaker—(See moving-coil speaker.)

dynamo—Direct-current generator.

dynamometer—Meter that is deflected by the force between a fixed and moving coil.

dynode—Secondary-emission electrode of silver-magnesium or copper-beryllium alloy, used in photomultiplier tubes. *Electron Tubes*.

E—1. Potential difference (d.-c. or rms value). 2. Exa- (S.I. unit prefix). 3. Einsteinium (also Es). 4. Erlag (unit of telephone traffic). 5. Electric field strength. 6. Modulus of elasticity.

e—1. Potential difference (instantaneous value). 2. Electron charge ($1.6021917 \times 10^{-19}$ coulomb) 3. Base of natural logarithms ($2.71828\ldots$).

E and I laminations—Silicon-steel plates used in transformer cores. *Transformers.*

ear—Organ for sensing sound. *Electroacoustics.*

EAROM—Electrically Alterable Read-Only Memory, a ROM (q.v.) whose contents may be erased by application of direct electronic erase signals, short-wave ultraviolet or x-radiation, and then reprogrammed by the user. *Computer Hardware.*

earphone—(See headphone.)

Earth—Third planet of solar system in distance outward from the Sun.

earth—Ground (British term).

earth ground—Connection to earth. Standard earth ground is a copper rod at least half an inch in diameter and eight feet long, driven into moist ground. If ground is dry (e.g., under a building), a means of moistening it (dripping water, for instance) must be provided. For many purposes underground water pipes may be used, provided there is a continuous metal connection to the main system, but caution is needed because so many nonconducting components are used in modern systems.

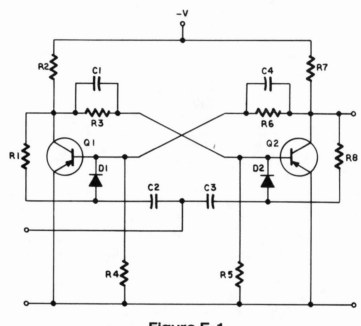

Figure E-1
Eccles-Jordan Multivibrator

From *Complete Guide to Reading Schematic Diagrams*, First Edition, page 230. ©1972. Reprinted with permission of Parker Publishing Company, Inc.

Eccles-Jordan multivibrator—Bistable multivibrator or flip-flop. (See Figure E-1.)

echo—1. Radar signal reflected from target. 2. Signal returned to talker using telephone after making one or more round trips between talker and listener. *Radar, Telephony.*

E C L—Emitter-coupled logic. *Microelectronics.*

eddy currents—Currents induced in a power transformer core by a varying magnetic field, representing lost power, are generally undesirable. However, they may be used to produce heat for cooking or for a metallurgical furnace. *Transformers.*

Edison battery—Storage battery consisting of cells containing a caustic potash solution as the electrolyte, with plates of nickel hydroxide and iron, mounted in a nickel-plated steel framework. Each cell has a voltage of 1.2 V.

Edison effect—(See thermionic emission.)

E-display—Rectangular radar display in which targets are shown as blips, their horizontal position representing range and their vertical position indicating elevation. *Radar.*

effective area of antenna—Effective area A_r of antenna is defined by:

$$A_r = G\lambda^2/4\pi$$

where G = antenna gain and λ = wavelength. *Antennas.*

effective height of antenna—Effective height of antenna h_e is given by:

$$h_e = \frac{\lambda}{\pi\sin(2\pi h/\lambda)} \sin^2(\pi h/\lambda)$$

for a vertical antenna where h is the actual height and $h \leqslant \lambda/4$. *Antennas.*

effective voltage—(See root-mean-square voltage.)

efficiency—Ratio of useful output to the input (usually power), expressed as a percentage.

EHF—Extremely high frequency (30-300 GHz).

E layer—Layer of ionosphere (110 km), important for daytime high-frequency propagation under 1000 miles. *Electromagnetic-Wave Propagation.*

electret—Material that retains its electric polarization after being subjected to a strong electric field, analogous to permanent magnetization of iron. Substances such as certain waxes, plastics, and ceramics have polarized molecules, randomly arranged that align themselves with a sufficiently strong electric field (approximately one megavolt/meter) and retain their alignment indefinitely. Practical use of this phenomenon has been in electrostatic microphones.

electrical equations—Electrical formulas commonly used in the field of radio and electronics. *Electrical Equations.*

ELECTRICAL EQUATIONS

The following equations are those required most often in the field of electronics. The meanings of the symbols used are listed below (numerical subscripts are added when the same symbol is used for more than one quantity in the same formula).

A = length of side adjacent to θ in the right triangle, in same units as other sides

B = susceptance, in siemens

C = capacitance, in farads

D = dissipation factor

d = thickness of dielectric (spacing of plates), in centimeters

dB = decibels

E = potential, in volts

F = temperature, in degrees Fahrenheit

f = frequency, in hertz

G = conductance, in siemens

H = length of hypotenuse (side opposite right angle) in right triangle, in same units as other sides

I = current, in amperes

K = dielectric constant; coupling coefficient: or temperature, in kelvins

L = self inductance, in henries

M = mutual inductance, in henries

N = number of plates; or number of turns

O = length of side opposite to θ in right triangle, in same units as other sides

P = power, in watts

p.f. = power factor

Q = figure of merit; or quantity of electricity stored, in coulombs

R = resistance, in ohms

S = area of one plate of capacitor, in square centimeters

X = reactance, in ohms

X_c = capacitive reactance, in ohms

X_L = inductive reactance, in ohms

Y = admittance, in siemens

Z = impedance

δ = 90-θ degrees

θ = phase angle, in degrees; angle, in degrees, in right triangle, whose sine, cosine, tangent, etc., is required.

λ = wavelength, in meters

π = 3.1416...

Admittance: (1) $Y = \dfrac{1}{\sqrt{R^2 + x^2}}$

(2) $Y = \dfrac{1}{Z}$

(3) $Y = \sqrt{G^2 + B^2}$

Average value: (1) Average value = $0.637 \times$ peak value

(2) Average value = $0.900 \times$ r.m.s. value

Capacitance: (1) Capacitors in parallel:

$C = C_1 + C_2 + C_3 \ldots$ etc.

(2) Capacitors in series:

$$C = \dfrac{1}{\dfrac{1}{C_1} + \dfrac{1}{C_2} + \dfrac{1}{C_3} \ldots \text{etc.}}$$

(3) Two capacitors in series:

$C = \dfrac{C_1 C_2}{C_1 + C_2}$

(4) Capacitance of capacitor:

$C = 0.0885 \dfrac{KS(N-1)}{d}$

(5) Quantity of electricity stored:

$Q = CE$

Conductance: (1) $G = \dfrac{1}{R}$

(2) $G = \dfrac{I}{E}$

(3) $G_{total} = G_1 + G_2 + G_3 \ldots$
(resistors in parallel)

(4) $I_{total} = EG_{total}$

(5) $I_2 = \dfrac{I_{total}G_2}{G_1 + G_2 + G_3 \ldots \text{etc.}}$ (current in R_2)

Cosecant: (1) $\csc \theta = \dfrac{H}{O}$

(2) $\csc \theta = \sec (90 - \theta)$

(3) $\csc \theta = \dfrac{1}{\sin \theta}$

Cosine: (1) $\cos \theta = \dfrac{A}{H}$

(2) $\cos \theta = \sin (90 - \theta)$

(3) $\cos \theta = \dfrac{1}{\sec \theta}$

Cotangent: (1) $\cot \theta = \dfrac{A}{O}$

(2) $\cot \theta = \tan (90 - \theta)$

(3) $\cot \theta = \dfrac{1}{\tan \theta}$

Decibel: (1) $dB = 10 \log \dfrac{P_1}{P_2}$

(2) $dB = 20 \log \dfrac{E_1}{E_2}$ (source and load impedance equal)

(3) $dB = 20 \log \dfrac{I_1}{I_2}$ (source and load impedance equal)

(4) $dB = 20 \log \dfrac{E_1 \sqrt{Z_2}}{E_2 \sqrt{Z_1}}$ (source and load impedances unequal)

(5) $dB = 20 \log \dfrac{I_1 \sqrt{Z_1}}{I_2 \sqrt{Z_2}}$ (source and load impedances unequal)

Frequency: (1) $f = \dfrac{3 \times 10^8}{\lambda}$

(2) $f = \dfrac{1}{2\pi\sqrt{LC}}$

Impedance: (1) $Z = \sqrt{R^2 + X^2}$

(2) $Z = \sqrt{G^2 + B^2}$

(3) $Z = \dfrac{R}{\cos \theta}$

(4) $Z = \dfrac{X}{\sin \theta}$

(5) $Z = \dfrac{E}{I}$

(6) $Z = \dfrac{P}{I^2 \cos \theta}$

(7) $Z = \dfrac{E^2 \cos \theta}{P}$

Inductance: (1)   Inductors in series: $L = L_1 + L_2 + L_3 \ldots$ etc.

(2)   Inductors in parallel:

$$L = \dfrac{1}{\dfrac{1}{L_1} + \dfrac{1}{L_2} + \dfrac{1}{L_3} \ldots \text{ etc.}}$$

(3)   Two inductors in parallel: $L = \dfrac{L_1 L_2}{L_1 + L_2}$

(4)   Coupled inductances in series with fields
       aiding: $L = L_1 + L_2 + 2M$

(5)   Coupled inductances in series with fields
       opposing: $L = L_1 + L_2 - 2M$

(6)   Coupled inductances in parallel with fields aiding:

$$L = \cfrac{1}{\cfrac{1}{L_1 + M} + \cfrac{1}{L_2 + M}}$$

(7)   Coupled inductances in parallel with fields opposing:

$$L = \cfrac{1}{\cfrac{1}{L_1 - M} + \cfrac{1}{L_2 - M}}$$

(8)   Mutual induction of two r-f coils with fields interacting:

$$M = \frac{L_1 - L_2}{4}$$

where L_1 - total inductance of both coils with fields aiding
where L_2 - total inductance of both coils with fields opposing

(9)   Coupling coefficient of two r-f coils inductively coupled so
       as to give transformer action:

$$K = \frac{M}{\sqrt{L_1 L_2}}$$

(Other coil formulas are given under *Coil Data.*)

Meter formulas: (1)   Ohms per volt $= \dfrac{1}{I}$

(I = full-scale current in amperes)

(2)   Meter resistance: $R_{meter} = R_{rheostat}$

(The meter is connected in series with a battery and a rheostat, and
the rheostat is adjusted until the meter reads full scale. A second
rheostat is then connected in parallel with the meter and adjusted
until the meter reads half scale. The resistance of the second
rheostat will equal that of the meter.)

(3)   Current shunt: $R = \dfrac{R_{meter}}{N - 1}$

where N is the new full-scale reading divided by the original full-
scale reading (both in the same units).

(4)   Voltage multiplier:

$$R = \frac{\text{Full-scale reading required}}{\text{Full-scale current of meter}} - R_{\text{meter}}$$

where reading is in volts and current in amperes.

Ohm's law formulas for d-c circuits:

(1) $I = \dfrac{E}{R}$

(2) $I = \sqrt{\dfrac{P}{R}}$

(3) $I = \dfrac{P}{E}$

(4) $R = \dfrac{E}{I}$

(5) $R = \dfrac{P}{I^2}$

(6) $R = \dfrac{E^2}{P}$

(7) $E = IR$

(8) $E = \dfrac{P}{I}$

(9) $E = \sqrt{PR}$

(10) $P = I^2R$

(11) $P = EI$

(12) $P = \dfrac{E^2}{R}$

Ohm's law formulas for a-c circuits:

(1) $I = \dfrac{E}{Z}$

(2) $I = \sqrt{\dfrac{P}{Z \cos \theta}}$

(3) $I = \dfrac{P}{E \cos \theta}$

(4) $Z = \dfrac{E}{I}$

(5) $Z = \dfrac{P}{I^2 \cos \theta}$

(6) $Z = \dfrac{E^2 \cos \theta}{P}$

(7) $E = IZ$

(8) $E = \dfrac{P}{I \cos \theta}$

(9) $E = \sqrt{\dfrac{PZ}{\cos \theta}}$

(10) $P = I^2 Z \cos \theta$

(11) $P = IE \cos \theta$

(12) $P = \dfrac{E^2 \cos \theta}{Z}$

Peak value:

(1) Peak value = $1.414 \times$ r.m.s. value

(2) Peak value = $1.570 \times$ average value

Peak-to-peak value:

(1) P-P value = $2.828 \times$ r.m.s. value

(2) P-P value = $3.140 \times$ average value

Phase angle:

$\theta = \text{arc tan } \dfrac{X}{R}$

Power factor:

(1) p.f. $= \cos \theta$

(2) $D = \cot \theta$

Q (figure of merit):

(1) $Q = \tan \theta$

(2) $Q = \dfrac{X}{R}$

Reactance:

(1) $X_L = 2\pi fL$

(2) $X_C = \dfrac{1}{2\pi fC}$

Resistance:

(1) Resistors in series: $R = R_1 + R_2 + R_3 \ldots$ etc.

(2) Resistors in parallel:

$$R = \dfrac{1}{\dfrac{1}{R_1} + \dfrac{1}{R_2} + \dfrac{1}{R_3} \ldots \text{etc.}}$$

(3) Two resistors in parallel: $R = \dfrac{R_1 R_2}{R_1 + R_2}$

Resonance:

(1) $f = \dfrac{1}{2\pi \sqrt{LC}}$

(2) $L = \dfrac{1}{4\pi^2 f^2 C}$

(3) $C = \dfrac{1}{4\pi^2 f^2 L}$

Right triangle:

(1) $\sin \theta = \dfrac{O}{H}$

(2) $\cos \theta = \dfrac{A}{H}$

(3) $\tan \theta = \dfrac{O}{A}$

(4) $\csc \theta = \dfrac{H}{O}$

(5) $\sec \theta = \dfrac{H}{A}$

(6) $\cot \theta = \dfrac{A}{O}$

Root-mean-square value:

(1) R.m.s. value = $0.707 \times$ peak value

(2) R.m.s. value = $1.111 \times$ average value

Secant:

(1) $\sec \theta = \dfrac{H}{A}$

(2) $\sec \theta = \csc (90 - \theta)$

(3) $\sec \theta = \dfrac{1}{\cos \theta}$

Sine:

(1) $\sin \theta = \dfrac{O}{H}$

(2) $\sin \theta = \cos (90 - \theta)$

(3) $\sin \theta = \dfrac{1}{\csc \theta}$

Susceptance:

(1) $B = \dfrac{X}{R^2 + X^2}$

(2) $B = \dfrac{1}{X}$

(3) $B = B_1 + B_2 + B_3 \ldots$ etc.

Tangent:

(1) $\tan \theta = \dfrac{O}{A}$

(2) $\tan \theta = \cot (90 - \theta)$

(3) $\tan \theta = \dfrac{1}{\cot \theta}$

Temperature:

(1) $C = 0.556F - 17.8$

(2) $F = 1.8C + 32$

(3) $K = C + 273$

Transformer ratio:

$\dfrac{N_p}{N_s} = \dfrac{E_p}{E_s} = \dfrac{I_s}{I_p} = \sqrt{\dfrac{Z_p}{Z_s}}$

(subscript p = primary; subscript s = secondary)

Wavelength:

$\lambda = \dfrac{3 \times 10^8}{f}$

Fourier analysis

The Fourier series expansion of the function $f(t)$ is:

$$f(\omega t) = \frac{a_0}{2} + \sum_{n=1}^{\infty} (a_n \cos n\, \omega t + b_n \sin n\, \omega t)$$

where:
$$a_n = \frac{1}{\pi}\int_{-\pi}^{+\pi} f(\omega t)\cos n\ \omega t d(\omega t)$$

$$b_n = \frac{1}{\pi}\int_{-\pi}^{+\pi} f(\omega t)\sin t d(\omega t)$$

If $e(\omega t)$ is a periodic voltage of amplitude E_{pk}, its Fourier series is:

$$e(\omega t) = E_o + \sum_{n=1}^{\infty} E_n\cos(n\omega t + \phi_n) = E_{pk}f(\omega t)$$

where:
$$E_o = \frac{a_o}{2}\ E_{pk}$$

$$E_n = \sqrt{(an^2 + bn^2)}$$

Half-wave rectified sine wave:

$$e(\omega t) = \frac{1}{\pi}\ E_{pk} + \frac{E_{pk}\sin\omega t}{2} - \frac{E_{pk}}{\pi} \times \sum_{n=1}^{\infty} \frac{1}{(4n^2 - 1)\pi}\ \cos 2n\omega t$$

Full-wave rectified sine wave:

$$e(\omega t) = \frac{2}{\pi}\ E_{pk} - \frac{2}{\pi}\ E_{pk} \sum_{n=1}^{\infty} \frac{1}{(4n^2 - 1)\pi}\ \cos 2n\omega t$$

Square wave:

$$e(\omega t) = E_{pk} \sum_{n=1}^{\infty} \frac{4}{(2n - 1)\pi}\ \sin(2n - 1)\omega t$$

Triangular wave:

$$e(\omega t) = E_{pk} \sum_{n=1}^{\infty} \frac{8}{(2n - 1)^2\pi^2}\ \cos(2n - 1)\omega t$$

Sawtooth wave:

$$e(\omega t) = E_{pk} \sum_{n=1}^{\infty} \frac{2}{n\pi}\ \sin n\ \omega t$$

Rectangular pulse train with duty cycle d:

$$e(\omega t) = dE_{pk} + E_{pk} \sum \frac{2}{n\pi}\ \sin n\ \pi d \cos n\ \omega t$$

electric charge—(See charge.)

electric discharge lamp—Lighting device consisting of a glass envelope within which a gas is ionized and made to glow by an applied voltage. Neon, argon, xenon, and helium are examples of gases used. Sodium and mercury vapors are also used for fluorescent and ultraviolet discharge lamps.

electric eye—1. Photoelectric cell (q.v.). 2. Cathode-ray tuning indicator tube used in some electron-tube radios.

electric field—Region around an electric charge in which an electric force is exerted on another charge; force per unit positive charge (symbol E).

electricity—Phenomenon of positively and negatively charged particles of matter at rest and in motion, individually as well as in great numbers.

electric motor—Device that converts electrical energy into mechanical energy, employing electromagnetic phenomena to do so.

electroacoustics—Science concerned with the electrical production, transmission, reception, and the effects of sound. *Electroacoustics.*

ELECTROACOUSTICS

The Nature of Sound

There are two ways of thinking about sound. It can be thought of as a purely mechanical phenomenon of pressure waves propagated through a medium, in which case it should be considered as *vibration*. Or, it can be thought of as *heard sound*, received by the ear and analyzed by the brain, a phenomenon of great significance to man, who is immersed in sound stimuli that influence all his activities.

The science of heard sound is called acoustics. It has much to do with how clearly sounds are heard. When electrical means are used to extend the range of hearing, store sound, or otherwise process it, the technology used is termed electroacoustics. This article covers the acoustical aspects of the reproduction of both speech and music.

The normal human ear is a remarkably sensitive organ, and it has been estimated that as little as 10^{-18} joule, or about 10^{-25} kilowatt-hour of sound energy, is enough to excite it and produce recognition of the sound. At the same time it is rugged enough to withstand sound intensity which is 10^{14} times greater than the minimum level.

Sound intensity is usually measured in decibels. The response of the human ear is logarithmic, so that if the output of a 5-watt amplifier were increased to 15 watts the change in sound amplitude would be more noticeable than it would be if the output of a 50-watt amplifier were increased to 100 watts. Although the change from 5 watts to 15 watts is a change of only 10 watts and the change from 50 watts to 100 watts is a change of 50 watts, the change from 5 to 15 watts has tripled the power, whereas the change from 50 to 100 watts has only doubled it. Expressed in decibels (dB), the first amplifier's increase of power is given by:

$$dB = 10 \log_{10} \frac{15}{5} = 10 \log_{10} 3 = 4.77$$

and the second's by:

$$dB = 10 \log_{10} \frac{100}{50} = 10 \log_{10} 2 = 3.01$$

Table I shows the intensity of some familiar sounds.

The velocity of sound varies with the medium. For example, the velocity of sound in dry air is 331 m/s, (1086 f/s), in water 1505 m/s (4937 f/s), and in iron 5170 m/s (16,962f/s). However, the velocity in a gas is much affected by the temperature of the gas (since this also affects its pressure), and so the velocity at some temperature other than 0°C is given by a formula:

$$V_t = V_o \sqrt{1 + t/273}$$

where V_o is the velocity at 0°C and t the Celsius temperature. For example, V_o for dry air is 331 m/s, so its velocity at 20°C will be:

$$V_t = 331 \sqrt{1 + 20/273} = 343 \text{ m/s}$$

Table I—Intensity Levels

Type of Sound	Intensity Level in dB	Intensity in W/cm^2
Threshold of painful sound	130	1000
Airplane, 1600 rpm, at 18 feet	121	126
Subway, local station, express passing	102	1.58
Noisiest spot at Niagara Falls	92	0.158
Average automobile, at 15 feet	70	10^{-3}
Average conversational speech at 40 inches	70	10^{-3}
Average office	55	3.16×10^{-5}
Average residence	40	10^{-6}
Quiet whisper at 60 inches	18	6.3×10^{-9}
Reference level	0	10^{-10}

Generally speaking, audible sound waves spread out in all directions from their source. Each note uttered by Shelley's skylark compresses the air around it, and this compression wave expands like a soap bubble, its radius growing at a rate equal to the speed of sound in air. However, on striking a surface, the wave is reflected, refracted, or absorbed according to the properties of the surface and the medium of which it is the boundary. In the case of a solid object with a flat surface (e.g., the wall of a room), it acts like another source of sound, and waves move outward from it in the same way that light is reflected by a mirror. The reflected sound-wave motion, moreover, can interfere with the original wave motion. It can reduce or increase the amplitude of the original wave. Accordingly, the sound level will vary in different parts of a room. This effect is more pronounced when the sound waves come from two identical sources (two loudspeakers). They cancel or reinforce each other, depending upon their location, the listener's position and the acoustical properties of the room. In large rooms, such as auditoriums, concert halls, churches and the like, a sound takes an appreciable time to go from the source to a reflecting surface and then to the listener. Consequently, he hears two sounds: the direct sound and an echo. This is called reverberation. A sharp sound, such as a choirboy dropping his hymn book, continues to echo back and forth for some time (to his embarrassment). The time it takes to die away is called the reverberation time.

For satisfactory reception of speech a relatively short reverberation time is obviously essential; otherwise successive utterances overlap. For best listening conditions, reverberation time should be about one second. For music the optimum reverberation time is 1.7 seconds (organ music 2 seconds). These figures are for an average frequency of 500 Hz. Reverberation time varies inversely with frequency, so that higher-pitched sounds die away faster, lower-pitched sounds more slowly. In a complicated sound pattern this results in certain frequencies being accentuated. For this reason concert halls are always constructed so that opposite surfaces are not parallel, and sound-absorbing materials are used to reduce the intensity of reflected sounds. The shape of a room also has a marked effect upon its acoustic properties. The most advantageous ratio for height, width, and length is when their proportions are 1: 2-1/3 : 2-2/3.

Sound absorption is provided by porous materials such as carpets, draperies, glass-fiber blankets, and so on, into which the molecules in the air carrying the sound energy can dissipate. When sound absorbing materials are placed within a room, reverberation time is greatly reduced. Consequently, sounds seem to come from their point of origin rather than from everywhere. A nonreverberant space seems more comfortable and conversation is easier.

Refraction takes place when sound passes from one medium to another, and is similar to what happens to a ray of light on passing from air to glass or water. Diffraction is the property sound waves have of flowing around obstacles in their path, and is greater with low frequencies. High-frequency sound "casts a shadow," so it is considerably attenuated by obstacles.

The foregoing characteristics of sound must all be considered in the acoustic treatment of enclosed spaces, whether the space is an auditorium or a room for listening to high-fidelity recordings. To summarize:

1. Diffusion is highly desirable, and both absorption and room geometry should be employed to enhance it.

2. Useful reflecting surfaces should be retained. Absorption should not be placed on ceilings as a general rule.

3. Absorption should be placed on the floor (carpeting) and behind the listener (drapes).

4. Too high a reverberation time is detrimental to intelligibility, but too little requires a much higher power output to overcome the deadening effect of too much absorption.

The Reproduction of Sound

The fundamental voice frequency range lies between 100 and 250 hertz, with the fundamental pitch of men's voices at the lower frequency end, women's voices at the upper end. However, overtones and voice inflection can produce frequencies up to 8000 hertz. The greatest amount of speech energy in any single octave is found between 300 and 600 hertz for male voices and 555 and 1110 hertz for female voices. The dynamic range of the human voice is 56 decibels, from the lowest audible whisper to the loudest possible shout. Most people, however, have a dynamic range of 40 dB.

A faithful reproduction of music requires an audio frequency range very much larger. The range for an orchestra is generally considered to be between 30 and 16,000 Hz. The very high-frequency notes also contain harmonic components that can extend well above 16,000 Hz. Consequently any high-fidelity system worthy of the name must have a frequency range of 30 to 20,000 Hz at least. Many have wider. But it is not enough that the amplifier has this range. The range of the *system* is that of the component with the smallest range, assuming no other losses (cabling, etc.). All components (cartridge, tape head, preamplifier, tuners, amplifier, and loudspeakers) must have adequate range, and that goes for the recording, too.

In a typical musical performance, the highest energy content is generally in the frequency range between 500 and 2000 Hz. This is in the musical tone range above "middle C" (261.63 Hz). The frequency range of various musical instruments is shown in Figure E-2. The fundamental tones of most of them do not extend much above 4000 Hz. However, for life-like reproduction it is necessary to reproduce the harmonic components which, as mentioned above, extend to 16,000 Hz or higher. It is these harmonics which give each instrument its characteristic sound. A large orchestra will also have a dynamic range of some 70 dB, which means that if the lowest level of sound it may produce is taken as 1 unit, then the maximum (presumably the finale of the "1812 Overture") will be 10,000,000 units.

High-fidelity phonograph records are long-playing microgroove disks that today are practically standardized at 12 inches diameter and a frequency response of 15 to 20,000 hertz. The speed of rotation of these records is 33-1/3 r.p.m. Smaller, 7-inch

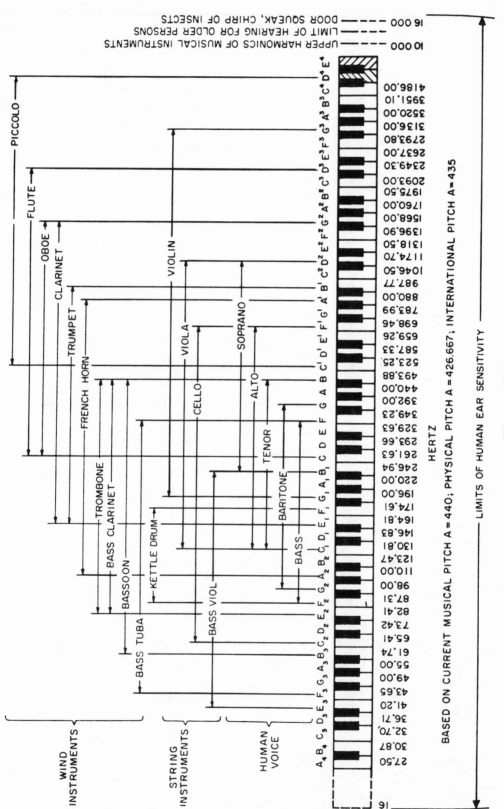

Figure E-2

Audio Frequency Spectrum

From *Complete Guide to Electronic Test Equipment and Troubleshooting Techniques*, by John Douglas Young, page 227.©1976. Reprinted with permission of Parker Publishing Company, Inc.

records rotate at 45 r.p.m. but because of their short playing time are used mainly for light music. It must also be recognized that 16-inch transcriptions are still in limited use, and 10-inch records also. Records with a speed of 78.26 r.p.m. are obsolete, but most record-players have provision for playing them. (Further details are given in the article on *Recording*.)

At one time the level of sound for different frequencies varied considerably among records of different manufacturers, so a special input circuit was required for each type. These circuits were called equalization circuits, and equalization controls will be found on older amplifiers with settings labeled "ORTHO," "NMB," "AES," etc.

All records today are standardized to the same characteristic curve, called the RIAA (Recording Industry Association of America), so this control will seldom be seen. If a foreign record with a nonstandard characteristic curve is played, adjustment of the treble and bass controls may be necessary.

Tape recordings are made on reel-to-reel tapes, cassettes, and cartridges. The preferred speed for all is 7-1/2 inches per second. Reel-to-reel tapes may be up to 7200 feet long (1.0-mil base on 14-inch reel). Other speeds (15, 3-3/4, and 1-7/8 i.p.s.) may also be used on reel-to-reel tapes. Cassettes and cartridges, however, are standardized at 7-1/2 i.p.s. Frequency response and dynamic range are generally 20-20,000 Hz and 65 dB for good quality tape. The chief problem with tape is background noise called hiss, resulting from residual magnetism. This may become noticeable at low sound levels, so a system to counteract it is frequently used. This system, the Dolby system, operates by boosting low passages in the recorded signal at the tape hiss frequencies prior to recording, and lowering them to their original levels during playback. The lowering process automatically reduces any noise that was introduced in recording. (Further details are given in the article on *Recording*.)

FM broadcasting is discussed in the article on *Broadcasting*. It has a frequency range from 50 to 15,000 hertz and a much higher dynamic range and freedom from noise compared to AM broadcasting. It borders on high fidelity and is satisfactory for most purposes to most listeners. It must be stressed, however, that only a high quality tuner is able to take advantage of this dynamic range and produce an undistorted output. The noise level of the tuner itself must be low so as not to be noticeable during quiet passages (see preceding remarks about Dolby).

An additional technical step is taken in FM broadcasting to render it less susceptible to noise. The noise components are more dominant in the higher frequencies, so the broadcasting station amplifies the latter more than the lower frequencies. This is called pre-emphasis. In the FM receiver, a circuit is included in the output of the demodulator to restore the correct amplitude relationship between the high and low audio frequencies. This network is called a de-emphasis circuit.

Microphones used to make recordings must also have the frequency range of high fidelity. This seems obvious, but is frequently overlooked in home systems. Very few microphones cover the full range. 100-10,000 Hz is a very common frequency range. It is also important that the impedance of the microphone be matched to the amplifier. The impedance can be of various values from 50 to 50,000 ohms.

The cartridge and stylus (pickup) of the record player tone arm have similar requirements to those of a microphone. The diamond in use today is generally made with an elliptical tip. Since a stereo record groove has a bottom radius of not greater than 0.0002 inch the maximum lateral diameter of the stylus is limited to 0.0004 inch. The longitudinal diameter is usually 0.0007 inch. The cartridge in hi-fi pickups is magnetic. Although one stylus is used, it induces electric currents in two coils, thereby converting the two sound tracks in the record groove to two signals for stereo reproduction.

The processing of the various inputs (AM, FM, phono, or tape) is done in the preamplifier. This also contains the circuits for recording on tape. The signals fed to the preamplifier are at very low levels, so its circuits are designed for maximum voltage amplification. Two or three stages are usually provided to build up the signal to a high enough level to drive the main amplifier.

The main audio amplifier is made up of one or more stages of voltage amplification, a driver stage, and a final power amplifier stage. Since nearly all amplifiers have to handle stereo recordings there are really two independent but otherwise identical amplifiers operating in parallel.

In the main amplifier, the two problems that are most likely to trouble the designer are the prevention of harmonic and intermodulation distortion (see *Distortion*). Pushpull output circuits and negative feedback, together with carefully biasing, are used to minimize these problems. Modern audio amplifiers fortunately do not require output transformers, so the difficulty of obtaining a flat frequency response over the full range, which made design of these transformers very tricky, is no longer a consideration.

The loudspeakers of a high-fidelity system are as important as any other components. Single speakers are generally not practical for extended range. Most units today consist of three speakers in one enclosure. The largest, called a "woofer," has a cone diameter of 10-15 inches, a massive permanent magnet, and a voice coil capable of handling considerable power. The woofer reproduces frequencies from 800 Hz down to 20 Hz. The second speaker is frequently a "midrange horn" speaker covering the range from 800 Hz to 8000 Hz. The third is a "tweeter," a small speaker for the frequencies from 8000 Hz to 25,000 Hz. To divide the audio signal according to the frequency range of each speaker, a cross-over network is used, which is a filter network to provide easy paths for the three bands of frequencies to their speakers. In stereo systems two identical speakers are used, placed approximately six feet apart in the average room.

Most of the speakers used for high-fidelity reproduction have their three component speakers mounted on a board which is called a baffle and forms the front face of the speaker enclosure. This is a closed box, lined with sound absorbing material, to ensure that sound waves from the rear of the woofer cannot get around to the front. It is called an infinite baffle, to distinguish it from other types that have ports or vents that allow sound to emerge from behind the speaker in phase with that from the front. These are intended to enhance the low-frequency response and are called by names such as "base reflex," "acoustic labyrinth," and so on. However, a large woofer does not need any enhancement, if the amplifier has the power it requires.

Much of what has been said applies also to public address systems, which include permanent (as installed in large auditoriums, schools, and so on); semi-permanent (set up in places that are not used frequently enough for a permanent installation); and temporary or portable systems (as used by rock groups or mobile PA systems installed in trucks). These systems consist basically of a microphone, an amplifier, and one or more speakers. However, the amplifier and speakers are "heavy-duty" types, especially if intended for use out-of-doors. The amplifier is of higher power (on the order of 100 watts, driving, say, four 25-watt speakers).

Successful speech and music reinforcement systems require the designer to coordinate reverberation time, speaker directivity, and distance from the sound source to the farthest listener so that he can still understand the words of the speaker. These factors in turn determine the amplifier power required.

The correct place to aim the center of a single loudspeaker in an auditorium is at the last seat, not the middle. It should also be located as high as possible, so that the sound waves strike the rear wall at an angle from above and are reflected downward, to minimize the echo effect. A speaker should never be aimed directly at a wall, so that the sound is reflected straight back to the speaker, if it can be avoided.

Human Hearing Characteristics

The preceding discussion outlined the requirements of the electrical portions of good sound reproduction. However, no outline can be complete without taking into consideration the characteristics of human hearing.

The human ear shows a greater sensitivity to high-frequency sounds than to lower ones. The peak sensitivity lies somewhere in the 2000 to 4000-hertz range, as shown in Figure E-3. High aural sensitivity in the 500 to 7000-hertz range is important in the intelligibility of speech sounds.

The variation in sensitivity of the human ear to different frequencies is not the same for all sound levels. At very high sound levels the ear has almost equal sensitivity to all frequencies. As the level decreases, sensitivity at the lower and upper ends of the sound spectrum diminishes.

Sound intensity is a measure of the energy in the sound wave as indicated by an instrument with a flat response at all audio frequencies. Loudness is the sensation in the ear and mind of the amplitude of the sound energy and is not proportional to the sound intensity at all frequencies. Each curve in Figure E-3 shows the intensity levels required to sound equally loud at all frequencies to the average human ear. The curve marked 0 dB represents the intensity of a just audible 1000-hertz tone. This curve intersects the vertical line for 400 Hz at the horizontal line representing 10 dB, which means that the intensity of sound at 400 Hz must be 10 dB greater (or 10 times stronger) to become just audible to the average human ear. On the other hand, if the frequency

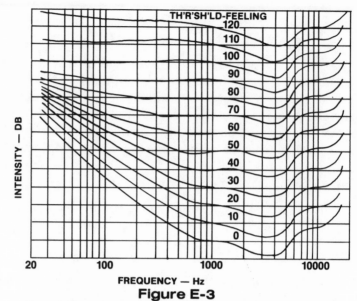

Figure E-3

*Sound Intensity Levels Required for Equal Loudness
(Fletcher-Munson Curves)*

is increased to a value between 3000 and 4000 Hz its intensity can be much lower and still sound the same.

These curves indicate very clearly that as the output of an amplifier is reduced the apparent loudness of the low notes will be reduced more rapidly than the middle frequencies. In fact the intensity of the sound must be comparatively very high to hear low notes at all. For this reason, an ordinary volume control is not used in a high-fidelity amplifier, but is replaced by a loudness control which boosts the bass and also the high frequencies, as it is turned down, so that the sound level at various frequencies is adjusted to agree with the applicable loudness curve of Figure E-3. A typical loudness control network is illustrated in Figure E-4.

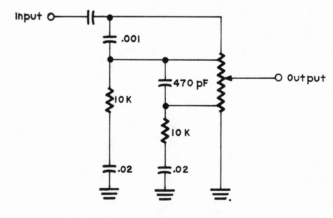

Figure E-4

Loudness Control

electrocardiograph (ECG or EKG)—Device consisting of a type of galvanometer which provides a graphic recording of the electric current generated by the heart muscle during a heart beat. The recording (electrocardiogram) provides information from which it can be determined if the heart action is normal or otherwise.

electrochemistry—Branch of chemistry concerned with the relation between electricity and chemical change. It comprises such things as the chemical action in batteries, electrolysis, and ionization.

electrocution—Method of execution used in 24 states in the USA, in the Philippines and Taiwan, in which a heavy current is made to flow through the condemned person.

electrode—Metal conductor used as either of two terminals of an electrically conducting medium, to conduct current into or out of the medium. The electrode from which electrons emerge is called the cathode, the electrode that receives electrons is the anode.

electroencephalograph (EEG)—Instrument used to record electrical activity in the brain. The recording is called an encephalogram and is used for studying how the brain works and for tracing connections between different parts of the central nervous system.

electroluminescence—Production of light by the flow of electrons, as within certain crystals, without simultaneous generation of heat, as in incandescence. There are two types. In Destriau luminescence, thermal agitation results in the ionization of atoms in the phosphor, which emit light on recombination with the ejected electrons. In the other type, charge injection (as when a voltage is applied to a p-n junction) creates holes, which are then filled by electrons with consequent emission of light.

electrolysis—Process in which an electric current is passed through a substance to effect a chemical change. Hydrogen and oxygen are produced by electrolysis of water. (See also electroplating.)

electrolyte—Substance that conducts electric current as a result of a dissociation into positive and negative ions, e.g., acids, bases, and salts, which dissociate in such solvents as water and alcohol.

electrolytic capacitor—Capacitor in which the dielectric is an oxide deposited electrolytically on an aluminum or tantalum plate. The electrolyte is the other plate. *Capacitors.*

electromagnet—Device consisting of a core surrounded by a wire coil through which an electric current is passed, imparting the power of attracting iron to the core while the current is flowing.

electromagnetic radiation—Propagation of energy through space by means of varying electric and magnetic fields. *Electromagnetic-Wave Propagation.*

electromagnetic spectrum—The entire distribution of electromagnetic waves, according to their frequency or wavelength, from long radio waves to gamma rays. Although it is convenient to designate different wavebands with names according to their characteristics, there are no precise boundaries between any of these contiguous ranges. (See Figure E-5.)

electromagnetic-wave propagation—Any of the waves of the electromagnetic spectrum propagated in free space at the uniform velocity of 2.99793×10^8 meters per second. *Electromagnetic-Wave Propagation.*

ELECTROMAGNETIC-WAVE PROPAGATION

The manner in which electromagnetic waves used in radio are generated is discussed in the article on *Antennas*. Electromagnetic fields are generated around the antenna by the alternating current flowing in it. With each change of direction of the antenna current the then-existing field is detached and repelled from the antenna. Successive fields of opposite polarity follow each other away from the antenna, so that the field potential at a fixed point fluctuates between alternating positive and negative values as the fields pass. The distance between two successive positive (or two successive negative) peaks is the wavelength. Since all waves travel at the same velocity c (the velocity of light), the number that pass the fixed point in each second is greater for shorter wavelengths, according to the formula:

$$f = \frac{c}{\lambda} \tag{1}$$

where f is the frequency in hertz, c is the velocity of light (approximately 3×10^8 m), and λ is the wavelength in meters. Since the waves are generated by the current in the antenna, they are an exact facsimile of the signal applied to the antenna, and upon encountering another antenna they induce currents in it that are identical with those in the first antenna, although much smaller.

The polarization of the field is such that the electrostatic lines of force always remain parallel to the antenna which generated them, while the magnetic lines of force, which were formed as circles around the antenna, are normal to the electrostatic lines. The field is also most intense in directions that are normal to the antenna, and least intense in directions that are collinear with it. (The exception is microwaves—see *Waveguides and Resonators.*)

Frequency (hertz)		Wavelength (meters)		Radioactive and Radio Waves	Ultraviolet, Visible, Infrared and Audio Waves
100 EHz	($\times$ 10^{18})	3 pm	($\times$ 10^{-12})	Gamma rays	
10 EHz	"	30 pm	"	(hard)	
1 EHz	"	300 pm	"	X rays	
100 PHz	($\times$ 10^{15})	3 nm	($\times$ 10^{-9})	(soft)	Ultraviolet rays
10 PHz	"	30 nm	"		
1 PHz	"	300 nm	"		
100 THz	($\times$ 10^{12})	3 μm	($\times$ 10^{-6})		Visible light rays
10 THz	"	30 μm	"		Infrared rays
1 THz	"	300 μm	"	12	
100 GHz	($\times$ 10^{9})	3 mm	($\times$ 10^{-3})	EHF - 11	
10 GHz	"	30 mm	"	SHF - 10	
1 GHz	"	300 mm	"	UHF - 9	
100 MHz	($\times$ 10^{6})	3 m	($\times$ 10^{0})	VHF - 8	
10 MHz	"	30 m	"	HF - 7	
1 MHz	"	300 m	"	MF - 6	
100 kHz	($\times$ 10^{3})	3 km	($\times$ 10^{3})	LF - 5	
10 kHz	"	30 km	"	VLF - 4	
1 kHz	"	300 km	"	VF - 3	Audio waves
100 Hz	($\times$ 10^{0})	3 Mm	($\times$ 10^{6})	ELF - 2	
10 Hz	"	30 Mm	"		
1 Hz	"	300 Mm	"		

Figure E-5
Electromagnetic Spectrum

Radio waves (and this term includes television and radar) are exactly the same as other electromagnetic waves, which include light, infrared, ultraviolet, x-rays, and gamma rays. Their different characteristics arise from their interaction with matter, and this interaction depends upon their frequency. Molecules may be widely separated and existing practically independently, as in space and in the upper atmosphere, or they may be interlocked in a solid. The spaces between them may be large enough for small wavelength radiation, such as x-rays, to penetrate and pass through, while longer waves such as light, radio, and so on are stopped. Radio waves of different frequencies also behave differently as they encounter obstructions in their path.

The greatest obstruction is the Earth itself. Radio waves penetrate the ground to a small extent only. To travel any distance they have to have a means of getting around the curvature of the Earth. If it were not for the atmosphere this would eliminate long-distance transmission, except by satellite.

The atmosphere at great heights above the ground is very thin, merging imperceptibly into space itself. At these levels it is exposed to the full force of the sun's radiation. High-energy ultraviolet rays strike atmospheric molecules and knock electrons out of their atoms. Although recombination takes place continuously, the liberation of other electrons maintains an electron density that varies with the intensity of the radiation. Since this increases with sunspot activity (and some other phenomena to be discussed later) it is not always the same, and disappears gradually during the hours of darkness.

There are actually several ionized layers in the atmosphere, as shown in Figure E-6. The F_2 layer, during the day, is between 250 and 400 kilometers above the Earth. Because it is the most exposed layer it stores more solar energy than the lower layers, and since its molecules are more dispersed the recombination rate is slower, so it remains ionized for many hours after dark. The F_1 layer below it exists only during daylight.

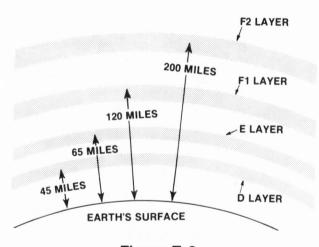

Figure E-6
Ionized Layers in the Upper Atmosphere

Lower yet is the E layer. Its ionization density corresponds closely with the elevation of the sun. Its molecules are not as thinly scattered, and it has another phenomenon, called sporadic E, in which irregular cloud-like areas of unusually high ionization occur, up to more than 50 percent of the time on certain days or nights. Sporadic-E ionization is ascribed to visible and subvisible-wavelength bombardment of the atmosphere.

The lowest layer, extending down to 50 kilometers above the surface, is the D layer. It exists only during daylight hours, and its ionization density corresponds with the elevation of the sun.

The ionization of these layers increases with height, and the ability of radio waves to pass through these layers varies according to their wavelength, only the shortest wavelengths being able to penetrate the upper layers. Each layer has a *critical frequency*, which is that frequency above which waves can pass through. Waves at or below the critical frequency are reflected back to earth just as if the ionized layer were a solid object.

Very low-frequency waves (3-30 kHz) and low-frequency waves (30-3000 kHz) have wavelengths from 1 to 100 km, so they have no chance of passing through the ionized layers. However, since the ground and the ionosphere act very much like a waveguide to the lower frequencies, they can travel a long way, and follow the curvature of the Earth. Since the conductivity of sea water is very high, their range over the ocean is even greater. Consequently, maritime and aircraft stations use these frequencies for navigational purposes, although satellite navigation stations are now being constructed which will have greater coverage at much higher frequencies.

Medium frequency waves (0.3-3 MHz) include the standard AM broadcast band, and in this range both surface waves and waves reflected from the ionosphere (sky waves) are important. During the day the sky waves are absorbed by the D and E layers, so propagation is almost entirely by the surface wave (ground wave) for distances between 25 and 100 miles. At night, when the D and E layers have disappeared sky wave reflection takes place from the F layer also. This is illustrated in Figure E-7. The fading area is the region where both ground and sky waves are received, but because of the different lengths of the paths traveled they may not always have the same phase, especially at sunset, when electron density and layer height are changing rapidly. This effect also varies seasonally and with changes in sunspot activity.

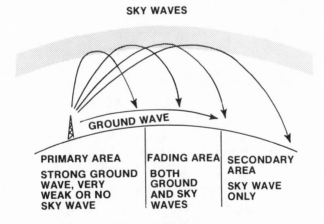

Figure E-7
Night Time Reception Conditions for Broadcast Stations

High frequency waves (3-30 MHz) are the most interesting and variable part of the radio spectrum, since this segment contains the bands most used for short-wave listening and, because it uses sky waves to a much greater extent, may be received

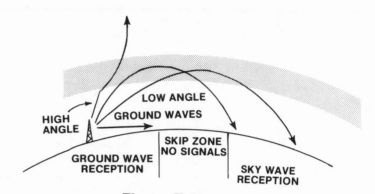

Figure E-8
Reception of High Frequency Signals Showing Skip Zone

over very great distances if conditions are favorable. The angle at which the wave encounters the reflecting layer determines the distance, as shown in Figure E-8.

The short-wave band can be divided into three sections, as follows:

3-10 MHz: For daytime operation, the low end of this band, between 3 and 7 MHz, is excellent for transmission or reception up to a few hundred miles. The night time range can be a thousand or more miles. At the high end of this band, between 7 and 10 MHz, the daytime range can often extend to almost a thousand miles, while at night, especially in winter, coverage can be thousands of miles. In this range the E layer dominates in the daytime, while at night signals are reflected by the F layer.

10-20 MHz: This range of frequencies is used by most of the international broadcast stations. The frequency is high enough to penetrate the E layer, so is reflected from the F layer during daylight hours. By using a low wave angle at the transmitter and by controlling the direction of transmission, signals can be beamed to other parts of the world over great distances. This is enhanced during periods of sunspot activity, but during a sunspot minimum the night time performance is more limited because the F layer is thinner and radiation escapes into space instead of being reflected.

20-30 MHz: This section is suitable for both local and long-distance work. Local communication is very good by ground wave up to about 20 miles and is used largely by the lower-frequency CB channels (26.96-27.23 MHz). Long distance communication is limited to times when the F layer is most dense, which is when sunspot activity is greatest. At other times it is poor or impossible.

Very high-frequency waves (30-300 MHz) and ultra high-frequency waves (0.3-3 GHz) normally are not reflected by the ionosphere, although low channel television signals (54-88 MHz) may be reflected by a very intense F layer during exceptional sunspot activity. This is most unusual, however. Sporadic E propagation can also give a longer-than-usual range.

As a general rule, though, propagation at these frequencies is very nearly in a straight line, much the same as a beam of light. It is not entirely straight because of the refractive index of the atmosphere. This decreases with height, so that waves near the ground travel more slowly and are bent downwards slightly. This extends the distance the wave can travel beyond the "line-of-sight" distance by about one-third, under average conditions in temperate climates. It does vary, however, with weather and atmospheric conditions, such as temperature, barometric pressure, and water vapor content in the air. The effect is more pronounced with VHF signals, less so with UHF; the higher the frequency the more nearly radio waves approach light in their behavior. Another effect of the lower atmosphere, as opposed to the ionosphere, on VHF, UHF and SHF waves, is called tropospheric scatter, in which weak but reliable fields are propagated several hundred miles beyond the horizon.

Signals above 30 MHz are readily reflected by objects on the ground, such as tall buildings, hills, or the ground itself. The reflected signal travels a longer path than the direct one, and so may arrive out of phase with it. The receiver then receives two identical signals one after the other. This will cause "ghosts" in a television picture, or degraded audio in a radio. Reflection from airplanes causes a television picture to flutter.

Television and FM stations transmit horizontally polarized waves, whereas AM stations send out vertically polarized waves. This is because at TV and FM frequencies, a vertical antenna would also pick up a lot of noise from electrical apparatus at these frequencies (or harmonics of them). By using a horizontal antenna the noise is reduced by about 20 dB.

Earth-space communication must pass through the Earth's atmosphere, so the best frequencies are those that do so with the least attenuation. Such frequencies are in bands called windows. One window extends from 10 MHz to 10 GHz. Another lies between 1 THz and 1 PHz (in the infrared and optical regions of the electromagnetic frequency spectrum).

electromechanical device—Any device using electrical energy to produce a mechanical effect.

electrometer—Instrument designed to measure electrical potential difference by means of electrostatic forces between charged bodies, such as plates or fibers. More recent models use a specifically designed d.c. amplifier tube with an input impedance of 10^{12} ohms. This is more rugged but not as sensitive as the older models.

electromotive force (e.m.f.)—Energy per unit electric charge imparted by an energy source (battery or generator), usually expressed in volts.

electron—Subatomic particle carrying a single basic charge of electricity. *Nuclear Physics.*

electron accelerator—Device to provide a beam of electrons to bombard an x-ray target, from which a beam of x-rays is emitted, e.g., as in a betatron. *Nuclear Physics.*

electron-beam evaporation—Method of depositing films of conductive, resistive, and dielectric material on a substrate by evaporation by electron beam in a vacuum, used in the manufacture of thin-film hybrid integrated circuits. *Microelectronics.*

electron-coupled oscillator—Oscillator using a pentode or tetrode electron tube, in a Hartley or Colpitts resonant circuit, to provide isolation between the resonant circuit and the output to avoid frequency shifts with varying loads. (See Figure E-9.)

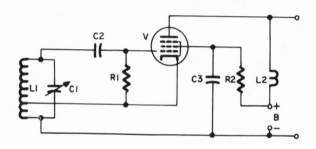

Figure E-9
Electron-Coupled Oscillator

From *Complete Guide to Reading Schematic Diagrams,* First Edition, page 82. ©1972. Reprinted with permission of Parker Publishing Company, Inc.

electron emission—Emission of electrons from a cathode surface by thermionic (or primary) emission, bombardment by other electrons (secondary emission), action of a high electric field (field emission), or the incidence of photons (photoemission). *Electron Tubes.*

electron gun—Electrode structure that produces and may control, focus, and deflect a beam of electrons, as in a television picture tube. At one end of the gun is the cathode, a metal cylinder of nickel coated with oxides of barium and strontium, heated by an internal coil so that electrons are emitted. These are drawn by a positive voltage through several electrodes that give the beam the required shape and direction. (See Figure E-10.)

electronic—Operated, operating, produced, or done by the action of electrons.

electronic flash unit—Device for producing a brilliant flash from a xenon-filled tube by the discharge of a capacitor through the tube. Mostly used in photography.

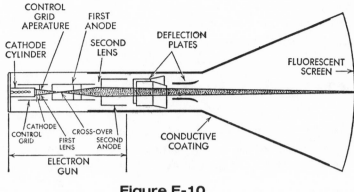

Figure E-10
Electron Gun

electronic organ—Keyboard musical instrument, producing sound by means of
electronic oscillators, and simulating the sound of the pipe organ.

electronics—Rapidly becoming the largest industry in the world, electronics is the key
technology of the last quarter of the twentieth century. Originally coined to describe
the behavior of free electrons in evacuated containers, it now includes their activity in
gases, liquids and solids, or wherever they appear. In fact, the electron is so universal
and fundamental as a constituent of matter that it is virtually impossible to limit the
scope of the subject or the potential of the industry.

electron image intensification—Tubular multipliers formed into a parallel array of
small-diameter elements, or a thin-film or porous supported layer having the side of
primary incidence made electrically conductive, are used in multiplier tubes
employed for amplifying very weak currents produced in camera tubes under low
light conditions. *Electron Tubes.*

electron microscope—Microscope using a beam of electrons instead of a beam of light
to illuminate the object of magnification. *Electron Microscope.*

ELECTRON MICROSCOPE

Limitations of Optical Microscope

Because of diffraction, the image of a point source of light formed by a perfect lens
appears as a bright disk surrounded by fainter concentric light and dark rings. This
pattern is called an Airy disk. Two point sources close together will produce Airy disks
that overlap. If the overlap is considerable the two disks are not seen as separate; they
are not resolved.

The shortest distance between two disks that allows them to be seen separately is
called the resolved distance, which is equal to the wavelength of the light divided by

the numerical aperture (equivalent to a camera's f-number). Since the numerical aperture of the best optical microscope is of the order of 1.6, the smallest resolved distance using visible light is about 0.3 micrometer. This allows bacteria to be seen, but not viruses.

Principle of the Electron Microscope

In 1924, Louis de Broglie demonstrated that electrons in a beam might be regarded as behaving like light, but with a wavelength one hundred-thousand times smaller. Soon afterwards the first electron microscopes were constructed, using magnetic or electrostatic fields instead of glass lenses, but the specimens examined were often severely damaged by heat. These and other problems have now largely been overcome, and the modern instrument is easy to operate, reliable, and capable of resolving objects that are less than 0.3 nanometer apart.

Transmission Electron Microscope

Figure E-11 illustrates the arrangement and operation of a transmission electron microscope, which has three essential systems: (1) an electron gun and condenser system that produces the electron beam and focuses it on the specimen being examined; (2) an image-producing system with an objective lens, movable specimen stage, intermediate and projector lenses, which focusses the electrons passing through the specimen to form a real, highly magnified image; and (3) the image-recording system, consisting of a fluorescent screen for viewing and focussing the image, and a camera for permanent records. A vacuum system and regulated electronic power supplies are also required.

The electron gun and condenser system. The cathode is a heated V-shaped tungsten filament with its point toward the condenser lenses. Electrons are emitted from the point and pass through an aperture in the shield surrounding the filament. Both filament and shield are at a high negative potential. The anode, which is a disk with an axial hole, is at a high positive voltage, so the electrons are strongly accelerated as they approach the anode and pass through the hole. These potentials must be well regulated to ensure uniform and satisfactory operation.

The condenser system focusses the beam so that it impinges on the specimen as a small spot. This is desirable to avoid unduly overheating it and damaging it by irradiation.

The image-producing system. The specimen is mounted on a copper grid carried in a small holder on a movable specimen stage, controlled by a micropositioning system working through vacuum seals, so that it can be moved around for examination of all parts. The objective lens provides the first magnification stage, some 20-200 times, and has a means of adjusting the coil current to improve the focus. Further magnification takes place in the intermediate and projection lenses, and the final image on the fluorescent screen is from 1,000 to 50,000 times the size of the specimen. (The maximum magnification possible with an optical microscope is about 1500 times.)

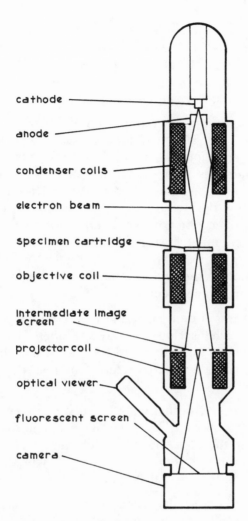

cathode

anode

condenser coils

electron beam

specimen cartridge

objective coil

intermediate image screen

projector coil

optical viewer

fluorescent screen

camera

Figure E-11
Transmission Electron Microscope

Image-recording system. The image is formed by the electron beam on a fluorescent screen and is viewed through a low-power optical microscope. However, this is used mainly for quick observation and focussing, as the screen image is deficient in brightness and contrast, and the screen is grainy. Therefore, after final adjustments have been made, it is photographed. The contrast of the negative can be enhanced, and enlargement to any desired size can then be done by normal photographic methods.

Scanning Electron Microscope. Certain disadvantages in the transmission electron microscope led to the development of the scanning electron microscope. Chief among these are: (1) the specimen has to be placed in a vacuum, so living organisms cannot be examined, and inanimate ones may be distorted; (2) specimens must not exceed 0.1 micrometer in thickness or electrons even accelerated by 60 kilovolts will not be able to pass through; (3) three-dimensional views are not possible; (4) the electron image is formed by heavy atoms in the specimen. If it does not have them, contrast is poor and heavy metals or their salts must be used to prepare the specimen; this in turn is an obstacle to analyzing its chemical composition.

The scanning electron microscope overcomes some of these limitations by scanning the surface of the specimen with a beam of focussed electrons. The action of the beam electrons is to stimulate emission of secondary electrons from the surface of the specimen, which emission varies from point to point according to the brightness of the surface. The secondary electrons emitted are collected and passed to a scintillator crystal, which produces tiny flashes of light as the electrons strike it. The flashes are detected by a photomultiplier tube which converts them to electrical signals, which are then used to form a very much enlarged image on a television picture tube, using standard television techniques. There is a point-by-point correspondence between the brightness of each location in the picture and the number of secondary electrons emitted from that location on the specimen.

No elaborate specimen-preparation technique is required, and large and bulky objects can be handled. The surface of the specimen is made conductive by evaporating a 10-20-nanometer-thick metal film on to it in a vacuum, or by spraying it with an antistatic spray, to obtain a sharp picture. Contrast is adjusted by varying the potential across the surface.

Electron-probe microanalyzer. Another scanning technique using an electron beam is called an electron-probe analyzer. This time the scanning beam excites the x-ray emission characteristic of the elements of which the specimen is composed. By measuring their intensity and studying the displayed television picture, valuable chemical information about the specimen is obtained.

High-voltage electron microscope. By using accelerating voltages in excess of 100 kilovolts (the maximum usually found in conventional transmission instruments) the theoretical resolving power is increased, thicker specimens can be studied, heating and irradiation damage is decreased, and improvements in the sharpness of the image result. Voltages up to 1,200,000 volts are used, although there are problems with voltage regulation, and x-ray emission is produced that could be dangerous to the operator without adequate protection.

Field-emission microscope. In this instrument a fine wire with a rounded tip is mounted in a cathode-ray tube. A high-potential electric field causes electrons to be emitted from the tip. The electrons then form an image of the metal tip on the screen, the magnification being proportional to the ratio of the radius of curvature of the screen to that of the tip. This is useful for studying the structure of the metal forming the tip, but the metals are limited to heat-resistant elements such as tungsten, platinum and molybdenum, because of the high current at the tip.

Proton-scattering microscope. In this microscope a beam of protons from an ion gun is used instead of electrons. This beam is made to bombard a target consisting of a crystal of a metal such as copper or tungsten. The pattern of scattered protons excites a fluorescent screen to provide information on the orientation of the crystal. A variant of the ion-beam microscope employs an ion field in a manner similar to that of the field-emission instrument to observe defects in the crystal lattice, radiation or chemical damage, and the like.

electron multiplier—Electron tube that amplifies a corpuscular or photon emission by means of the secondary electron emission produced by it. *Electron Tubes.*

electron tube circuits—Circuits employing electron tubes as the active element because of higher voltage-handling capability than semiconductor devices, e.g., high-power radio transmitters, radar pulse generators; and special applications, such as television cameras, cathode-ray oscilloscopes, and so on. *Electron Tube Circuits.*

ELECTRON-TUBE CIRCUITS

Applications

Transistors have largely replaced electron tubes in low-power applications. They are, however, still used in the high-power stages of radio and similar transmitters, and as pulse generators for radar and other pulse service equipment. (See article on *Electron Tubes.*) Electron-tube circuits are divided into amplifiers and oscillators.

Classification of Amplifier Circuits

1. Amplifiers are classified as follows, according to the conditions under which they are operated:

Class A:	Biased so that plate current flows continuously throughout the electrical cycle.
Class AB:	Biased so that plate current flows during appreciably more than half but less than the entire electrical cycle.
Class B:	Biased so that plate current flows during only half of the electrical cycle.
Class C:	Biased so that plate current flows during appreciably less than half of the electrical cycle.

If positive grid current flows during some portion of the cycle the subscript 2 may be added; if it does not, the subscript 1 may be used.

2. They are also classified according to their external circuits. In this method, the circuit containing the amplifier tube is regarded as a four-terminal network with two input and two output terminals. One input and one output terminal are common, as shown in Figure E-12.

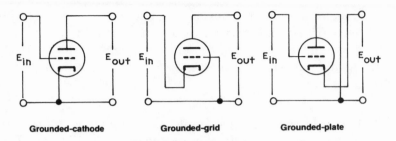

Figure E-12
Amplifier Circuit Classification

Amplifiers Pairs

A *grounded-cathode, grounded plate* pair is shown in Figure E-13. This circuit provides the gain and phase reversal of a grounded-cathode stage with a low source impedance at the output. It is especially useful in feedback circuits or for amplifiers driving a low or unknown load impedance. The direct coupling is an advantage for pulse work.

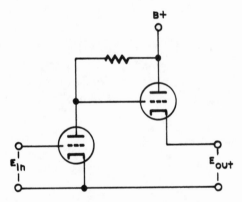

Figure E-13
Grounded-Cathode, Grounded-Plate Amplifier

A *grounded-plate, grounded-grid* pair, shown in Figure E-14, can be used as a differential amplifier. In this circuit, if both inputs are identical there will be no output, but if there is a difference it will be amplified. It is, therefore, able to suppress common-mode signals. If one input is grounded it becomes a phase inverter.

A *grounded-cathode, grounded-grid* pair, often called a *cascode amplifier,* is shown in Figure E-15. It has characteristics resembling a pentode, but does not require screen current. V_2 isolates V_1 from the output load, drastically reduces capacitive feedback without introducing partition noise. This makes it valuable as an r.-f. amplifier, and it was much used in TV tuners.

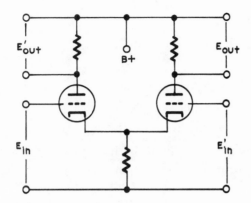

Figure E-14
Grounded-Plate, Grounded-Grid Amplifier

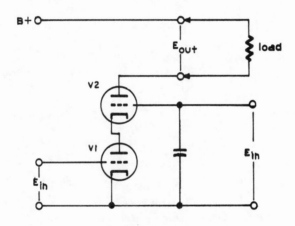

Figure E-15
Grounded-Cathode, Grounded-Grid, or Cascode Amplifier

Cathode-follower Data

A cathode-follower circuit (see Figure E-12) has the following characteristics:

> High-impedance input, low-impedance output
> Input and output have one side grounded
> Good wideband frequency and phase response
> Output in phase with input
> Voltage gain less than one
> Power gain obtained
> Input capacitance reduced

Relaxation Oscillators

These comprise multivibrators and blocking oscillators. There are three types of multivibrator:

    Monostable, having one resting state

    Bistable, having two resting states

    Astable, having no resting state

A *monostable multivibrator* is typified by the Schmitt trigger (Figure E-16), which is commonly used for pulse generation. An input signal results in a pulse output. Between inputs the circuit returns to its resting state.

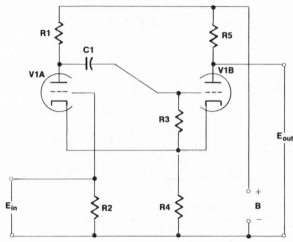

Figure E-16
Monostable Multivibrator

A *bistable multivibrator* is also called a flip-flop and is suitable for binary counters and frequency dividers. As shown in Figure E-17, the circuit is cross-connected and operates so that one tube is saturated while the other is cut off. A positive input signal to the grid of the cutoff tube turns it on, and the resultant negative-going potential at its plate coupled to the grid of the other tube, turns that tube off. The circuit remains in that state until reset by a positive signal at the other input.

An *astable multivibrator* (Figure E-18) is free running and requires no inputs. Its repetition rate $f = 1/t$, where t is the period in seconds. This is given approximately by

    $t = 2[R_1 C_1 \log_e(E_b - E_m/E_x)]$

    where E_b = the plate-supply voltage
             E_m = the minimum alternating voltage on the plate
             E_x = the cutoff voltage corresponding to E_b
             R_1 = the resistance value of R_1 in ohms
             C_1 = the capacitance of C_1 in farads

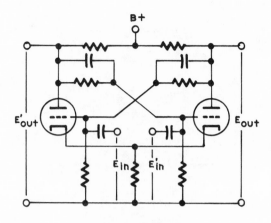

Figure E-17
Bistable Multivibrator

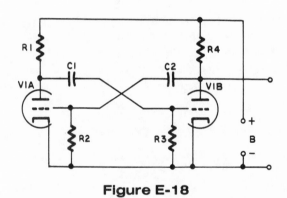

Figure E-18
Astable Multivibrator

A *blocking oscillator* may be free running or driven. A typical driven type (such as would very likely be used in an electron-tube television receiver) is shown in Figure E-19. In this circuit, when power is first applied an increasing plate current flows through the upper winding of T1, and because the magnetic field is building up it induces a voltage in the lower winding that couples through C1 to make the grid positive. Grid current flows and charges C1 negatively until the tube is cut off. C1 now discharges through R1 and R2. Just before the tube would resume conducting of its own accord a sync pulse arrives that triggers it into conduction, and the cycle is repeated. In the absence of a sync signal the oscillator will free run. Frequency is determined by the time constant of C1 and R1, R2.

electron tube—Evacuated glass or metal envelope in which the conduction of electricity takes place through a vacuum or gas; used to amplify weak radio or audio signals, generate electrical oscillations, produce x-rays, and so on. *Electron Tubes.*

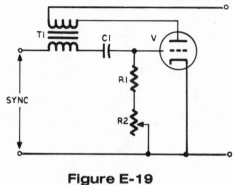

Figure E-19
Blocking Oscillator

ELECTRON TUBES

Electron Emission

An electron tube is an evacuated glass or metal envelope in which a stream of electrons is generated and controlled in such a manner as to perform a useful function in an electronic circuit. Electron tubes include both vacuum tubes in which the air or gas has been exhausted from the enclosure, and gas tubes, in which a small amount of gas has been introduced into the enclosure.

The generation of the electron stream is performed by one of four processes of emission from the surface of the cathode electrode:

Thermionic emission

Secondary emission

Field emission

Photoemission

Thermionic emission is obtained by heating the emitter substance until its electrons have sufficient thermal energy to escape from the surface. There are two ways of doing this: direct heating and indirect heating.

In direct heating, a current is passed through a filament of the emitter substance. This may be tungsten, thoriated tungsten (tungsten containing dissolved thorium), or tantalum. Thermionic emission occurs in the range 1677-2327°C, depending on which substance is used.

In indirect heating, a current is passed through a heater wire which is inside a nickel sleeve coated with oxides of barium, calcium, and strontium. The oxide coating emits electrons in the range 727-877°C. This is more efficient than direct heating and has a further advantage that alternating current can be used in the heater without introducing hum. However, indirect heating cannot be used with very high plate voltages, as in transmitter tubes, because oxide coatings would be apt to disintegrate. Such tubes are generally provided with thoriated tungsten filaments.

Secondary emission occurs when the surface of a solid is bombarded by charged particles. If the bombarding particles are electrons (the usual case), they are called *primary electrons*. On striking the solid with sufficient energy, they excite the surface electrons to an energy level that enables them to escape as *secondary electrons*. These form the bulk of the secondary emission, which also includes a small percentage of reflected primaries. The most common materials used for secondary emission are silver-magnesium and copper-beryllium alloys, although the highest emission is obtained from single-crystal insulators having low electron affinity. Conversely, emission is lowest from porous carbon deposits because the surface structure physically traps the electrons.

Field emission results when an electric field of sufficient magnitude affects the surface of a metal. This may be done intentionally, as in the field-emission microscope, but has to be guarded against in the design of very-high-voltage electron tubes. While field emission is a factor in some cold-cathode gas tubes it is used very little in high-vacuum tubes.

Photoemission is the phenomenon where electrons are emitted from a surface which is struck by protons of visible light. Such surfaces consist of various compounds designated by S numbers in which cesium or rubidium are common elements. The emitted electrons are called photoelectrons, and their energy (expressed in volts) is proportional to the wavelength of the incident light and the temperature.

Electron Currents and Beams

A current of energetic electrons within an evacuated vessel is a characteristic element of many technical devices of electronics engineering. The electrons, usually produced from a thermionic cathode as explained above, are controlled by electrical and magnetic fields generated by voltages applied to various electrodes within the vacuum vessel or by currents in external coils. Since the electron is a negatively charged particle it is attracted by positive charges, repelled by negative ones, and deflected by magnetic fields. It is also extremely light (mass $= 9.10908\,(13) \times 10^{-31}$kg), so that it can be accelerated easily. Many electron-tube devices contain an electron gun, a succession of apertures through which the electrons are accelerated by electrostatic fields into a well-defined beam of a few kilovolts energy. If much higher energies are required ($>10^7$ eV), the electrons are made to travel in bunches along a tube or round and round in a magnetic field while being continuously accelerated by a synchronous electric field. Each electron then suffers a vast relativistic increase of mass and travels very nearly with the velocity of light. (In the Tektronix 7104 oscilloscope cathode-ray tube the electron beam can move laterally across the microchannel plate *at speeds exceeding the velocity of light.*)

Since the flow of electrons is always outward from the cathode, the simplest thermionic vacuum diode is a rectifier of alternating current. This current may be controlled by the voltage applied to a grid through which it passes on its way to the anode. Such an electron tube is a triode, the prototype electronic valve device, as shown in Figure E-20. Further grids are added to control space charge (the build-up of

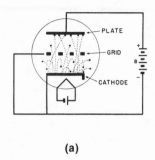
(a)

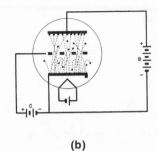
(b)

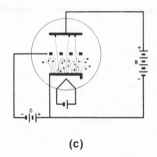
(c)

Figure E-20
Principle of the Triode Electron Tube

(a) *When there is no grid bias, an average number of electrons flows to the anode (plate), and the rest form a space charge between the cathode and the grid.*

(b) *Making the grid positive greatly increases the number of electrons moving to the anode.*

(c) *Making the grid negative reduces the number of electrons moving to the anode. If it is made sufficiently negative, they cannot flow at all.*

electrons in a region of the tube giving rise to fields opposing the entry of more electrons), and secondary emission from the anode (caused by the impacts of primary electrons). To obtain amplification or oscillation at very high frequencies, the electrons are formed into a beam and travel in bunches through a succession of resonant cavities in a straight line or bent around a magnetic field.

Electrode Dissipation

After the electron stream has given up the useful component of its energy, the remainder is dissipated as heat in some suitable part of the tube. There are five methods used for heat removal:

Radiation cooling

Water cooling

Force-air cooling

Evaporative cooling

Conduction cooling

Radiation cooling is used for smaller, low-power tubes. In this method the part of the tube from which the heat is dissipated is allowed to reach a temperature at which heat is radiated to the surroundings.

Water cooling requires that the tube have an additional structure through which water circulates. High-power tubes need a flow of water in gallons per minute to prevent an excessive rise of temperature, so a pump is necessary.

Forced-air cooling is achieved by forcing a stream of air past a suitable radiator, and so requries suitable ducting and a blower.

Evaporative cooling is a method in which a tube has a specially designed anode that is immersed in a boiler filled with distilled water. The heat from the anode boils the water, and the steam rises up through a condenser which turns it back into water. This water runs down into the boiler by gravity, so no pump is needed in this system, or, at most, only a very small one, compared to water cooling. Since the method exploits the latent heat of steam it is much more compact. Heat can be dissipated at a rate of up to 500 watts/centimeter^2.

Conduction cooling is obtained by means of a heat sink. Heat is conducted from the anode via a ceramic insulator to the heat sink since the anode is at a high potential. Beryllia is a suitable ceramic for use with a typical anode temperature of up to 250°C, but as it is a toxic substance it must be handled carefully.

Noise in Tubes

Noise in tubes can have several causes, among which are:

> Shot effect
> Partition noise
> Flicker effect
> Collision ionization
> Induced noise
> Miscellaneous noise

Shot effect noise arises as a result of random variations in the number and velocity of electrons emitted by the cathode. It sets a limit to the minimum signal that can be amplified and is the cause of snow on a television screen and hiss in a radio receiver.

Partition noise is caused by random fluctuations as the electron stream divides between the electrodes in multicollector tubes.

Flicker effect is the outcome of random variations in cathode activity and temperature. It is important only at low frequencies.

Collision ionization results from electrons colliding with gas molecules. The electrons liberated by the collisions and collected by the anode appear as noise in the anode circuit. Further noise is caused by the positive ions disrupting the space charge.

Induced noise occurs at high frequencies in miniature tubes because the electron stream may induce current in the electrode leads. This is more likely to be encountered above 15 megahertz.

Miscellaneous noise includes hum, leakage microphonics, secondary emission, and so on.

Electron Tube Symbols and Formulas for Operation Up to and Including VHF

A — gain of stage

μ — amplification factor

r_p — dynamic anode resistance, in ohms

g_m — mutual conductance, in siemens

E_p — anode voltage

E_g — grid voltage

I_p — anode current, in amperes

I_k — cathode current, in amperes

R_L — anode load resistance, in ohms

E_s — signal voltage

Δ — change or variation in value

Amplification factor:

$$\mu = \frac{\Delta E_p}{\Delta E_g} \text{(with } I_p \text{ constant)}$$

Dynamic anode resistance:

$$r_p = \frac{\Delta E_p}{\Delta I_p} \text{(with } E_g \text{ constant)}$$

Mutual conductance:

$$g_m = \frac{\Delta I_p}{\Delta E_g} \text{(with } E_p \text{ constant)}$$

Gain of stage:

$$A = \mu \left(\frac{R_L}{R_L + r_p} \right)$$

Voltage output across R_L:

$$E_o = \mu \left(\frac{E_s R_L}{r_p + R_L} \right)$$

Power output across R_L:

$$P_o = R_L \left(\frac{\mu E_s}{r_p + R_L} \right)^2$$

Maximum power output across R_L, when $R_L = r_p$:

$$P_o = \frac{(\mu E_s)^2}{4 r_p}$$

Maximum undistorted power output across R_L, when $R_L = 2r_p$:

$$P_0 = \frac{2(\mu E_s)^2}{9 r_p}$$

Cathode biasing resistor value:

$$R_k = \frac{E_g}{I_k}$$

Operation Above VHF

It is obvious that even electrons must take some time to travel from the cathode to the anode, although at low and medium frequencies this time is an insignificant fraction of a radio-frequency cycle. At ultra-high frequencies such a fraction becomes very significant, and therefore reduced cathode-grid and grid-anode spacing are used, resulting in smaller tubes. The electrodes are smaller also to bring down interelectrode capacitance, and the external connections are kept short and carefully dressed. Even so, the conventional space-charge control tube suffers from such serious losses in performance in the microwave region (>1 GHz) that it is virtually useless.

Microwave Tubes

Microwave tubes generally function on the basis of the modulation of the *velocity* of the electron stream rather than the modulation of its density. There are two principal types:

>Linear beam
>
>Crossed-field

Linear Beam Tubes include the klystron, the traveling-wave amplifier, and the backward wave oscillator.

A *klystron* consists of the elements shown in Figure E-21. The electron gun creates a beam of electrons that passes through the two cavity resonators and the drift space between them. A cavity resonator is a region bounded by conducting walls within which resonant electromagnetic fields may be excited, so it has a specific frequency, and a field of this frequency is produced by applying a radio-frequency signal to it. The electron beam is alternately accelerated and decelerated by the successive half-cycles of the field, the amounts of acceleration and deceleration being proportional to the magnitude of the input signal voltage. The drift space is a region substantially free of externally applied alternating fields, in which the bunches of accelerated electrons catch up with the decelerated ones, so that the electron beam becomes a stream with fluctuating current density. This beam now traverses the output cavity resonator and the variations in density induce amplified voltage waves in the output circuit.

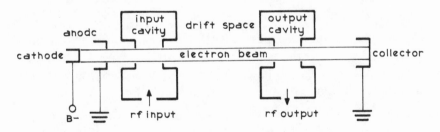

Figure E-21
Two-Cavity Klystron

A *reflex klystron* has only one cavity resonator. After passing through it and being bunched, the electron beam is repelled by a negative reflector potential of several hundred volts, and re-enters the cavity resonator in the opposite direction. In this way one cavity is used for both input and output, but the operation is basically the same.

Reflex klystrons with powers from 10 to 100 milliwatts are used as local oscillators, while those with outputs from 500 milliwatts to 2 watts are suitable for antenna testing and pumping cryogenically cooled masers. Those with output power up to 10 watts are used for frequency modulators in microwave links.

Klystrons with more than two cavities, called multicavity klystrons, are capable of output powers of around 10 kilowatts for continuous-wave frequencies up to 5 gigahertz, for use in UHF television transmitters and so on. Pulsed klystrons also have been designed for outputs of 30 megawatts at 3 gigahertz for nuclear particle acceleration and the like.

Traveling-wave tubes also have an electron gun, which beams a stream of electrons as in Figure E-22. Spaced closely around the beam is a slow-wave device (typically a helix as shown), to which the external radio-frequency input is coupled. This generates an electromagnetic wave that travels along the tube at approximately the same velocity as the electrons in the beam. As it does so it interacts with them to rearrange them, with the result that direct-current energy is transferred from the electron beam to become amplified radio-frequency output power.

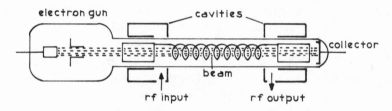

Figure E-22
Traveling-Wave Tube

Traveling-wave tubes have continuous-wave powers ranging from tens of kilowatts in the UHF region to over 100 watts at 10 gigahertz. Several megawatts of pulsed power have been achieved at 3 gigahertz.

Backward-wave oscillator tubes are similar to TWT's, but the wave on the helix travels in the opposite direction, against the electron-beam direction. With a beam current of sufficient magnitude, oscillation takes place. Power output ranges from 100 mW at 1 GHz to 25 mW at 12 gigahertz.

Electron-beam parametric amplifiers are devices consisting of an electron gun, an input coupler and an output coupler, as in Figure E-23. As the electron beam traverses the first coupler it interacts with the input signal, and as it traverses the second coupler

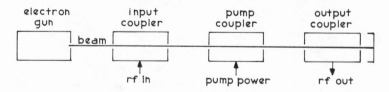

Figure E-23
Electron-Beam Parametric Amplifier

it interacts with a pump frequency that is commonly twice the frequency of the input signal. This serves to boost the energy of the modulated electron beam, provided the two signals are correctly phased, and so achieves amplification. Its greatest advantage, however, is its low-noise behavior. In this respect it resembles the maser (q.v.).

Crossed-Field Tubes include magnetrons, carcinotrons, and crossed-field amplifiers. They generate more power than linear beam tubes.

Magnetrons are high-vacuum tubes used as oscillators. The construction of a typical magnetron is shown in Figure E-24. It is a diode, with a cathode positioned like an axle in the center of a wheel, the wheel being the anode. Several "spokes" divide the wheel structure into a number of cavities. A high direct-current potential of several thousand volts is applied between the cathode and anode, and a flow of electrons is set up from the cathode to the anode. Surrounding the anode is an array of permanent magnets that provide a magnetic field. This field deflects the electrons sideways so that they take a circular path around the cathode. The interaction between the rotating electron field and the cavities generates oscillations, the frequency being determined by the resonant frequency of the separate cavities. Continuous-wave magnetrons have power outputs ranging from 1 kW at 3 GHz to a few watts at 30 GHz, and with pulsed power from 3 MW at 3 GHz to 100 kW at 30 GHz.

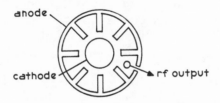

Figure E-24

Magnetron Oscillator (circumferential magnet structure not shown)

Carcinotrons are structured so that they look like magnetrons, but their method of operation is like that of a backward-wave oscillator. The electron beam follows a circular path between a slow-wave structure (which replaces the magnetron cavities) and a "sole," which takes the place occupied by the cathode in a magnetron. The sole is negatively charged, so the electrons have more the form of a beam. After leaving the cathode they make one circuit before being absorbed by the collector. As they do so an interaction with the slow-wave structure takes place that is similar to that in the BWO. The carcinotron has continuous-wave capability of several hundred watts through the X-band (5.2 through 10.9 GHz) and several milliwatts even beyond 300 GHz.

Crossed-field amplifiers are similar to magnetrons in their structure and in the use of a magnetic field; and to carcinotrons in having a circular slow-wave structure for the radio-frequency signal. Thus both the electron stream and radio-frequency signal follow synchronous circular paths around the cathode, the electron stream being the inner path. Voltage peaks in the r-f circuit wave form "spokes" in the electron stream, so if it were visible it would look like a rotating toothed wheel. The interaction between the beam and the circuit wave results in a growing of the latter, and thus gain. This device operates as a saturated amplifier so cannot be used for amplifying a-m signals. Other disadvantages are low gain, limited bandwidth, and high noise. However, it is capable of very high peak output power with relatively high efficiency.

Gas Tubes

Gas tubes are electron tubes containing gas at low pressure. Collisions therefore occur between the electrons traveling from the cathode to the anode and gas atoms, dislodging electrons so that the gas atoms become positive ions. These positive ions neutralize the space charge, with the result that an abundance of electrons becomes available for conduction.

Hot-cathode gas tubes produce electrons by thermionic emission. Ionization takes place when a sufficient voltage is applied between the anode and cathode. The current the tube can handle depends mostly on the emission capability of the cathode.

Cold-cathode gas tubes produce electrons by a combination of secondary emission and field emission. When a high enough voltage is applied between anode and cathode, electrons are emitted from the cathode and flow toward the anode. Collisions with gas atoms produce positive ions which bombard the cathode, freeing more electrons. There are two types of cold-cathode gas tube: glow discharge tubes and arc discharge tubes.

Glow discharge tubes require a drop of several hundred volts between anode and cathode, and the current is of the order of tens of milliamperes.

Arc discharge tubes are typified by the *ignition*. The cathode is a pool of mercury. A small electrode called an igniter touches the surface of the mercury. An external voltage applied between the igniter and the cathode provides an arc which vaporizes the mercury. The vaporized mercury then is able to conduct a heavy current of thousands of amperes at voltage drops of tens of volts.

Transmit-receive tubes are used in radar to isolate the receiver from the antenna when a pulse is being transmitted. Normally the tube conducts signals from the antenna to the receiver, but when a high-power signal comes from the transmitter it ionizes the gas in the tube, which detunes the structure, reflecting all the transmitter power to the antenna. (See Figure E-25.)

Light-Sensing and Emitting Tubes

Image orthicons are the camera tubes most widely used in commercial television. As shown in Figure E-26, this camera tube is housed in a cylindrical glass envelope with an enlarged section at one end. This end has an optically flat glass plate, coated on the inside with a light-sensitive material called a photocathode. The optical image is focussed through the glass on to the coating, where it liberates electrons by

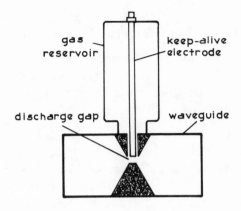

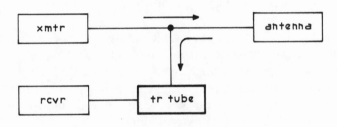

Figure E-25
Transmit-Receive Tube

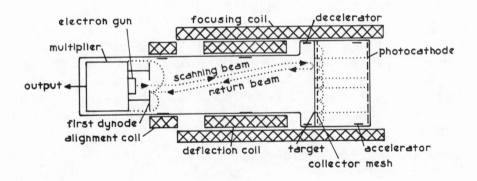

Figure E-26
Image Orthicon Camera Tube

photoemission. The number of electrons emitted at each point of the photocathode corresponds to the amount of light in the image at that point.

The liberated photoelectrons are accelerated toward the surface of a thin semiconducting target and focussed on it magnetically. They strike this target with sufficient energy to liberate a larger number of secondary electrons. The latter are collected by a mesh screen that lies parallel and close to the target.

Where a larger number of photoelectrons strikes the target, the loss of secondary electrons is larger, making that area more positive. Conversely where a smaller number of electrons strikes it, that area of the target is less positive. Because the target is extremely thin, these potentials appear on both sides of it.

At the other end of the tube is an electron gun. The electron beam from this gun is made to scan the rear surface of the target by the magnetic fields generated by the alignment, focussing, and deflection coils surrounding the tube. The electrons in this beam have a low velocity, so that only negligible secondary emission occurs. If the area on the target is positive (bright area) it absorbs electrons until it is neutralized. If the area is not positive (dark area) the beam electrons are turned back and focussed into a five-stage electron multiplier. The number of electrons returned from each point in the scanning pattern corresponds to the brightness of the image at that point. These electrons then enter the electron multiplier.

This consists of a number of secondary-emitting dynodes operated at progressively higher potentials and terminated by an electron-collecting electrode. Electrons striking the first dynode cause the emission of a larger number of secondary electrons, which are accelerated to the second dynode, where the process is repeated, and so on. The amplified electron current at the final stage of the electron multiplier is then passed out of the image orthicon to external circuits.

The typical color television camera contains three image orthicon tubes, with an optical system that casts an identical image on the sensitive surface of each tube. This optical system has semitransparent color-selective mirrors arranged so that each image is red, blue, or green as required to form the television color signals.

Vidicons are small camera tubes, adaptable to portable cameras. Like the image orthicon, the vidicon is housed in a cylindrical glass tube with an electron gun at one end and a transparent conducting film on the inside of the glass face plate at the other. The optical image is focussed on a thin layer of photoconductive material deposited on the gun side of the conducting film. This photoconductive layer is scanned by the electron beam from the gun, and the electrons charge it uniformly to the same potential as the cathode.

However, the charge leaks through the layer at various rates depending upon the intensity of light falling on its outer surface, so an electrical image is formed of corresponding positive potentials with respect to the cathode. Electrons flow from the cathode to the photoconductive layer, through it to the conducting film, and then out to the external circuits. Since the value of the current varies according to the "leakiness" of each picture element scanned, a picture signal is obtained.

Photodiodes are vacuum tubes containing only a photocathode and an anode. Light falling on the photocathode, which is coated with a material containing cesium, causes photoemission, and the photoelectrons are collected by the anode, which has a positive charge. Photodiodes are used in flying-spot scanning systems, in which the light source is a cathode-ray tube. The electron beam in the CRT is deflected in the standard scanning pattern, so a raster appears on the screen. The light from the spot producing the raster is focussed by a lens on the surface of, for example, a motion-picture film, and passes through to excite the photodiode, which produces an electrical signal corresponding to the transparency of the film at each point.

Cathode-ray tubes usually have a large glass, metal or ceramic envelope, with a neck at one end and a flared or cone-shaped portion at the other. The latter has a flat face of glass, coated on its inner surface with a phosphor, and the neck encloses an electron gun. Cathode-ray tubes are mostly used for:

Picture displays

Waveform displays

Character displays

Radar displays

A typical television *picture tube* is shown in Figure E-27. It is a highly evacuated, funnel-shaped structure. At the wide end is the viewing screen, which is a coating of fluorescent material called a phosphor (see Table I on page 215). On the inner side of this coating there is also a thin film of aluminum through which the bombarding electrons pass without hindrance. The aluminum provides a mirror surface that prevents the backward-emitted light from being lost in the interior of the tube, and reflects it forward to the viewer.

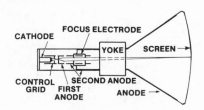

Figure E-27
Monochrome Picture Tube

At the opposite end of the tube is an electron gun. This supplies electrons by thermionic emission from a cathode, then forms them into a beam and accelerates them toward the screen, where in the absence of deflection they would produce a small brilliant spot in the center.

The first electrode next to the cathode is the control electrode, which consists of a flat metal disk or a cup with a hole through which the electrons pass. The picture signal is applied between the cathode and the control electrode. When the signal makes the control electrode less negative with respect to the cathode it permits more electrons to pass, and vice versa. This controls the brightness of the spot on the screen. A brightness control is also provided, which adjusts the average voltage between the cathode and the control electrode, and thus the overall brightness of the picture.

As the electron beam passes from the neck of the picture tube into the flared portion it is deflected by two sets of coils that are placed on the outer surface of this part of the tube. Currents in these coils generate magnetic fields that build up and collapse in such a way as to force the electron beam to sweep across the screen, and also from top to bottom. The horizontal field causes the spot to traverse the screen 15734.264 times per second.

The vertical field moves it from top to bottom 59.94 times per second. These rates are such that each horizontal sweep of the spot traces a line that is slightly below its predecessor, until 525 lines have been scanned (actually two sets of 262½ interlaced; see *Broadcasting*). During its travel, the brightness of the spot varies (explained above) in accordance with picture content. The scanning process is controlled by synchronizing pulses that keep it locked to the scanning process in the television camera from which the picture is orginating. Because of the high speed of the spot, coupled with persistence of vision and duration of phosphorescence, the viewer sees a steady uniform image on the screen.

In a color picture tube there are three electron guns, which produce three separate beams. These are deflected simultaneously by the standard scanning process just described. One beam is controlled by the red primary-color signal, one by the blue, and one by the green. The viewing screen is composed of three sets of individual phosphor patterns. Each set is made of a different phosphor that glows red, blue, or green when excited by its corresponding electron beam. The three beams approach the screen from slightly different angles and pass through apertures in a mask or grille lying directly behind it. In older models the mask (called a "shadow mask") contained about 200,000 precisely located holes, each aligned with three different phosphor color dots. Because of the different paths of the beams, only electrons from the "red" gun hit the "red" dot, electrons from the "blue" gun hit the "blue" dot and electrons from the "green" gun hit the "green" dot. The mask was, however, not altogether satisfactory. It limited brightness and required a high degree of precision in its manufacture and in the adjustment of the receiver. It has since been superseded by a different type of color tube, the Trinitron, in which the shadow mask is replaced by a metal grille having vertical slits extending from the top to the bottom of the screen. The three electron beams pass through the slits to the color phosphors, which are in the form of vertical stripes aligned with the slits. The grille directs the majority of the electrons through the slits, resulting in a much brighter picture. The electron guns of the Trinitron tube share a common focussing mechanism, thus keeping all three beams in fine focus simultaneously. Because the colored elements are so small (they can be seen only with a magnifying glass), the three colored images merge in the mind of the viewer to give him a full-color rendition of the image.

The sizes of picture tubes are usually given by the diagonal measurement across their screens. In the case of a rectangular screen this means the diagonal of a circular screen with the same area. Sizes range from 3 to 27 inches. Newer tubes have a deflection angle of 114°, allowing for shallower overall depth relative to screen size.

In the *oscilloscope*, waveforms are viewed with a rather different type of cathode-ray tube. This tube is longer in relation to its screen dimension, employs static deflection instead of magnetic, and is usually monochrome (see Figure E-28).

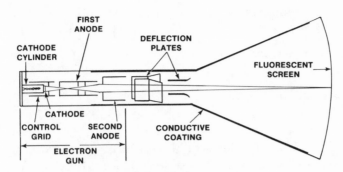

Figure E-28
Oscilloscope Cathode-Ray Tube

The electron gun is similar to that in a picture tube, but the signal is not applied to the control electrode, which is used only to vary the brightness. On leaving the gun the electron beam passes between two pairs of deflection plates, which change its direction by means of electrostatic fields created between them by the voltages applied to the plates. This system does not bend the beam as much as in the picture tube (hence the greater length), but can be employed over a great range of frequencies, whereas the magnetic deflection coils used with a picture tube operate at fixed frequencies.

One pair of plates deflects the beam horizontally. A sawtooth voltage waveform sweeps the beam at an even rate from left to right at a speed selected by the operator, so that it crosses the screen in a precise period of time. The signal whose waveform is to be viewed is applied, after suitable amplification, to the other pair of plates. The potentials on these plates correspond from instant to instant with the amplitude of the signal, and the electron beam is deflected upward and downward accordingly. By choosing a convenient horizontal sweep rate, the spot on the screen displays the amplitude variations of the signal with respect to a time base.

To facilitate analysis of the displayed waveform a graticule of vertical and horizontal lines is provided on the screen. On older models this was scribed on a separate filter mounted over the screen, but modern precision instruments have it etched in the plane of the phosphor to avoid parallax error. The divisions of the graticule are scaled to agree with the settings of the sweep and vertical attenuator controls of the oscilloscope.

Although CRT's generally fit this description, additional refinements are required to display extremely high-frequency signals. In a standard CRT the electron beam delivers sufficient primaries to the screen to produce a bright trace at the lower sweep speeds, but at the higher sweep speeds required to display fast pulses and similar phenomena the beam moves so fast that the number of electrons delivered to each point of the line it traces is very much reduced. The trace is dimmer, even at maximum brightness, and this limits the highest frequency that can be displayed. One method of overcoming this problem is to increase the accelerating voltage to obtain a stronger beam, but the increased velocity of the electrons makes it harder to deflect them, so a smaller screen, much longer CRT, and much higher deflection voltages are required. The following devices are used in CRT's for very high frequencies:

Post-deflection acceleration

Differential deflection plates

Scan-expansion lens

Electron multiplier

Post-deflection acceleration allows the beam to be accelerated moderately by the electron gun, so that it is reasonably flexible as it passes between the deflection plates. It then enters the flared portion of the envelope. This has a narrow ribbon of aquadag in the form of a spiral deposited on it, which acts like a continuous voltage divider to provide an accelerating voltage increasing all the way to the screen. As a result the beam electrons' velocity is greatly increased *after* they have been deflected.

Differential deflection plates are a means of deflecting the beam with an electric field that moves in synchronization with it, so that it is under its influence longer, and so may be deflected more for a given deflecting voltage. Earlier types consisted of a lumped-parameter delay line, with a succession of separate plates and inductors. More recently a helical slow-wave structure has been employed (similar to that in a traveling-wave tube).

A *scan-expansion lens* consists of a number of lens-shaped plate elements, each connected to a different voltage, that increases the deflection of the passing electron beam about four times, after it has been deflected by the deflection plates themselves, so that the deflection plates do not need extremely high potentials to cover the full screen height and width.

Electron multiplication is achieved by placing a microchannel plate just behind the screen. This is a louvered grille, with the louvers spaced 25 micrometers apart. They are coated with a material that gives off secondary electrons when struck by the primaries. The more numerous secondaries are further accelerated by the high voltage on the aluminized screen and thereby provide a display that is sufficiently bright for practical use even at sweep speeds as high as 200 picoseconds/centimeter.

A *storage CRT* produces a visual display of controllable duration. In addition to the standard electron gun (called a writing gun), one or more flood guns are included that deposit low-velocity electrons all over the screen. These do not cause luminescence and very little secondary emission; therefore the screen acquires an overall negative charge. Where high-energy electrons from the writing gun strike the screen, secondary emission depletes it of electrons, so the visible display has a positive charge. This is more attractive to flood-gun electrons so they strike the written area with greater energy and keep it glowing. Since there are two stable states that the phosphor may be in—unwritten and negative, or written and positive—this CRT is called a bistable storage tube. There are other types which do not have a visible display, in which the image is stored on a barrier grid as an electrostatic charge that is subsequently read by scanning, as in a vidicon.

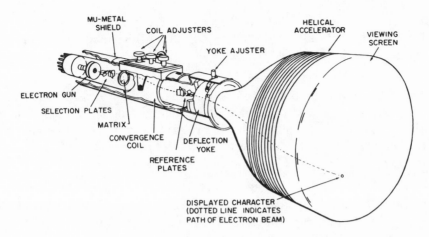

Figure E-29
Charactron Display Tube

Character displays are provided by a cathode-ray tube such as a CHARACTRON. In this CRT the electron beam is deflected by the selection plates (see Figure E-29) and then passes through the proper character in the matrix. A 6-bit code selects the desired character from 64 within the matrix, and a 20-bit signal selects the position on the screen where the character is to be shown. This tube has a display rate of 20,000 characters per second.

Radar displays. (See article on *Radar.*)

TABLE I
Characteristics of Some Commonly Used Phosphors

P Number	Chemical Composition	Color	Persistence	Use
P1	Zn_2SiO_4: Mn	Yellow-green	Medium	Oscilloscope
P2	ZnS: Cu	Yellow-green	Medium	Oscilloscope
P4	ZnS: Ag + ZnCdS: Ag	White	Medium-short	Monochrome TV
P7	ZnS: Ag on ZnCdS: Cu	White	Medium-short	Oscilloscope
P11	ZnS: Ag(Ni)	Blue	Medium-short	Photographic
P15	ZnO: Zn	Green	Short	Flying-spot scanner
P22B	ZnS: Ag	Blue	Short	Color TV
P22G	Zn_2SiO_4: Mn	Green	Medium	Color TV
P22R	$Zn_3(PO_4)_2$: Mn	Red	Medium	Color TV
P31	ZnS: Cu	Green	Medium-short	Oscilloscope

electronvolt (eV)—Unit of energy of charged particles accelerated by electric fields, equal to 1.6×10^{-19} joule, and is kinetic energy acquired by a particle bearing one unit of electric charge (1.6×10^{-19} coulomb) that has been accelerated through a potential difference of one volt. *Nuclear Physics.*

electrophoresis—Movement of electrically charged particles in a fluid under the influence of an electric field. Used to apply coatings to elements employed in electron tubes.

electrophorus—Apparatus consisting of an insulated disk of resin, shellac, ebonite, etc., on the surface of which static electricity is generated by friction, and a similar metal disk with an insulated handle that is brought near so it can pick up an induced charge, to be conveyed where required.

electroplaque—Flattened cell found in electric organ of electric eel, electric catfish, etc. Large numbers of electroplaques are arranged in series and parallel to build up voltage and current-carrying ability of the organ. In the electric eel the ultimate discharge is from 600-1000 volts at one ampere, obtained from electroplaques that generate a voltage of about 150 millivolts each.

electroplating—Deposition of an adherent metal coating by electrolysis. The most commonly used metals for plating are silver, tin, copper, chromium, nickel, zinc, and cadmium. The object to be plated is the cathode, the metal for plating is the anode, both being immersed in an electrolytic solution which deposits metal on the cathode and withdraws it from the anode by ionic action, when a suitable current flows from cathode to anode. (See Figure E-30.)

electroscope—Instrument for detecting the presence of an electric charge, usually consisting of a pair of thin gold leaves suspended from an electrical conducting rod mounted vertically in a sealed glass jar. The upper end of the rod extends out of and above the glass jar and is the terminal. A potential applied to this terminal causes the gold leaves to be charged similarly, so that they stand apart at an angle ("like charges repel"). (See Figure E-31.)

electrostatic—Pertaining to static electricity (q.v.).

electrostatic deflection—Method of deflecting an electron beam by passing it between oppositely charged metal plates, as in a cathode-ray tube. *Electron Tubes.*

electrostatic generator—Apparatus for producing heavy electrical discharges at very high voltage, the voltage thus generated being used especially for accelerating charged particles, as in a Van de Graff generator. *Nuclear Physics.*

element—Fundamental material of matter, of which there are 105 known elements that exist separately and in compounds with other elements. Elements are composed of atoms that are characteristic for each element, and their properties depend upon the number and arrangement of the subatomic particles (protons, neutrons, and electrons) within each atom. Elements arranged according to the configuration of

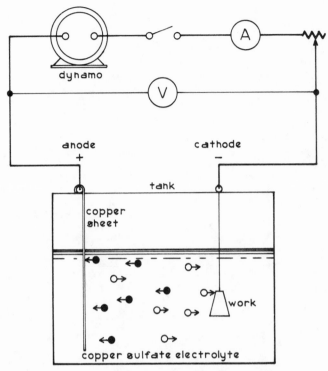

Figure E-30
Electroplating Copper

Figure E-31
Electroscope

From *Complete Guide to Electronic Test Equipment and Troubleshooting Techniques*, page 38.©1976. Reprinted with permission of Parker Publishing Company, Inc.

IA	IIA	IIIB	IVB	VB	VIB	VIIB	VIII	VIII	VIII	IB	IIB	IIIA	IVA	VA	VIA	VIIA	O
1 H 1.0080 1																	2 He 4.0026 2
3 Li 6.939 2 1	4 Be 9.012 2 2											5 B 10.81 2 3	6 C 12.011 2 4	7 N 14.007 2 5	8 O 15.999 2	F 18.998 2 7	Ne 20.183 2 8
11 Na 22.99 2 8 1	12 Mg 24.31 2 8 2											13 Al 26.98 2 8 3	14 Si 28.09 2 8 4	15 P 30.974 2 8 5	16 S 32.06 2 8 6	17 Cl 35.453 2 8 7	18 Ar 39.948 2 8 8
19 K 39.102 2 8 8 1	20 Ca 40.08 2 8 8 2	21 Sc 44.96 2 8 9 2	22 Ti 47.90 2 8 10 2	23 V 50.94 2 8 11 2	24 Cr 52.00 2 8 13 1	25 Mn 54.94 2 8 13 2	26 Fe 55.85 2 8 14 2	27 Co 58.93 2 8 15 2	28 Ni 58.71 2 8 16 2	29 Cu 63.54 2 8 18 1	30 Zn 65.37 2 8 18 2	31 Ga 69.72 2 8 18 3	32 Ge 72.59 2 8 18 4	33 As 74.92 2 8 18 5	34 Se 78.96 2 8 18 6	35 Br 79.909 2 8 18 7	36 Kr 83.80 2 8 18 8
37 Rb 85.47 2 8 18 8 1	38 Sr 87.62 2 8 18 8 2	39 Y 88.90 2 8 18 9 2	40 Zr 91.22 2 8 18 10 2	41 Nb 92.91 2 8 18 12 1	42 Mo 95.94 2 8 18 13 1	43 Tc (99) 2 8 18 13 2	44 Ru 101.07 2 8 18 15 1	45 Rh 102.90 2 8 18 16 1	46 Pd 106.4 2 8 18 18	47 Ag 107.87 2 8 18 18 1	48 Cd 112.40 2 8 18 18 2	49 In 114.82 2 8 18 18 3	50 Sn 118.69 2 8 18 18 4	51 Sb 121.75 2 8 18 18 5	52 Te 127.60 2 8 18 18 6	53 I 126.90 2 8 18 18 7	54 Xe 131.30 2 8 18 18 8
55 Cs 132.90 2 8 18 18 8 1	56 Ba 137.34 2 8 18 18 8 2	57 La 138.91 2 8 18 18 9 2	72 Hf 178.49 2 8 18 32 10 2	73 Ta 180.95 2 8 18 32 11 2	74 W 183.85 2 8 18 32 12 2	75 Re 186.2 2 8 18 32 13 2	76 Os 190.2 2 8 18 32 14 2	77 Ir 192.2 2 8 18 32 15 2	78 Pt 195.09 2 8 18 32 17 1	79 Au 196.97 2 8 18 32 18 1	80 Hg 200.59 2 8 18 32 18 2	81 Tl 204.37 2 8 18 32 18 3	82 Pb 207.19 2 8 18 32 18 4	83 Bi 208.98 2 8 18 32 18 5	84 Po (210) 2 8 18 32 18 6	85 At (210) 2 8 18 32 18 7	86 Rn (222) 2 8 18 32 18 8
87 Fr (223) 2 8 18 32 18 8 1	88 Ra (226) 2 8 18 32 18 8 2	89 Ac (227) 2 8 18 32 18 9 2	104 Rf (261) 2 8 18 32 32 10 2	105 Ha (260) 2 8 18 32 32 11 2													

58 Ce 140.12 2 8 18 20 8 2	59 Pr 140.91 2 8 18 21 8 2	60 Nd 144.24 2 8 18 22 8 2	61 Pm (145) 2 8 18 23 8 2	62 Sm 150.35 2 8 18 24 8 2	63 Eu 151.96 2 8 18 25 8 2	64 Gd 157.25 2 8 18 24 9 2	65 Tb 158.92 2 8 18 27 8 2	66 Dy 163.50 2 8 18 28 8 2	67 Ho 164.93 2 8 18 29 8 2	68 Er 167.26 2 8 18 30 8 2	69 Tm 168.93 2 8 18 31 8 2	70 Yb 173.04 2 8 18 32 8 2	71 Lu 174.97 2 8 18 32 9 2
90 Th 232.04	91 Pa (231) 2 8 18 32 20 9 2	92 U 238.03 2 8 18 32 21 9 2	93 Np (237) 2 8 18 32 22 9 2	94 Pu (244) 2 8 18 32 24 8 2	95 Am (243) 2 8 18 32 25 8 2	96 Cm (245) 2 8 18 32 25 9 2	97 Bk (247) 2 8 18 32 26 9 2	[illegible] (249) 2 8 18 32 28 8 2	99 Es (254) 2 8 18 32 29 8 2	100 Fm (252) 2 8 18 32 30 8 2	101 Md (256) 2 8 18 32 31 8 2	102 No (254) 2 8 18 32 32 8 2	103 Lw (257) 2 8 18 32 32 9 2

Figure E-32
(See page 219.)

Figure E-32
Periodic Table of the Elements

- *The letter in the circle is the symbol of the element.*
- *The number above the symbol is the element's atomic number.*
- *The number below the symbol is the element's atomic weight (numbers in parentheses are the atomic weights of the most stable isotopes of those elements).*
- *The figures on the left are the numbers of electrons in each atomic shell, starting with the innermost.*
- *Elements with atomic numbers 58 through 71 are a very closely related series called the lanthanons, or rare-earth elements; therefore they are all listed in Group IIIB, following lanthanum.*
- *Elements with atomic numbers 90 through 103 are a very closely related series called the actinons, or transuranium elements; therefore they are listed in Group IIIB, following actinium.*

their electrons obey the periodic law, which is illustrated by the periodic table. (See Figure E-32.)

ELF—Abbreviation for extremely low frequency (30-300 Hz). *Frequency Data.*

e.m.f.—Abbreviation for electromotive force.

EMI—Abbreviation for electromagnetic interference.

emission—1. Production of electrons in an electron tube by thermionic, secondary, field, or photoemission. *Electron Tubes.* 2. Designation consisting of bandwidth in kilohertz and symbol to show type of modulation of the main carrier of a radio transmission. *Frequency Data.*

emission-type tube tester—Instrument that measures the emission of electrons from the cathode of an electron tube. Usually the indicating meter reads only "good" or "bad," according to whether a satisfactory level of emission exists or not.

emitter—The region of a bipolar transistor that, when forward biased, injects charge carriers into the base region. These are electrons in an npn unit, holes in a pnp unit, which therefore become minority carriers in the base. *Transistors.*

emitter-coupled logic (ECL)—High-speed bipolar logic gate resembling a differential amplifier. Also abbreviated CML (current-mode logic). *Microelectronics.*

emitter follower—Common-collector transistor circuit. *Transistor Circuits.*

end effect—Capacitance added to wire antenna by the insulators, compensated for by shortening the length by about five percent (half-wave dipole for frequency not exceeding 30 MHz). *Antennas.*

energy—Equivalent of or capacity for doing work.

energy band—One of the bands or ranges of allowed energies, called energy levels, available to an electron in a solid material. Two important bands of allowed energy are the valence band and the conduction band. *Semiconductors.*

enhancement mode—Manner of operation of an insulated-gate field-effect transistor (IGFET) where increasing the gate voltage enhances conduction. This transistor is normally nonconducting until the gate voltage V_G reaches the threshold voltage V_T. *Transistors.*

envelope—1. Glass, metal, or ceramic container of an electron tube. *Electron Tubes.* 2. Outline of an amplitude-modulated wave.

ephemeris time—Time scale based on the second defined as 1/31 556 925.9747 of the tropical year for 12^h ephemeris time January 0, 1900 (= December 31, 1899). The tropical year is the interval between two consecutive vernal equinoxes. Unfortunately this interval decreases by about 5.3 milliseconds per year, so this second must be defined in relation to the tropical year at a particular time. The second is presently defined as the duration of 9 192 631 770 periods of radiation corresponding to the transition between hyperfine levels of the ground state of the cesium-133 atom, this being also the duration of the ephemeris second as defined above. *Frequency Data.*

epitaxial—Having the same structure as an underlying layer. Used to describe the process of growing a crystal layer upon another crystal so that the new crystal has the same crystalline structure as the one on which it is grown.

epoxy resin—Polyether resin comprising a class of thermosetting materials formed by the polymerization (chaining of molecules) of suitable epoxides such as ethylene oxide. These resins are used for adhesives and coatings.

EPROM—Electronically Programmable Read-only Memory (see ROM).

equalizing pulses—In the composite television signal, the six pulses before and after the vertical synchronizing pulse interval that maintain proper interlace. *Broadcasting.*

equal-loudness contours—Contours showing relation between sound intensity levels and frequencies that seem of equal loudness to an average listener. Also called Fletcher-Munson curves. *Electroacoustics.*

equation—In mathematics, an expression of equality between two quantities, as shown by the equal symbol (=). *Electrical Equations.*

erase head—Head on tape recorder that erases recording on a tape by subjecting it to an alternating magnetic field as it runs by the head. See also bulk eraser. *Recording.*

erlang (E)—International dimensionless unit of traffic intensity, equal to one traffic unit, 30 equated busy-hour calls (EBHC), or 36 ces/hour (a ces is a 100-second call or 100 call-seconds of traffic). *Telephony.*

error-correcting code—Communication code used to detect errors in wire transmission. For example, in the commonly used Moore ARQ code each permissible code character contains exactly 3 ones and 4 zeros, in various combinations. Any combination that does not contain 3 ones and 4 zeros is an error and triggers an automatic request-repeat (ARQ). Of course, this code fails if the error pattern consists of 3 ones and 4 zeros, but the percentage of times this happens is small enough to be negligible.

error signal—(See feedback.)

Esaki diode—(See tunnel diode.)

escape velocity—Velocity required for a body to escape from a gravitational center of attraction without further acceleration. For the Earth, escape velocity is approximately 7 miles per second, for the Moon 1.5 miles per second.

etching—Chemical erosion with etchants (acids or other corrosive agents) that produces a design on a surface by eating away selected exposed portions of the surface. *Microelectronics.*

ether—Universal substance believed during the 19th century to be the medium for transmission of electromagnetic waves. Einstein's special theory of relativity in 1905 led to the abandonment of this hypothesis. No evidence for the existence of the ether has ever been found, but strong evidence against its existence was provided by the Michelson-Morley experiment of 1881. *Electromagnetic-Wave Propagation.*

eutectic—Term used to describe that mixture of two substances which has the lowest melting point, and in which both substances liquefy simultaneously. Solder containing 63 percent of tin and 37 percent of lead is eutectic, melting sharply at 361°F. Solders containing tin and lead in other proportions pass through an intermediate range of plasticity, where one substance is only partially melted, before reaching the melting point.

eV—Abbreviation for electronvolt (q.v.).

exclusive OR gate—Logic gate which gives a 0 output when the inputs are the same, but a 1 output when they are different. *Computer Hardware.*

exosphere—Outermost region of the atmosphere (q.v.), extending from between 300 and 600 miles to approximately 3000 miles above the Earth's surface.

expanded-scale meter—Segmental meter, used for monitoring a.c. line voltage, in which a non-linear bridge circuit does not allow current to flow through the movement until the voltage reaches 100 V a.c. Between this voltage and 130 V a.c. the meter reads normally. (See Figure E-33.)

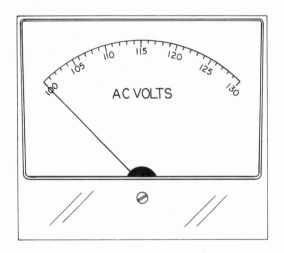

Figure E-33
Expanded-Scale Meter

exponent—Name for a superscript that denotes the number of like terms, multiplied together, that make up the number that is their product. For example, $5 \times 5 \times 5$, which make up 125, may be written 5^3, and called "5 cubed," or "5 to the power of 3." The superscript 3 is the exponent.

exponential function—Transcendental function in the form $y = a^x$. The most important exponential function is $y = \epsilon^x$, in which the constant ϵ (2.7182818 . . .) is the base. A well-known instance of its use is in the RC charge equation:

$$e = E(1 + \epsilon^{-t/RC})$$

exposure meter—Device that measures scene brightness to assist a photographer in selecting the proper combination of shutter speed and lens aperture for his camera. Exposure meters are either photovoltaic or photoconductive. The latter type is more rugged, but requires a battery. (See Figure E-34.)

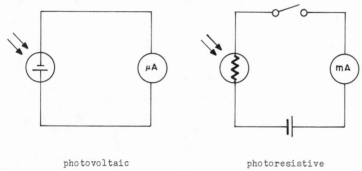

photovoltaic   photoresistive

Figure E-34
Exposure Meter Circuits

extremely high frequency (EHF)—Frequency band between 30 and 300 GHz.

extremely low frequency (ELF)—Frequency band between 30 and 300 Hz.

extrinsic semiconductor—Semiconductor after modification by dopants, traps, dislocations, mechanical stress, or electrical fields. *Semiconductors.*

F—1. Class designation letter for protective device (e.g., a fuse). 2. Symbol for Fahrenheit (q.v.). 3. Symbol for farad. *S.I. Units.* 4. Symbol for chemical element fluorine.

f—1. Symbol for focal length. 2. Symbol for frequency. 3. Symbol for prefix femto- (10^{-15}). *S.I. Units.*

F−—(See A−)

F+—(See A+)

FAA—Abbreviation for Federal Aviation Administration (q.v.).

Fabry-Perot interferometer—In a semiconductor injection laser a resonant cavity consisting of two parallel mirrors facing each other and perpendicular to the plane of the junction. *Lasers.*

facsimile transmission—Electrical transmission of pictures, maps, and other printed material by the technique of scanning, in which a photoelectric cell scans the subject matter line by line, converting the light and dark portions into corresponding electric current. At the receiver, this varying current is used to reproduce the original image.

fade—Gradual lowering of the signal amplitude, instead of cutting it off sharply. Used in broadcasting to "fade in" a program. (Fade in is the opposite of fade out.)

fading—Decrease in intensity of an incoming radio signal due to changes in the ionosphere, precipitation, or other factors. *Electromagnetic-Wave Propagation.*

Fahrenheit temperature scale—Scale based on 32° for the freezing point of water and 212° for the boiling point of water, commonly used in the U.S. Most other countries use the Celsius scale (q.v.). *Conversion Factors.*

fall time—Time of decay from 90 percent to 10 percent of pulse amplitude. (See pulse.)

fan-in—The number of inputs available to a gate.

fan marker—In the Instrument Low-Approach System (ILS), one of the markers that informs the pilot of his distance from the runway. *Navigation Aids.*

fan-out—Number of gates that a given gate can drive within a given logic family.

farad—The electrical capacitance that exists between two conductors when the transfer of an electric charge of one coulomb from one conductor to the other changes the potential difference between them by one volt. *S.I. Units.*

faraday—Unit of quantity of electricity, used in the study of electrochemical reactions, and equal to the amount of electricity that liberates one gram equivalent of new substance from an electrode. 1 faraday = 9.649×10^4 coulombs.

Faraday effect—The rotation of the plane of polarization of a light beam by a magnetic field.

Faraday's law of induction—The magnitude of the electromotive force (emf) induced in a circuit is proportional to the rate of change of the magnetic flux that cuts across the circuit. If the rate of change of magnetic flux is expressed in webers per second, the induced emf is given in volts.

Faraday's laws of electrolysis—1. The amount of chemical change produced by current at an electrode-electrolyte boundary is proportional to the quantity of electricity used. 2. The amounts of chemical changes produced by the same quantity of electricity in different substances are proportional to their equivalent weights. (See also faraday.)

far-field region—The outer part of the field in front of a paraboloidal antenna, starting about five times the extension of the near field (given by area of antenna aperture/2λ) from the antenna, where the power density begins to decrease in proportion to the inverse square of the distance from the antenna. Also called the Fraunhofer region. *Antennas.*

FCC—Abbreviation for Federal Communications Commission (q.v.).

F-display—In radar, a display in which vertical position of a signal gives elevation error and horizontal position gives azimuthal error. Aiming the radar antenna in accordance with this information moves the signal to the center of the screen.

Federal Aviation Administration—U.S. government agency that operates navigation aids and traffic control systems for both military and civil aircraft. *Navigation Aids.*

Federal Communications Commission—U.S. government agency that regulates communications by telephone, telegraph, radio, and television.

feedback—1. Transfer of a voltage from the output of an amplifier to its input. If the signal is fed back in phase with the input signal it is called positive, direct, or regenerative feedback because it adds to the voltage of the input. If the signal fedback is 180° out of phase with the applied signal it is called negative, inverse, or degenerative feedback because it subtracts from the input voltage.   2. Feedback control of a system is by comparing the output with a reference and generating a correcting signal. *Feedback.*

FEEDBACK

Since there is more energy in the output of an amplifier than there is in its input, it is easily possible to take a part of the output energy and insert it into the input circuit. When this is done the amplifier is said to have *feedback*.

There are two types of feedback. If the voltage that is inserted into the input is 180° out of phase with the signal voltage, the feedback is called *negative, inverse,* or *degenerative.* On the other hand, if the voltage is fed back in phase with the input signal, the feedback is called *positive, direct,* or *regenerative.* With negative feedback the voltage that is fed back opposes the signal voltage; this decreases the amplitude of the input voltage. With a smaller signal voltage, of course, the output is also smaller. The effect of negative feedback, then, is to reduce the amount of amplification.

Negative Feedback

The circuit in Figure F-1 gives degenerative feedback. Resistor R is in series with the regular plate register R_p, and thus is part of the load for the tube. Therefore, part of the

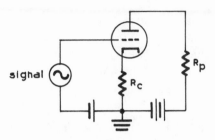

Figure F-1
Negative Feedback (current feedback)

output voltage will appear across R_c. However, R_c also is connected in series with the grid circuit, and so the output voltage that appears across R_c is in series with the signal voltage. In this circuit, the output voltage across R_c opposes the signal voltage and the actual a.c. voltage between the grid and the cathode therefore is equal to the difference between the two voltages.

The advantage of using negative feedback is that the amplification becomes more independent of amplifier characteristics. This means that the frequency-response characteristics of the amplifier become flat—that is, amplification tends to be the same at all frequencies within the range for which the amplifier is designed. Also, any distortion generated in the plate circuit of the tube tends to "buck itself out" when some of the output voltage is fed back to the grid. Amplifiers with negative feedback are therefore comparatively free of harmonic distortion. These advantages, secured at the expense of voltage amplification, are worthwhile if the amplifier otherwise has enough gain for its intended use. Alternatively, additional stages can sometimes be added.

The circuit shown in Figure F-2 can be used to give either negative or positive feedback. In this case the secondary winding of a transformer is connected back into the grid circuit to insert a desired amount of feedback voltage. Reversing the connections of the secondary winding will reverse the phase of the voltage fed back.

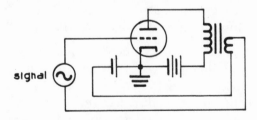

Figure F-2
Negative or Positive Feedback (voltage feedback)

Positive Feedback

Positive feedback increases the amplification because the fed-back voltage adds to the original signal voltage and the resulting larger voltage on the grid causes a larger output voltage. It has the opposite characteristics to negative feedback; the amplification tends to be greatest at one frequency (depending upon the circuit arrangement) and harmonic distortion is increased. If the energy fed back becomes large enough a self-sustaining oscillation will be set up at one frequency; in this case all the signal voltage on the grid is supplied from the plate circuit; no external signal is needed. It is not even necessary to start the oscillation; any small irregularity in the plate current—and there are always some such irregularities—will be amplified and thus give the oscillations an opportunity to build up. Oscillations obviously would be undesirable in an audio-frequency amplifier, and for that reason (as well as for the others mentioned above) positive feedback is never used in a.f. amplifiers. Positive feedback finds its use in oscillators at both audio and radio frequencies.

Feedback of both types is an important characteristic of operational amplifiers (q.v.).

feedthrough capacitor—Feedthrough insulator that provides capacitor between the feedthrough conductor and the chassis or shield through which it is fed.

female connector—Connector that receives a male connector; a socket or receptacle.

femto—Prefix meaning 10^{-15}. *S.I. Units.*

Fermi level—Any energy level having the probability that it is exactly one-half filled with electrons. *Semiconductors.*

ferrimagnetism—Type of weak magnetism occurring in ferrites. The best known example is lodestone, or magnetite (Fe_3O_4). For every four oxygen atoms in this mineral there are three iron atoms, but as two of the iron atoms pair off in opposite directions, only one is left to align with the nonpaired iron atoms in other molecules to form an external magnetic field. Ferrimagnetism is entirely disrupted above the Curie point (q.v.), which is about $570°\,C$ for magnetite, but revives when the temperature is lowered beneath it.

ferrite—Ceramic-like chemical compound of iron oxide and a metal, which exhibits weaker magnetism than iron, but is a much poorer conductor of electricity; thus, it is useful in applications where ordinary ferromagnetic material would introduce too great a loss of electric energy, especially at high frequencies. Examples of such use include loopstick antennas, ferrite cores for inductors, and so on.

ferroelectricity—Property of certain nonconducting crystals, or dielectrics, such as barium titanate and Rochelle salt, that is analogous to ferromagnetism. These crystals are called polar, because their basic structural unit is a tiny electric dipole in which chemical forces keep the centers of positive and negative charge separated. These dipoles are aligned in domains, which can be aligned also by a strong electric field in a manner resembling that of ferromagnetic domains (q.v.).

ferromagnetic domain—Microscopic region within a ferromagnetic material (e.g., iron), in which each atom, behaving as a tiny magnet, aligns itself spontaneously in the same direction. In an unmagnetized piece of iron the domains are randomly oriented with respect to each other, so their fields weaken each other. But under the influence of an external field they align themselves with it. If they remain aligned after the removal of the external field, the iron has been permanently magnetized. *Transformers.*

ferromagnetism—Property of certain materials that produces a strong magnetic field. The magnetism is caused by alignment patterns of their atoms that act as elementary electromagnets. (See also ferromagnetic domain.) Ferromagnetic substances include iron, nickel, and cobalt. Ferromagnetism is disrupted above the Curie point (q.v.), which is 690°C for silicon steel (iron 96%, silicon 4%), commonly used for transformer cores. *Transformers.*

FET—Abbreviation for field-effect transistor (q.v.).

fiber optics—Technique that uses optical-fiber waveguides. *Opto-electronics.*

field—1. Region in space in which a given force (q.v.) has effect.   2. Either of two sets of 262½ lines that are transmitted alternately and interlaced to form the television raster. *Broadcasting.*   3. In computer programming, a region of a statement or line: e.g., label field, operator field, operand field, comment field.

field coil—Coil of insulated wire wound around an iron core to produce a magnetic field as, for instance, the field coil or field winding in an electric generator.

field-effect transistor (FET)—Majority-carrier unipolar three-terminal device in which field modulation of current is the active process. In the junction field-effect transistor (JFET) the gate electrode is a reverse-biased pn junction where the depletion layer extends into a conducting channel and effectively removes carriers from it. The later and more important insulated-gate field-effect transistor (IGFET) has a metal gate electrode insulated from the conducting channel by a thin insulator through which an electric field modifies its conductivity. *Transistors.*

field emission—The escape of electrons from a metal surface subjected to an electric field of sufficient magnitude. *Electron Tubes.*

field strength—Intensity of energy at a given point in a field: e.g., the strength of radio signals at some point in the transmission field of the antenna.

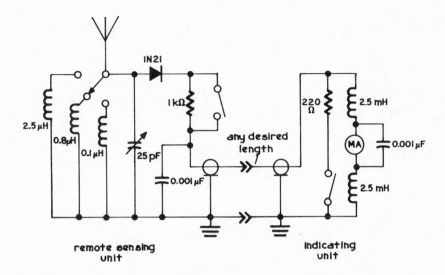

Figure F-3

Field Strength Meter (tuning coils are for 28 MHz, 50 MHz,
and 144 MHz; other values can be substituted)

field strength meter—Used for determining the radiation pattern of an antenna. (See Figure F-3.)

filament—Fine electrical conductor, usually a metallic thread of tungsten or other metal of high melting point, used in incandescent lamps and some electron tubes. In the latter the filament provides a source of electrons, emitted either directly from the hot filament, or from a cathode heated by the filament.

filament transformer—Transformer used to supply current for electron-tube filament, especially where a heavier-than-usual current is required, as in a transmitter power output stage.

filament winding—Secondary winding on a power transformer to provide electron-tube filament current.

file—A block of records stored in a computer memory device.

film capacitor—1. Capacitor with plastic film dielectric. *Capacitors.* 2. Capacitor formed on a hybrid integrated-circuit chip by a film-deposition process. *Microelectronics.*

filter—Arrangement of electronic components that allows some frequencies to pass and blocks others. *Active Filters, Passive Filters.*

filter capacitor—Capacitor used in a filter circuit; usually an electrolytic capacitor used in a power supply circuit. *Rectifiers and Filters.*

fine-tuning control—In the tuner of a television receiver, a small variable capacitor for making fine adjustment to the oscillator frequency. Its manual control is usually an outer ring concentric with the channel selector.

fire-control radar—Radar used in gunnery control.

first detector—Mixer or converter stage in a superheterodyne receiver.

five-bit code—Teleprinter code consisting of five-bit code groups.

five-layer device—Semiconductor device with four junctions. *Semiconductors.*

fixed capacitor—Capacitor with a fixed value of capacitance as opposed to a variable capacitor. *Capacitors.*

fixed decimal point—On a calculator, adding machine, or cash register, where the decimal point is a fixed number of places from the right, commonly two to divide dollars from cents. (See also floating decimal point.)

fixed memory—Read-only memory (ROM). *Computer Hardware.*

fixed resistor—Resistor with a fixed value of resistance as opposed to a rheostat or potentiometer. *Resistors.*

flag—In a computer, a one-bit memory used to indicate zero, carry, sign, parity, and the like. Also called status flag, status bit, or condition code. *Computer Hardware.*

flange—Projecting rim or collar on end of waveguide section by which it is attached to a microwave component or another waveguide. *Waveguides and Resonators.*

flash, electronic—Artificial light source used to provide external light for photographers. (See electronic flash unit.)

flashbulb—Bulb made of glass with plastic coating, containing wire or shredded foil of aluminum with a few percent magnesium (shredded zirconium in small bulbs) in an atmosphere of oxygen, and ignited by a battery (although some newer flashbulbs do not require a battery). Each flashbulb is good for only one flash (flashcubes and other multiple devices contain several flashbulbs).

flash tube—Tube containing xenon gas, which emits a burst of high-intensity light when flashed by a high-voltage pulse from the discharge of a capacitor. (See electronic flash unit.)

flat frequency response—If an amplifier, or other equipment, has a constant-amplitude output for all frequencies of a constant-amplitude input, it is said to have a flat frequency response within whatever tolerance is specified (usually expressed in decibels).

flat pack—Integrated circuit package consisting of a thin rectangular capsule with pins on each side lying in the plane of the capsule. (See Figure F-4.)

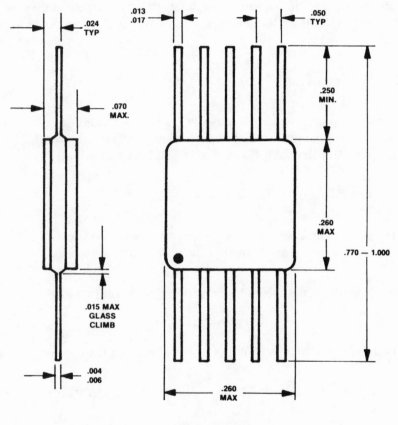

Figure F-4
Flat Pack

F-layer—Region of the ionosphere between 175 and 400 kilometers above the Earth's surface. In the daytime it consists of the lower (F_1) and upper (F_2) layers, which merge at night. *Electromagnetic-Wave Propagation*.

Fleming's rules—1. If the thumb, first, and second fingers of the right hand are extended at right angles to each other, the second finger indicates the direction (conventional) of the flow of current induced in a conductor that cuts lines of magnetic force that run in the direction of the first finger when the conductor is moving in the direction of the thumb. The same relationships in a motor are shown by the left hand.   2. If the fingers of the right hand are placed around a conductor in which the direction (conventional) of current flow is indicated by the direction of the thumb, the fingers indicate the direction of the lines of force of the magnetic field.

Fleming valve—Original name of the electron-tube diode (invented by Sir John Ambrose Fleming in 1904). *Electron Tubes.*

Fletcher-Munson curves—Set of curves of intensity versus frequency that show the real level of sounds that to the ear seem equally loud over the audible range. *Electroacoustics.*

flexible disk—(See floppy disk.)

flip-chip—Standard planar silicon transistor, diode, or integrated circuit having metallic pellets deposited in small holes in oxide over lead areas. This permits facedown bonding after the die is flipped over and located correctly on the substrate.

flip-flop—(See bistable multivibrator.)

flipover cartridge—Phonograph cartridge with two styli, one for microgroove, the other for 78-r.p.m. records, that is turned over to change styli.

floating decimal point—Calculator characteristic that displays the decimal point in its correct position automatically.

floating ground—Common return that is not connected to earth ground.

flood gun—Electron gun used in a storage cathode-ray tube in addition to the writing gun. *Electron Tubes.*

floppy disk—Flexible plastic disk resembling a 45-r.p.m. record used in a computer disk memory system. *Computer Hardware.*

flowchart—Graphic representation of a computer operation indicating the various steps to be taken as the problem moves through the computer. Also used to delineate a manufacturing process. Alternate name, flow diagram. *Computer Software.*

flow soldering—(See wave soldering.)

fluctuating wave—Like pulsating waves (q.v.), fluctuating waves are confined to one side of the zero current or voltage axis; but, unlike pulsating waves, they never touch zero. The chief difference is the d.c. content of the fluctuating wave, hence it is often called a composite wave. (See Figure F-5.)

fluorescence—Emission of a light photon or other radiation by an atom as a result of transition from the excited level to the ground level, after having previously been raised to the excited level by absorption of radiant energy. Interval between absorption and emission is approximately 10^{-8} second. Not to be confused with phosphorescence (q.v.), which persists from 10^{-3} second to days or years, depending on the circumstances.

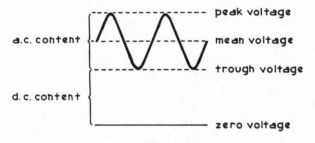

Figure F-5
Fluctuating Wave

fluorescent lamp—Sealed glass tube filled with argon gas and containing a small amount of mercury. When an electric discharge is established in the tube, ultraviolet radiation is produced which causes a phosphor coating on the inside wall of the tube to fluoresce with a bright white glow. A 40-watt fluorescent lamp gives as much light as a 150-watt incandescent bulb and is much cooler in operation.

fluorescent material—(See phosphor.)

fluorescent screen—Image-forming device used in cathode-ray tube or fluoroscope. *Electron Tubes.*

fluorine—Pale greenish-yellow gas with an irritating odor, the most reactive chemical element, used as an oxidizer in rocket fuels. Symbol, F; atomic weight, 18.9984; atomic number, 9.

fluoroscope—Fluorescent screen used to display images formed by x-rays or gamma rays. Used for diagnosis of disease, inspection of manufactured articles, welded joints, etc.

flutter—Deviation from correct pitch (frequency) in a recording resulting from deviation from the mean speed of the medium exceeding 0.1%. Flutter is the name for cyclic deviations occurring at a relatively high rate (e.g., 10 Hz). Cyclic deviations at, say, once per revolution of a turntable are termed "wow." *Recording*.

flux—1. Total amount of radiation passing through a unit area per unit time. If measured in terms of its heating ability, it is expressed in watts; if measured in terms of stimulation to the standard photopic human eye, it is expressed in lumens; it may also be expressed in photons per second.   2. Substance used in soldering to remove oxide film from the metal surfaces so that the intermetallic solvent process (soldering) can take place. Fluxes are resin (rosin), organic, or chloride ("acid").   3. Product of the average component of magnetic induction perpendicular to any given surface in a magnetic field by the area of that surface, expressed in webers.

flux density—Measure of the magnitude and direction of a magnetic field, or concentration of the magnetic flux lines that perpendicularly intersect a unit cross-sectional area. Magnetic flux is expressed in teslas. (One tesla=10,000 gauss, the c.g.s. unit.) The Earth's magnetic field has an average flux density of 50 μT, or 0.5 gauss.

flyback—In a sawtooth wave, the portion of the waveform that returns from the peak amplitude to the reference or zero level.

flyback power supply—Television circuit that uses the flyback portion of the horizontal sawtooth waveform to generate a high d.c. potential for application to the picture-tube anode.

flyback transformer—Transformer used in a flyback power supply. Also called a horizontal output transformer.

flying spot scanner—1. Television pickup device used to transmit images from film transparencies, either still or motion pictures, in which a scanning light-spot on a CRT screen is focussed optically on the surface of the film. The intensity of the light passing through is converted by a photoelectric cell into electric signals.   2. Device that scans printed matter, converting the light reflected from black and white areas into electrical signals which are digitized for computer manipulation. (See Figure F-6.)

FM—Abbreviation for frequency modulation (q.v.).

f-number—Relative aperture of a camera lens, given by the ratio of focal length of the objective to the diameter of the entrance pupil and expressed as a simple ratio—e.g., 1/4.5. In a camera lens with focal length of 50 mm, this would give an entrance pupil diameter, or lens diaphragm opening, of 11 mm.

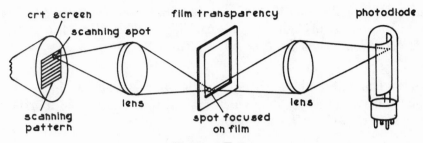

Figure F-6
Flying Spot Scanner

focal length—(See lens.)

focus—(See lens.)

focussing anode—Electrode in the focussing system of an electron gun (q.v.). Also called a focussing electrode. *Electron Tubes.*

focussing coil—Coil around neck of picture or camera tube using a variable magnetic field to focus the electron beam. *Electron Tubes.*

focussing magnet—Permanent-magnet assembly on neck of picture tube used to adjust focus of electron beam. *Electron Tubes.*

folded dipole—Antenna consisting of two simple dipoles parallel and close together, joined at each end and fed in the middle. *Antennas.*

folded horn—Horn type loudspeaker in which the horn is folded back over itself to make it more compact. (See Figure F-7.)

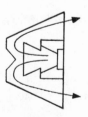

Figure F-7
Folded Horn

foldover—Television picture defect due to extreme nonlinearity in either the horizontal or the vertical deflection circuits.

footcandle—Unit of illumination on a surface that is everywhere one foot from a uniform point source of light of one candle, and equal to 1.076 391 0 = 10^1 lumens/meter^2. *Conversion Factors, S.I. Units.*

footlambert—Unit of luminescence equal to the luminescence of a perfectly diffusing surface that emits or reflects one lumen per square foot. One footlambert equals 3.426 259 candelas/meter^2. *Conversion Factors, S.I. Units.*

forbidden band—Range of energies that an electron cannot attain while within a solid material (in free space it may have any specified energy). Also called forbidden zone, or energy gap, its width determines whether the material is an insulator or a semiconductor (conductors have either no energy gaps or very small ones). *Semiconductors.*

force—Any action that tends to maintain or alter the position of a body or to distort it. Four fundamental forces of interaction among particles of matter are known: gravitational, electromagnetic, and the so-called strong and weak forces that bond components of atomic nuclei. The unit of force is the newton (q.v.).

forecast—A forecast of radio propagation conditions, broadcast from WWV (q.v.) in voice at 14 minutes after each hour, for propagation along paths in the North Atlantic area. *Frequency Data.*

fork oscillator—Oscillator in which the frequency of oscillation is determined by a tuning fork.

form factor—Diameter of a single-layer coil divided by its length. The approximate value of the low-frequency inductance (L) of the coil is

$$L = F\, n^2 d \text{ microhenries}$$

where F is the form factor, n = the number of turns, d = diameter of coil in inches. *Coil Data.*

forming voltage—In an electrolytic capacitor, the voltage at which the anodic oxide has been formed. The thickness of the oxide layer is proportional to this voltage. *Capacitors.*

Formvar—High-temperature (to 105°C) wire-insulating material (vinyl acetate) for wire used in transformer windings. Dimensions of Formvar-coated wire are very nearly the same as for enamelled wire.

FORTRAN (formula translation)—Programming system that includes the language and a compiler, which permits programs to be written in mathematical-type language. It is classed as a fundamental algorithmic and procedural "macrolanguage." *Computer Software.*

fortuitous distortion—Distortion resulting from random causes.

forward bias—External potential applied to a pn junction that lowers the potential barrier, allowing much larger forward current to flow. *Semiconductors.*

forward scatter propagation—Propagation by ionospheric or tropospheric scatter. *Electromagnetic-Wave Propagation.*

Foster-Seeley discriminator—(See discriminator.)

Fourier waveform analysis—Representation of a complex waveform by a Fourier series, in which the individual terms are sinusoidal functions. The complex waveform must meet the following requirements: (a) it must be single-valued and finite in the interval $-\pi$ to $+\pi$ (one single period of oscillation); (b) it must have a finite number of (or zero) discontinuities and a finite number of maxima and minima in the interval $-\pi$ to $+\pi$. *Electrical Equations.*

four-layer diode—Pnpn device with two terminals. Also called a Shockley diode. *Semiconductors.*

four-layer transistor—Pnpn device with three terminals; e.g., a thyristor. *Semiconductors.*

four-terminal oscillator—One of two classes of oscillators, the other being a two-terminal oscillator. *Transistor Circuits.*

Frahm frequency meter—(See vibrating-reed frequency meter.)

frame—1. In television broadcasting, one complete picture consisting of 525 horizontal lines; 30 frames are transmitted and displayed per second to create the illusion of continuous smooth motion. A television frame consists of two fields (q.v.). *Broadcasting.* 2. In motion pictures, a single picture on a length of motion picture film; 24 frames are projected per second.

Fraunhofer region—The region at a substantial distance in front of a paraboloidal antenna in which the power density begins to decrease in proportion to the inverse square of the distance from the antenna; also called the far field. (See also Fresnel region.) *Antennas.*

free-running multivibrator—(See astable multivibrator.)

free-running sweep—In an oscilloscope, an unsynchronized recurrent sweep.

free space—Space completely empty of matter, where there is nothing to reflect, refract, or absorb radiations; an artificial concept, useful for definitions in radiation theory. *Electromagnetic-Wave Propagation.*

free space dielectric constant—$e_v = 8.85 \times 10^{-14}$ F/cm.

free-space room—(See anechoic enclosure.)

F-region—Upper ionized layer of the atmosphere. *Electromagnetic-Wave Propagation.*

Freon—Trade name for members of a group of aliphatic organic compounds containing carbon and fluorine, and in some cases chlorine and hydrogen. Used as refrigerants, aerosols, solvents, fire extinguishers, etc.

frequency—The number of waves that pass a fixed point in one second, expressed in hertz, abbreviated Hz.

frequency counter—(See counter.)

frequency data—Frequency allocations, bandwidths, types of emission, time signals, etc. *Frequency Data.*

FREQUENCY DATA

International Frequency Allocations

| Band No. 12: | 300 gigahertz to 3 terahertz: no frequency allocated |
| Band No. 11: | 30 to 300 gigahertz (millimetric waves), extremely high frequency (EHF) |

GHz	Allocated for:
275	Amateur radio
250-275	Satellites, space research, radio astronomy
240-250	Amateur radio

GHz	Allocated for:
220-240	Satellites, space research, radio astronomy
200-220	Amateur radio
170-200	Satellites, space research, radio astronomy
152-170	Amateur radio
84-152	Satellites, space research, radio astronomy
71-84	Amateur radio
50-71	Satellites, space research
48-50	Amateur
40-48	Satellites
36-40	Mobile
35.2-36.0	Radio location
34.2-35.2	Radio location, space research
33.4-34.2	Radio location
32.3-33.4	Radio navigation
31.8-32.3	Radio navigation, space research
31.5-31.8	Space research
31.3-31.5	Radio astronomy
31.0-31.3	Fixed, mobile, space research

Band No. 10: 3 to 30 gigahertz (centimetric waves), superhigh frequency (SHF)

GHz	Allocated For:
29.50-31.00	Fixed satellite (Earth to space)
27.50-29.50	Fixed satellite (Earth to space), fixed, mobile
25.25-27.50	Fixed, mobile
24.25-25.25	Radio navigation
24.05-24.25	Radio location, amateur radio
24.00-24.05	Amateur radio
23.60-24.00	Radio astronomy
22.00-23.60	Fixed, mobile
21.20-22.00	Earth exploration satellite (space to Earth), fixed, mobile
19.70-21.20	Fixed satellite (space to Earth)
17.70-19.70	Fixed satellite (space to Earth), fixed, mobile
15.70-17.70	Radio location
15.40-15.70	Aeronautical radio navigation
15.35-15.40	Radio astronomy
14.50-15.35	Fixed, mobile
14.40-14.50	Fixed satellite (Earth to space), fixed, mobile
14.30-14.40	Fixed satellite (Earth to space), radio navigation satellite
14.00-14.30	Fixed satellite (Earth to space), radio navigation
13.40-14.00	Radio location
13.25-13.40	Aeronautical radio navigation
12.75-13.25	Fixed, mobile

GHz	Allocated for:
12.50-12.75	Fixed satellite (Earth to space), fixed, mobile (except aeronautical)
12.20-12.50	Broadcasting, fixed, mobile (except aeronautical)
11.70-12.20	Broadcasting, broadcasting satellite, fixed, fixed satellite (space to Earth) mobile, (except aeronautical)
11.45-11.70	Fixed satellite (space to Earth), fixed, mobile
11.20-11.45	Fixed, mobile
10.95-11.20	Fixed satellite (space to Earth), fixed, mobile
10.70-10.95	Fixed, mobile
10.68-10.70	Radio astronomy
10.60-10.68	Radio astronomy, fixed, mobile, radio location
10.55-10.60	Fixed, mobile, radio location
10.50-10.55	Radio location (CW only)
10.00-10.50	Radio location, amateur radio
9.80-10.00	Radio location, fixed
9.50-9.80	Radio location
9.30-9.50	Radio navigation, radio location
9.20-9.30	Radio location
9.00-9.20	Aeronautical radio navigation (ground-based radar), radio location
8.85-9.00	Radio location
8.75-8.85	Radio location, aeronautical radio navigation (doppler radar)
8.50-8.75	Radio location
8.40-8.50	Space research (space to Earth), fixed, mobile
5.925-8.400	Satellites, fixed, mobile
5.725-5.925	Radio location, amateur radio
5.670-5.725	Radio location, amateur radio, deep space research
5.650-5.670	Radio location, amateur radio
5.470-5.650	Maritime radio navigation, radio location
5.460-5.470	Radio navigation, radio location
5.350-5.460	Aeronautical radio navigation, radio location
5.255-5.350	Radio location
5.250-5.255	Radio location, space research
5.000-5.250	Aeronautical radio navigation
4.990-5.000	Radio astronomy
4.700-4.990	Fixed, mobile
4.400-4.700	Fixed satellite (Earth to space), fixed, mobile
4.200-4.400	Aeronautical radio navigation
3.700-4.200	Fixed satellite (space to Earth), fixed, mobile
3.500-3.700	Fixed satellite (space to Earth), fixed, mobile, radio location
3.400-3.500	Fixed satellite (space to Earth), radio location, amateur
3.300-3.400	Radio location, amateur radio
3.100-3.300	Radio location

Band No. 9:   300 to 3000 megahertz (decimetric waves), ultra-high frequency
                (UHF)

MHz	Allocated for:
2900-3100	Radio navigation (ground-based radar), radio location
2700-2900	Aeronautical radio navigation
2690-2700	Radio astronomy
2500-2690	Satellites, fixed, mobile (except aeronautical)
2300-2500	Radio location, fixed, mobile, amateur radio (2300-2450 MHz)
2290-2300	Space research (space to Earth), fixed, mobile
1790-2290	Fixed, mobile
1770-1790	Meteorological satellite, fixed, mobile
1710-1770	Fixed, mobile
1700-1710	Space research (space to Earth), fixed, mobile
1690-1700	Meteorological satellite (space to Earth), meteorological aids
1670-1690	Meteorological satellite (space to Earth), meteorological aids, fixed
1660-1670	Meteorological aids, radio astronomy
1645-1660	Aeronautical mobile satellite
1644-1645	Aeronautical mobile satellite, maritime mobile satellite
1636.5-1644	Maritime mobile satellite
1558.5-1636.5	Aeronautical radio navigation
1543.5-1558.5	Aeronautical mobile satellite
1542.5-1543.5	Aeronautical mobile satellite, maritime mobile satellite
1535-1542.5	Maritime mobile satellite
1525-1535	Space operations (telemetering), Earth exploration satellite, fixed, mobile
1429-1525	Fixed, mobile
1427-1429	Space operations (telecommand), fixed, mobile (except aeronautical)
1400-1427	Radio astronomy
1350-1400	Radio location
1300-1350	Aeronautical radio navigation, radio location
1215-1300	Radio location, amateur radio
960.0-1215	Aeronautical radio navigation
942.0-960.0	Fixed
890.0-942.0	Radio location, fixed
470.0-890.0	Broadcasting (television)
460.0-470.0	Meteorological satellite, fixed, mobile (Citizens Band: 462.5375-462.7375 MHz and 467

MHz	Allocated for:
450.0-460.0	Fixed, mobile
420.0-450.0	Radio location, amateur radio
410.0-420.0	Fixed, mobile (except aeronautical)
406.1-410.0	Radio astronomy, fixed, mobile (except aeronautical)
406.0-406.1	Mobile satellite (Earth to space)
403.0-406.0	Meteorological aids, fixed, mobile (except aeronautical)
402.0-403.0	Meteorological aids, meteorological satellite (Earth to space), fixed, mobile (except aeronautical)
401.0-402.0	Space operations (telemetering), meteorological aids, meteorological satellite (Earth to space), fixed, mobile (except aeronautical)
400.15-401.0	Space research (telemetering and tracking), meteorological satellite (maintenance telemetering), meteorological aids
400.05-400.15	Standard frequency satellite
399.9-400.05	Radio navigation satellite
335.4-399.9	Fixed, mobile
328.6-335.4	Aeronautical radio navigation (glide-path systems)
Band No. 8:	30 to 300 megahertz (metric waves), very high frequency (VHF)

MHz	Allocated for:
273.0-328.6	Fixed, mobile
267.0-273.0	Space operations (telemetering), fixed, mobile
225.0-267.0	Fixed, mobile (survival craft and equipment 243.0 MHz)
220.0-225.0	Radio location, amateur radio
216.0-220.0	Radio location, fixed, mobile
174.0-216.0	Broadcasting (television), fixed, mobile
150.05-174.0	Fixed, mobile (distress and calling-telephone-156.8 MHz)
149.9-150.05	Radio navigation satellite
148.0-149.9	Fixed, mobile
144.0-148.0	Amateur radio
138.0-144.0	Space research (space to Earth), radio location, fixed, mobile
137.0-138.0	Space research (space to Earth), space operations (telemetering and tracking), meteorological satellite
136.0-137.0	Space research (space to Earth)
117.975-136.0	Aeronautical mobile
108.0-117.975	Aeronautical radio navigation
88.0-108.0	Broadcasting (FM radio)
75.4-88.0	Broadcasting (television), fixed, mobile
74.6-75.4	Aeronautical radio navigation
73.0-74.6	Radio astronomy
54.0-73.0	Broadcasting (television), fixed, mobile
50.0-54.0	Amateur radio
38.25-50.0	Fixed, mobile

MHz	Allocated for:
37.75-38.25	Radio astronomy, fixed, mobile
30.01-37.75	Fixed, mobile
30.005-30.01	Space operations (satellite identification), fixed, mobile

Band No. 7:	3 to 30 megahertz (decametric waves), high frequency (HF)

MHz	Allocated for:
29.70-30.005	Fixed, mobile
28.00-29.70	Amateur radio
27.50-28.00	Meteorological aids, fixed, mobile
26.10-27.50	Fixed, mobile (except aeronautical) (Citizens Band: 26.96-27.23 MHz)
25.60-26.10	Broadcasting (international AM radio)
25.11-25.60	Fixed, mobile (except aeronautical)
25.07-25.11	Maritime mobile
25.01-25.07	Fixed, mobile (except aeronautical)
24.99-25.01	Standard frequency (WWV/WWVH)
23.35-24.99	Fixed, land mobile
23.20-23.35	Aeronautical fixed and mobile
22.72-23.20	Fixed
22.00-22.72	Maritime mobile
21.87-22.00	Aeronautical fixed and mobile
21.85-21.87	Radio astronomy
21.75-21.85	Fixed
21.45-21.75	Broadcasting (international AM radio)
21.00-21.45	Amateur radio
20.01-21.00	Fixed
19.99-20.01	Standard frequency (WWV/WWVH)
18.068-19.99	Fixed
18.052-18.068	Fixed, space research
18.03-18.052	Fixed
17.90-18.03	Aeronautical mobile
17.70-17.90	Broadcasting (international AM radio)
17.36-17.70	Fixed
16.46-17.36	Maritime mobile
15.45-16.46	Fixed
15.10-15.45	Broadcasting (international AM radio)
15.01-15.10	Aeronautical mobile
14.99-15.01	Standard frequency (WWV/WWVH)
14.35-14.99	Fixed
14.00-14.35	Amateur radio
13.36-14.00	Fixed
13.20-13.36	Aeronautical mobile
12.33-13.20	Maritime mobile
11.975-12.33	Fixed

MHz	Allocated for:
11.70-11.975	Broadcasting (international AM radio)
11.40-11.70	Fixed
11.175-11.40	Aeronautical mobile
10.10-11.175	Fixed
10.005-10.10	Aeronautical mobile
9.995-10.005	Standard frequency (WWV/WWVH)
9.775-9.995	Fixed
9.500-9.775	Broadcasting (international AM radio)
9.040-9.500	Fixed
8.815-9.040	Aeronautical mobile
8.195-8.815	Maritime mobile
7.300-8.195	Fixed
7.000-7.300	Amateur radio
6.765-7.000	Fixed
6.525-6.765	Aeronautical mobile
6.200-6.525	Maritime mobile
5.950-6.200	Broadcasting (international AM radio)
5.730-5.950	Fixed
5.450-5.730	Aeronautical mobile
5.250-5.450	Fixed, land mobile
5.060-5.250	Fixed
5.005-5.060	Fixed, broadcasting (international AM radio)
4.995-5.005	Standard frequency (WWV/WWVH)
4.850-4.995	Fixed, land mobile, broadcasting (international AM radio)
4.750-4.850	Fixed, broadcasting (international AM radio)
4.650-4.750	Aeronautical mobile
4.438-4.650	Fixed, mobile (except aeronautical)
4.063-4.438	Maritime mobile
4.000-4.063	Fixed
3.500-4.000	Fixed, mobile (except aeronautical), amateur radio
3.400-3.500	Aeronautical mobile
3.200-3.400	Fixed, mobile (except aeronautical), broadcasting (international AM radio)
3.155-3.200	Fixed, mobile (except aeronautical)
3.025-3.155	Aeronautical mobile

Band No. 6:	300 to 3000 kilohertz (hectometric waves), medium frequency (MF)

kHz	Allocated for:
2850-3025	Aeronautical mobile
2505-2850	Fixed, mobile
2495-2505	Standard frequency (WWV/WWVH)
2300-2495	Fixed, mobile, broadcasting (international AM radio)
2194-2300	Fixed, mobile

kHz	Allocated for:
2170-2194	Mobile (distress and calling - telephone - 2182 KHz)
2107-2170	Fixed, mobile
2065-2107	Maritime mobile
2000-2065	Fixed, mobile
1800-2000	Fixed, mobile (except aeronautical), radio navigation, amateur radio
1605-1800	Fixed, mobile, aeronautical radio navigation, radio location
535-1605	Broadcasting (domestic AM radio)
525-535	Mobile, broadcasting, aeronautical radio navigation
510-525	Mobile, aeronautical radio navigation
490-510	Mobile (distress and calling - telegraph - 500 kHz)
415-490	Maritime mobile (radiotelegraphy only)
405-415	Maritime radio navigation (radio direction-finding), aeronautical radio navigation
325-405	Aeronautical radio navigation, aeronautical mobile
Band No. 5:	30 to 300 kilohertz (kilometric waves), low frequency (LF)

kHz	Allocated for:
285-325	Maritime radio navigation (radio beacons), aeronautical radio navigation
200-285	Aeronautical radio navigation, aeronautical mobile
160-200	Fixed
130-160	Fixed, maritime mobile
70-130	Fixed, maritime mobile, radio navigation, radio location
Band No. 4:	3 to 30 kilohertz (myriametric waves), very-low frequency (VLF)

kHz	Allocated for:
20.05-70.00	Fixed, maritime mobile (WWVB at 60 kHz)
19.95-20.05	Standard frequency (WWV/WWVH)
14.00-19.95	Fixed, maritime mobile
10.00-14.00	Radio navigation, radio location
10.00	No allocations
Band No. 3:	300 to 3000 hertz: no frequencies allocated
Band No. 2:	30 to 300 hertz: no frequencies allocated

Notes

1. The foregoing internationally agreed frequency allocations apply in the Western Hemisphere (Region 2). Not all allocated frequencies are actually being used.
2. The expression "fixed" means point-to-point radio communication between fixed stations.

3. The expression "mobile" means radio communication between any type of craft or vehicle and another, or between it and a fixed station. If a particular type of craft or vehicle is specified, radio communication at that frequency is restricted to communication between two mobile stations of that type or between a mobile station and a fixed station.

4. Radio location means radar and tracking systems, whereas radio navigation means navigation aid.

5. Where a band is shared between amateur radio and another service, the other service is the primary service. Amateur operators may not cause harmful interference to it.

6. Meteorological aids include radar and radiosondes.

Microwave Bands

Band Letter	Frequency Range in Gigahertz
P	0.225 - 0.390
L	0.390 - 1.550
S	1.550 - 5.200
X	5.200 - 10.900
K	10.900 - 36.000
Q	36.000 - 46.000
V	46.000 - 56.000
W	56.000 - 100.000
C	3.900 - 6.200
K_1	15.350 - 24.500

(These bands may be divided into subbands designated by lower-case letter suffixes.)

Designation of Emissions

For classes of emissions see under the appropriate entry: e.g., "Class A1 emission."

Standard-Frequency Broadcasts

The standard-frequency and time stations of the National Bureau of Standards are WWV, WWVB, and WWVL at Fort Collins, Colorado, and WWVH at Kekaha, Kauai, Hawaii. WWV and WWVH broadcast continuously on 5, 10, 15, 20, and 25 MHz, giving time signals in Coordinated Universal Time, radio propagation forecasts, and geophysical alerts. WWVB broadcasts time signals in binary code on a frequency of 60 kHz. WWVL is an experimental VLF station whose frequency and schedule are variable.

Coordinated Universal Time (UTC) is based on the time kept by an atomic clock which has an accuracy of one part in 10^{12}. Owing to variations in the speed of rotation of the Earth, International Atomic Time (TAI) and Ephemeris Time (ET) do not always agree exactly, so UTC has to be corrected occasionally by adding or subtracting a "leap second." This keeps the difference to not more than −0.7 second. These adjustments are made at the end of the last minute of December 31 or June 30, as required, so these minutes may contain 59 or 61 seconds according to the sign of the adjustment.

The WWV/WWVH broadcasts consist of a pulse sent out at each second, starting with the first in the minute and accompanied by a continuous audio tone. This goes on for 45 seconds, when the tone ceases and the seconds are marked by ticks. The WWVH station then announces what the UTC time will be at the end of the minute; the WWV station does the same, but 7.5 seconds later. Precisely on the 60th second WWVH transmits a 1200-hertz tone for 0.8 second and simultaneously WWV transmits a 1000-hertz tone for 0.8 second. The audio tone is then resumed unless an announcement is to be made (propagation forecast or geophysical alert). The audio tone alternates between 600 and 500 hertz in successive minutes, and the 29th and 59th second pulses are always omitted. Both stations identify the beginning of each hour with a 1500-hertz tone for 0.8 second.

frequency distortion—(See distortion.)

frequency divider—Bistable multivibrator (q.v.), single or cascaded, that gives an output that is an integral submultiple of the input frequency.

frequency meter—(See vibrating-reed frequency meter.)

frequency modulation (FM)—Varying the frequency of a carrier signal in accordance with the amplitude of the information signal. The amplitude of the carrier does not change. *Broadcasting.*

frequency response—In sound systems, a measure of the ability of a device or system to handle all frequencies within a desired band equally. *Electroacoustics.*

frequency-shift keying (FSK)—Method of digital modulation in which the two binary states are represented by two different frequencies. *Modulation.*

frequency spectrum—(See electromagnetic spectrum.)

Fresnel region—Region in front of and close to a paraboloidal antenna in which the radiation is substantially confined within a cylindrical pattern having the same diameter as the antenna. This region extends to a distance equal to the area of the antenna aperture divided by twice the wavelength, where it merges into the transition zone. The latter is about four times the length of the Fresnel region (or near field), and eventually changes into the Fraunhofer region (q.v.).

friction error—The maximum percentage of the full-scale value that a meter pointer may move when the instrument is tapped after the pointer comes to rest.

fringe area—Area at a distance of 100 miles or more from a television transmitter, assuming flat open terrain between, where a high-gain antenna must be used for satisfactory reception.

frit—Semifused compounds used in ceramics for glazing, such as low-temperature glass, used to make joints of ceramics to metals, or for glass seals.

front end—The section of a receiver the selects the channel or tunes the station and converts the r.f. signal to the i.f.

front porch—In the composite television signal, the portion of the pedestal that precedes the synchronizing pulse. *Broadcasting.*

front-to-back ratio—Ratio of power gain of directional antenna when pointed toward the other station and power gain when the antenna is rotated 180 degrees to point in the opposite direction.

FSK—Abbreviation for frequency-shift keying (q.v.).

fuel cell—Electrochemical cell ("batttery") that differs from others in that the chemicals supplying the electrode processes are stored separately and supplied on demand. The cell continues to operate as long as these reactive chemicals are supplied. As they are usually gaseous, the electrodes must provide a stable interface between the reactants (gaseous), the electrolyte (liquid), and the conductors (solid). The simplest fuel cell is one that uses hydrogen and oxygen. Hydrogen is supplied to the inside of a porous-tube electrode that provides an interface between the gas and the electrolyte, so that hydrogen ions and electrons are produced. The ions travel through the electrolyte to the other electrode, which is supplied with oxygen in a similar manner. At the oxygen-electrolyte interface they absorb electrons from the electrode and combine with the oxygen to form water. Since the hydrogen electrode accumulates electrons while the oxygen electrode loses them, the former acquires a negative charge, the latter a positive charge, so that if they are connected by an external circuit a current will flow in it. A means is also provided for removing, by evaporation or wicking, the water formed in the reaction so as not to dilute the electrolyte. The latter is acidic or alkaline according to cell type.

full-adder—Digital circuit consisting of two half-adders (q.v.) that can add two bits, plus the carry from the previous stage, to give a sum and carry. *Computer Hardware.*

full scale value—Maximum value of a meter range. If the scale does not have zero at one end, the full scale value will be the total value from one end of the scale to the other, regardless of sign.

full-wave rectifier—Power-rectifier circuit that uses both positive and negative halves of the input wave. *Rectifiers and Filters.*

function—In mathematics, any rule f that to each element a in a set A assigns an element b in a set B. The element b is called the value of f at the point a, and it is denoted b = f(a); or b is a function of a, as in $b = a^2$, or b = a + 2.

functional block—Three-dimensional circuit in which the molecular arrangement of the material performs the desired electrical functions. A piezoelectric delay line is an example of a functional block.

function generator—Generator that supplies sine waves, square waves, triangle waves, etc., either singly or severally.

fundamental frequency—The first harmonic or lowest frequency of a complex wave.

fuse—Safety device, usually a short piece of wire of easily fusible metal, that melts when the current in the circuit in which it is installed exceeds its rating. To facilitate replacement of blown fuses, they are mounted in fuse holders, fuse clips, fuse boxes, etc. (See Figure F-8.)

fast-blow   **slow-blow**

Figure F-8
Fuses

fuze—Explosive device used in military and other equipment to fire munitions. Also used to separate stages of a launch vehicle (command fuze or squib).

G—1. Symbol for conductance. 2. Symbol for gravitational constant. 3. Symbol for metric prefix giga-. *S.I. Units.* 4. Class designation letter for electronic chopper, generator, ignition magneto, interrupter vibrator, oscillator, rotating amplifier, or telephone magneto.

g—Symbol for acceleration due to gravity at sea level, 40 degrees latitude, = 9.8017 m/s^2 (32.1578 ft/sec^2).

Ga—Symbol for chemical element gallium (q.v.).

GaAs—Symbol for gallium arsenide (q.v.).

gage—(See gauge.)

gain—In radio, sound reproduction, etc., an increase in signal power from one stage to another in an amplifying system. Expressed in decibels, or so many times; e.g., if the output signal of a stage has five times the power of the input signal, the stage is said to have a gain of 7 dB, or x 5. Used also of voltage gain, current gain, etc.

gain-bandwidth—Product of maximum voltage gain and the 3-db bandwidth of an amplifier.

gain control—Control to vary the gain of an amplifier; e.g., by varying the bias voltage.

galactic noise—Radio interference originating outside Earth or its atmosphere. *Radio Noise and Interference.*

galena—Lead sulfide, crystals of which were used as point-contact detectors in early crystal radio sets.

gallium—Silvery-white soft metal, with very low melting point (about 30°C [86°F]), best known for its compound gallium arsenide (q.v.). Symbol, Ga; atomic weight, 69.72; atomic number, 31.

gallium arsenide—Semiconductor material used for light-emitting diodes (q.v.). Symbol, GaAs. *Semiconductors.*

galvanic series—Two dissimilar metals in sea water comprise a galvanic cell, so that if they are connected by a conductor, current will flow and chemical action take place. This results in one metal being corroded because it is the anode. The galvanic series lists metals in order from most corroded (anodic) to most noble (cathodic). *Properties of Materials.*

galvanizing—Application of a zinc coating to iron and steel to protect them from rust, by either hot-dip process or electrolytic deposition. (See electroplating.)

galvanometer—Sensitive D'Arsonval-movement (q.v.) instrument for measuring small currents, usually with a small mirror attached to the moving coil (however, some have pointers); the angle through which it rotates provides a measure of the current flowing in the coil. (See also ballistic galvanometer.)

gamma ray—Radiation originating in atomic nuclei during radioactive decay. *Nuclear Physics.*

gang—Set of like devices arranged to work together; as variable capacitors sharing a common shaft, etc.

garnet—Name for several varieties of common silicate minerals. In addition to silicon and oxygen, garnets contain calcium, magnesium, iron, aluminum, and chromium. Garnet substrates are used in bubble memories and other microelectronic devices. *Computer Hardware, Micro-electronics, Lasers, Masers.*

gas laser—By far the most varied category of lasers. Gases used most commonly are helium-neon, argon, and carbon dioxide. *Lasers.*

gas masers—Ammonia and hydrogen masers are used as frequency standards. *Masers.*

gas tube—Electron tube containing gas at very low pressure (≈ 1 mm of mercury [133 Pa]), in which ionization takes place by collisions between moving electrons and gas atoms. *Electron Tubes.*

gate—1. A circuit having two or more inputs and one output, the output depending on the combination of logic signals at the inputs. *Computer Hardware.* 2. The control electrode of a field-effect transistor. *Transistors.*

gated-beam detector—FM demodulator circuit employing a gated-beam tube in which the i.-f. input is applied to the limiter grid, to give an electron current consisting of equal amplitude pulses at the i.f. frequency. These are accelerated by the accelerator grid and pass through the quadrature grid to the plate. The potential on this grid oscillates at the i.f. but lags it by 90 degrees, so the number of electrons reaching the plate varies according to the frequency. (See Figure G-1.)

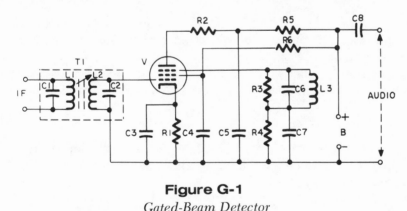

Figure G-1

Gated-Beam Detector

From *Complete Guide to Reading Schematic Diagrams*, First Edition, page 143.

gauge—1. Any device for measuring something. 2. A standard dimension, as of wire, etc.

gauss—Centimeter-Gram-Second (CGS) unit of magnetic flux density. The S.I. unit tesla (q.v.) is preferred. 1 tesla=10,000 gauss. *S.I. Units.*

gauss—Former unit of magnetic flux density, now replaced by the tesla (q.v.). 1 tesla - 10,000 gauss. *S.I. Units.*

Gaussian distribution—(See normal distribution.)

GCA—Abbreviation for ground-controlled approach (q.v.).

GCT—Abbreviation for Greenwich civil time or Universal Time (q.v.)

G-display—Type of radar display in which the blip appears as a spot with wings ("wingspot"). The vertical position indicates elevation error, the horizontal, azimuth error. The length of the wings is inversely proportional to the range. *Radar.*

Ge—Chemical symbol for germanium (q.v.).

Geiger counter—Radiation detector. *Nuclear Physics.*

generator—1. Signal generator (q.v.). 2. Electromechanical device to convert mechanical energy to electrical energy. 3. High-energy-particle accelerator. *Nuclear Physics.*

geometrical distortion—Cathode-ray tube distortion that causes vertical and horizontal lines near the edge of the screen to be bowed outward or inward. Also called "barrel" or "pincushion" distortion.

German silver—Name for a variety of Sheffield plate in which nickel alloyed with copper and zinc was substituted for copper as the base metal underlying the silver coating. The alloy is usually called nickel-silver, but contains no silver itself. Nickel-silver, often incorrectly called German silver, is commonly used for fuse wire. It was formerly called pai-t'ung, its Chinese name, the Chinese having used it from about 200 B.C.

germanium—Silvery-gray brittle metalloid, never found free in nature, used in the manufacture of semiconductor devices. Symbol, Ge; atomic weight, 72.59; atomic number, 32. *Semiconductors.*

getter—Small cup containing chemicals placed inside a vacuum tube (electron tube that is not a gas tube). After initial evacuation the tube is heated and the getter chemicals combine with any residual gas molecules, forming metal compounds that are deposited on the inner surface of the tube envelope (this is the origin of the silvery coating observed on the glass of some tubes).

giga-—S.I. prefix defined as one thousand million times, or 10^9. Symbol, G. *S.I. Units.*

gilbert—Former unit of magnetomotive force equal to $7.957\,747\,2 \times 10^{-1}$ ampere turns. *S.I. Units.*

gimmick—Small capacitance added to VHF circuit, made by twisting together two insulated wires, etc.

glass—Supercooled liquid made by fusing together silica, lime and soda, to form, when cooled, a hard, brittle, transparent substance with good insulating and dielectric properties.

glassivation—Deposition and melting of vitreous material on a semiconductor die to passivate and protect underlying device junctions. *Microelectronics.*

glide slope—Vertical guidance provided by ILS to aircraft approaching runway. *Navigation Aids.*

glitch—1. Random pulses (e.g., noise spikes) having sufficient amplitude to cross the threshold of a device and activate it as if it were a legitimate logic pulse. 2. Narrow horizontal bar that moves vertically on a television screen, caused by low-frequency interference.

glow discharge tube—1. Electron tube containing a cathode and an anode, and a gas at a pressure of one thousandth of an atmosphere or thereabouts. When a potential of several hundred volts is applied between the electrodes, ionization of the gas takes place and light is emitted, its color depending on the gas: neon, red; mercury vapor, bluish; helium in amber glass, gold; mercury vapor in yellow glass, green; and so on. 2. Glow lamps contain a high resistance filament in rarefied gas. The voltage drop across the filament causes the gas to glow faintly.

gm—Symbol for mutual conductance of an electron tube, expressed in siemens.

GMT—Abbreviation for Greenwich mean time (q.v.).

gold-bonded diode—Diode made by pressing a preformed gold whisker against an n-type germanium chip, then alloying it into the chip with controlled pulses of current. A pn junction is thought to be formed. *Semiconductors.*

gold-leaf electroscope—(See electroscope.)

goniometer—Inductive or capacitive device having a stator connected to a fixed antenna system and a rotor connected to a receiver or transmitter, allowing the pattern of the fixed array to be rotated as the rotor is rotated. Extensively used in direction finding and in omnirange beacons. *Navigation Aids.*

gram—One thousandth of a kilogram (q.v.) equal to 0.035 ounce, *S.I. Units.*

gram atomic volume—Volume in cubic centimeters occupied in the solid state by an element at its melting point when the element consists of the Avogadro number of atoms (6.0225×10^{23}).

graph—In analytic geometry, the graph of an equation is the locus of all points satisfying the equation. For example, the graph of $x^2 + y^2 = 1$ in Cartesian coordinates is a circle of radius one centered at the origin. (See Figure G-2.)

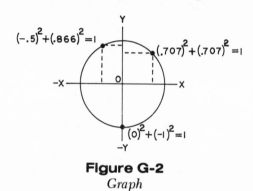

Figure G-2
Graph
$(x^2 + y^2 = 1)$

graphic symbols—Symbols used in schematic diagrams (q.v.) to indicate the components in a circuit and their interconnections. A graphic symbol is not an illustration of a part but a representation of its function. *Graphic Symbols.** (See Figure G-3.)

AMPLIFIER (A)

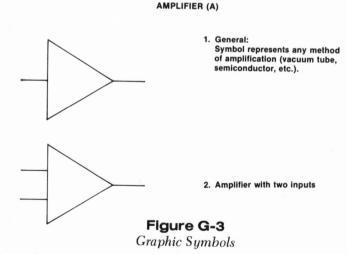

1. **General:**
Symbol represents any method of amplification (vacuum tube, semiconductor, etc.).

2. **Amplifier with two inputs**

Figure G-3
Graphic Symbols

---

*From *Complete Guide to Reading Schematic Diagrams*, Second Edition, Table I.

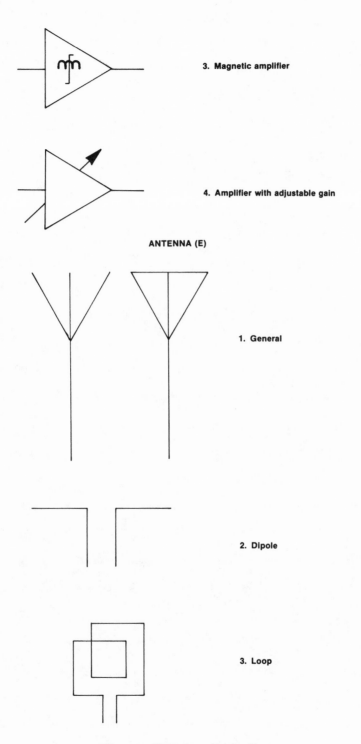

3. Magnetic amplifier

4. Amplifier with adjustable gain

ANTENNA (E)

1. General

2. Dipole

3. Loop

Figure G-3 (continued)
Graphic Symbols

AUDIBLE SIGNALING DEVICE (LS)

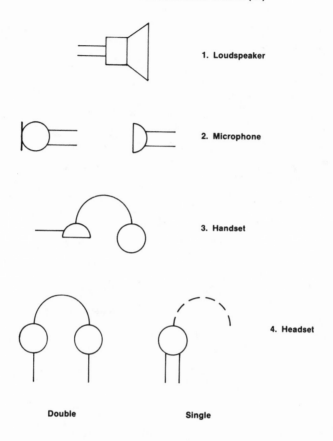

1. Loudspeaker

2. Microphone

3. Handset

4. Headset

Double

Single

BATTERY (BT)

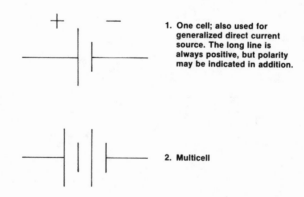

1. One cell; also used for generalized direct current source. The long line is always positive, but polarity may be indicated in addition.

2. Multicell

Figure G-3 (continued)
Graphic Symbols

CABLE, CONDUCTOR, WIRING (W)

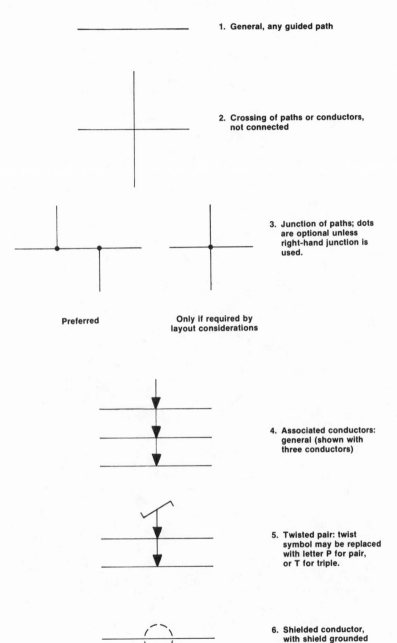

1. General, any guided path

2. Crossing of paths or conductors, not connected

3. Junction of paths; dots are optional unless right-hand junction is used.

Preferred

Only if required by layout considerations

4. Associated conductors: general (shown with three conductors)

5. Twisted pair: twist symbol may be replaced with letter P for pair, or T for triple.

6. Shielded conductor, with shield grounded

Figure G-3 (continued)
Graphic Symbols

CABLE, CONDUCTOR, WIRING (W)
(continued)

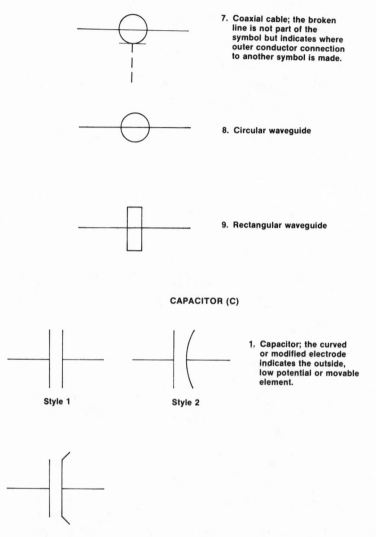

7. Coaxial cable; the broken line is not part of the symbol but indicates where outer conductor connection to another symbol is made.

8. Circular waveguide

9. Rectangular waveguide

CAPACITOR (C)

Style 1

Style 2

1. Capacitor; the curved or modified electrode indicates the outside, low potential or movable element.

Style 1 modified to identify electrode

Style 1

Style 2

2. Polarized capacitor

Figure G-3 (continued)
Graphic Symbols

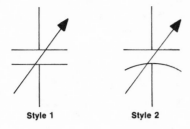

Style 1   Style 2

3. Adjustable or variable
   capacitor. If mechanical
   linkage of more than one
   unit is to be shown, the
   tails of the arrows are
   joined by a dashed line.

CIRCUIT BREAKER (CB)

1. General

CONNECTOR: FEMALE (J), MALE (P)

Female (jack)   Male (plug)

1. General

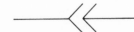

2. Male and female connectors
   engaged

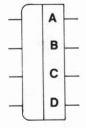

3. Alternative way of showing multiple connector.
   Plug is on the left,
   jack on the right.

Figure G-3 (continued)
Graphic Symbols

CONNECTOR
(continued)

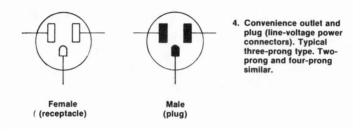

Female
((receptacle)

Male
(plug)

4. Convenience outlet and
   plug (line-voltage power
   connectors). Typical
   three-prong type. Two-
   prong and four-prong
   similar.

CRYSTAL UNIT, PIEZOELECTRIC (Y)

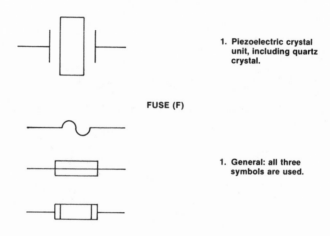

1. Piezoelectric crystal
   unit, including quartz
   crystal.

FUSE (F)

1. General: all three
   symbols are used.

GROUND, CIRCUIT RETURN (no class letter)

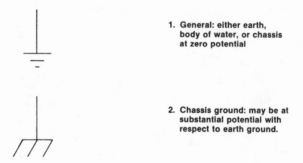

1. General: either earth,
   body of water, or chassis
   at zero potential

2. Chassis ground: may be at
   substantial potential with
   respect to earth ground.

Figure G-3 (continued)
Graphic Symbols

INDUCTOR (L)

 or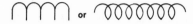

1. General: right-hand symbol is deprecated and should not be used on new schematics.

2. Magnetic core inductor

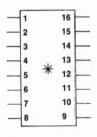

3. Tapped inductor

4. Adjustable inductor

INTEGRATED CIRCUIT (U)

```
 ┌──────────┐
─┤1       16├─
─┤2       15├─
─┤3       14├─
─┤4       13├─
─┤5   *   12├─
─┤6       11├─
─┤7       10├─
─┤8        9├─
 └──────────┘
```

1. General: unused pin connections need not be shown.

   The asterisk is not part of the symbol. It indicates where the type number is placed.

LAMP (DS)

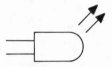

1. General; light source, general

2. Glow lamp, neon lamp (a.c. type)

Figure G-3 (continued)
Graphic Symbols

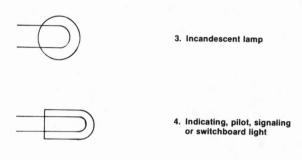

3. Incandescent lamp

4. Indicating, pilot, signaling or switchboard light

METER (M)

1. The asterisk is not part of the symbol. It indicates where to place a letter or letters indicating the type of meter:

A = ammeter   W = wattmeter
DB, VU = audio level meter
F = frequency meter
MA = milliammeter
OHM = ohmmeter   V = voltmeter

RELAY (K)

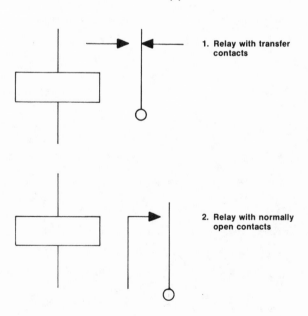

1. Relay with transfer contacts

2. Relay with normally open contacts

Figure G-3 (continued)
Graphic Symbols

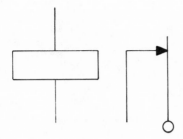

3. Relay with normally closed contacts

RESISTOR (R)

1. General

2. Tapped resistor

3. Resistor with adjustable contact (potentiometer)

4. Continuously adjustable resistor (rheostat)

t°

5. Thermistor

6. Photoconductive transducer (e.g., cadmium-sulfide photocell)

Figure G-3 (continued)
Graphic Symbols

**SEMICONDUCTOR DEVICE
DIODE (D OR CR)**

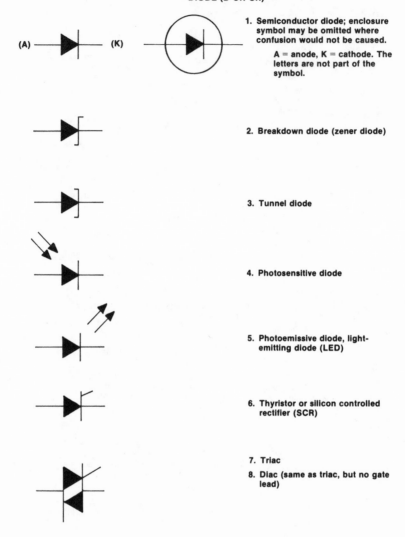

1. Semiconductor diode; enclosure
   symbol may be omitted where
   confusion would not be caused.

   A = anode, K = cathode. The
   letters are not part of the
   symbol.

2. Breakdown diode (zener diode)

3. Tunnel diode

4. Photosensitive diode

5. Photoemissive diode, light-
   emitting diode (LED)

6. Thyristor or silicon controlled
   rectifier (SCR)

7. Triac

8. Diac (same as triac, but no gate
   lead)

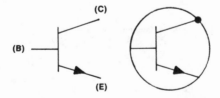

1. NPN transistor; enclosure
   symbol may be omitted where
   confusion would not be caused
   (unless an electrode is connected to it, as shown here).

   C = collector, E = emitter,
   B = base; these letters are
   not part of the symbol.

Figure G-3 (continued)
Graphic Symbols

SEMICONDUCTOR DEVICES
(continued)

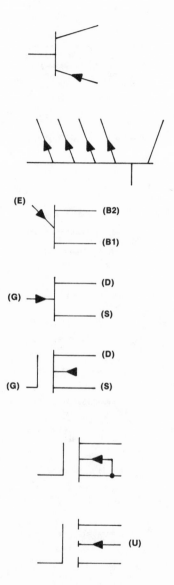

2. **PNP transistor**

3. **NPN transistor with multiple emitters (four shown in this example)**

4. **Unijunction transistor with N-type base. If arrow on emitter points in opposite direction base is P type.**

5. **Junction field-effect transistor (JFET) with N-channel junction gate.**

   G = gate, D = drain, S = source; these letters are not part of the symbol.

6. **Insulated-gate field-effect transistor (IGFET) with N-channel (depletion type), single gate, positive substrate.**

7. **Insulated-gate field-effect transistor (IGFET), with N-channel (depletion type), single gate, active substrate internally terminated to source.**

8. **Insulated-gate field-effect transistor (IGFET) with N-channel (enhancement type), single gate, active substrate externally terminated.**

   U = substrate; this letter is not part of symbol.

9. **Same as previous example, but with two gates**

Figure G-3 (continued)
Graphic Symbols

SEMICONDUCTOR DEVICES
(continued)

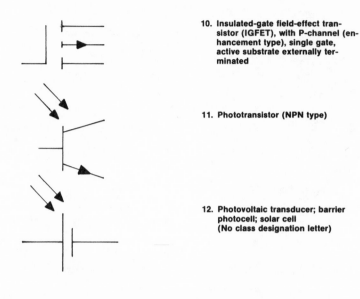

10. Insulated-gate field-effect transistor (IGFET), with P-channel (enhancement type), single gate, active substrate externally terminated

11. Phototransistor (NPN type)

12. Photovoltaic transducer; barrier photocell; solar cell
(No class designation letter)

SWITCH (S)

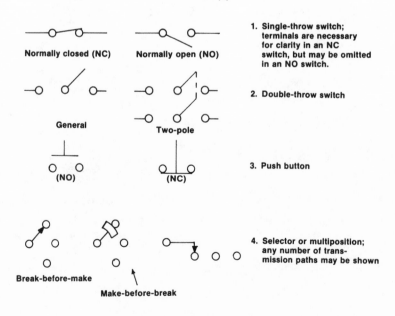

Normally closed (NC)   Normally open (NO)

General   Two-pole

(NO)   (NC)

Break-before-make

Make-before-break

1. Single-throw switch; terminals are necessary for clarity in an NC switch, but may be omitted in an NO switch.

2. Double-throw switch

3. Push button

4. Selector or multiposition; any number of transmission paths may be shown

Figure G-3 (continued)
Graphic Symbols

SWITCH (continued)

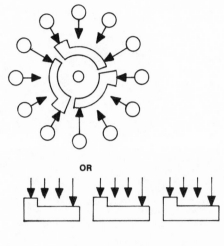

OR

5. **Rotary or wafer-type switch.** Viewed from end opposite control knob. For more than one section the first is the one nearest the control knob. With contacts on both sides front contacts are nearest control knob.

6. **Flasher; self-interrupting switch**

SYNCHRO (B)

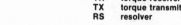

1. **General:** if identification is required, add appropriate letter combination from following list adjacent to symbol:

CDX	control-differential transmitter
CT	control transformer
CX	control transmitter
TDR	torque-differential receiver
TDX	torque-differential transmitter
TR	torque receiver
TX	torque transmitter
RS	resolver

2. **Synchro: control transformer; receiver; transmitter**

3. **Synchro: differential receiver; differential transmitter**

Figure G-3 (continued)
Graphic Symbols

SYNCHRO (continued)

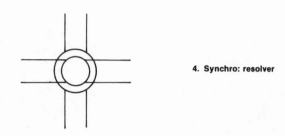

4. Synchro: resolver

PICKUP HEAD (PU)

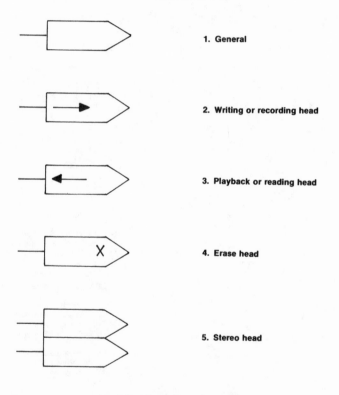

1. General

2. Writing or recording head

3. Playback or reading head

4. Erase head

5. Stereo head

PIEZOELECTRIC CRYSTAL UNIT (Y)

1. Also called quartz crystal unit

Figure G-3 (continued)
Graphic Symbols

TRANSFORMER (T)

1. General: international symbol on right

2. Magnetic core, non-saturating

3. Shielded transformer with magnetic core. A ferrite core is often shown by dashed lines, with arrow if tunable.

4. Magnetic core with electro-static shield between wind-ings. (Shield shown connected to frame.)

5. Saturating transformer

Figure G-3 (continued)
Graphic Symbols

TRANSFORMER (continued)

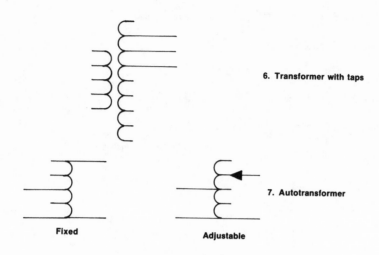

6. **Transformer with taps**

7. **Autotransformer**

Fixed   Adjustable

VACUUM TUBE (V)

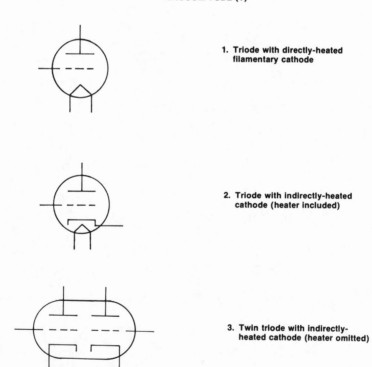

1. **Triode with directly-heated filamentary cathode**

2. **Triode with indirectly-heated cathode (heater included)**

3. **Twin triode with indirectly-heated cathode (heater omitted)**

Figure G-3 (continued)
Graphic Symbols

VACUUM TUBE (continued)

4. **Pentode with indirectly-heated cathode (heater omitted)**

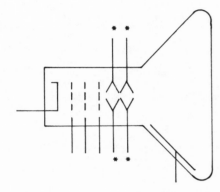

5. **Cathode-ray tube (CRT) with deflection plates (*). Same symbol without deflection plates is used for monochrome picture tube (single electron gun, magnetic deflection).**

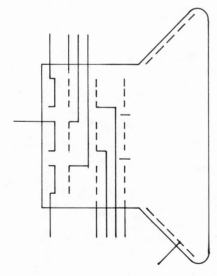

6. **Color picture tube with three electron guns and electromagnetic deflection**

Figure G-3 (continued)
Graphic Symbols

VACUUM TUBE (continued)

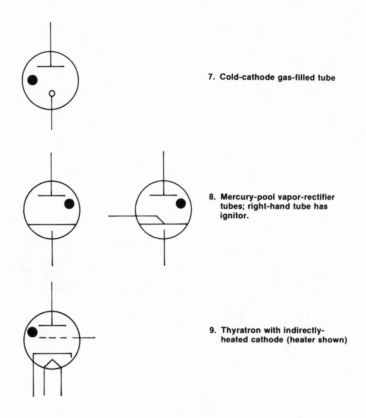

7. **Cold-cathode gas-filled tube**

8. **Mercury-pool vapor-rectifier tubes; right-hand tube has ignitor.**

9. **Thyratron with indirectly-heated cathode (heater shown)**

Figure G-3 (continued)
Graphic Symbols

graphite—Black to dark steel-gray form of carbon (q.v.), having a metallic luster, existing naturally as a mineral in soft platy or flaky masses, or synthesized from petroleum coke. Used for lubricants, crucibles, polishes, batteries, carbon brushes, and pencil lead.

graticule—Framework of rectangular vertical and horizontal lines, evenly spaced, either etched in the screen of an oscilloscope or scribed on a filter mounted in front of it, and dimensioned to agree with the settings on the vertical attenuator and time-base controls of oscilloscopes in which these have calibrated values, to enable the operator to make precise measurements of waveform parameters.

gravitation—Universal force of attraction between two masses. Any particle of matter attracts any other with a force varying directly as the product of the masses and inversely as the square of the distance between them (Newton), as long as they are confined to situations involving small relative speeds and weak fields (Einstein).

gravitational constant—Number expressing the force of gravity, as opposed to acceleration due to gravity (g), assumed to be constant and universal and equal to 6.673×10^{-11} N m^2 kg^{-2} approximately. *Constants.*

Greek alphabet—Writing system of Greece dating from c. 1000 B.C. Its letters are widely used as symbols in science and mathematics. *Greek Alphabet.*

GREEK ALPHABET

Name	Capital	Small	Designates
Alpha	A	α	Angles, coefficients, attenuation constant, absorption factor, area
Beta	B	β	Angles, coefficients, phase constant
Gamma	Γ	γ	Complex propagation constant (CAP), specific gravity, angles, electrical conductivity, propagation constant
Delta	Δ	δ	Increment or decrement*, determinant (CAP), permittivity (CAP), density, angles
Epsilon	E	ϵ	Dielectric constant, permittivity, base of natural logarithms, electric intensity
Zeta	Z	ζ	Coordinates, coefficients
Eta	H	η	Intrinsic impedance, efficiency, surface charge density, hysteresis, coordinates
Theta	Θ	θ	Angular phase displacement, time constant, reluctance, angles
Iota	I	ι	Unit vector
Kappa	K	κ	Susceptibility, coupling coefficient
Lambda	Λ	λ	Permeance (CAP), wavelength, attenuation constant
Mu	M	μ	Permeability, amplification factor, prefix micro-
Nu	N	ν	Reluctivity, frequency

GREEK ALPHABET (continued)

Xi	Ξ	ξ	Coordinates
Omicron	O	o	
Pi	Π	π	3.141 592 653 589 793 238 ...
Rho	P	ρ	Resistivity, volume charge density, coordinates.
Sigma	Σ	σ	Summation (CAP), surface charge density, complex propagation constant, electrical conductivity, leakage coefficient, deviation
Tau	T	τ	Time constant, volume resistivity, time-phase displacement, transmission factor, density
Upsilon	Υ	υ	
Phi	Φ	φ	Scalar potential (CAP), magnetic flux, angles
Chi	X	χ	Electric susceptibility, angles
Psi	Ψ	ψ	Dielectric flux, phase difference, coordinates, angles
Omega	Ω	ω	Ohms (CAP), solid angle (CAP), angular velocity

All use small letter unless capital (CAP) is indicated. An asterisk (*) indicates either capital or small may be used.

Greenwich mean time—Universal time scale based on the mean angle of rotation of Earth about its axis in relation to the Sun, taking the prime meridian passing through the Royal Greenwich Observatory, London, as the reference. Also called Universal Time (UT), "Z time" (from the suffix added to code time groups to indicate they are applicable to the Z, or 0, time zone): abbreviated GMT. *Frequency Data.*

grid—In an electron tube, an electrode that has one or more openings for controlling electrons or ions as they pass through it. Usually the term "grid" is understood to mean the control grid, located between the cathode and the anode, that varies the rate of flow of electrons from the one to the other in accordance to the potential applied to it. In some tubes a screen grid is placed between the control grid and the anode to reduce the electrostatic influence of the latter on the control grid, and a suppressor grid to capture secondary electrons. *Electron Tubes.* (See Figure G-4.)

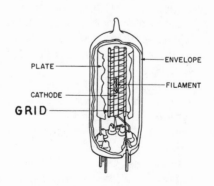

Figure G-4
Grid

grid battery—See C– (minus).

grid bias—D.C. potential applied between grid and cathode of an electron tube that determines its operating point. *Electron Tube Circuits.*

grid-controlled gas rectifier—Electron tube containing argon at low pressure, or mercury vapor, commonly used in heavy-duty power supplies. *Rectifiers and Filters, Electron Tubes.*

grid-dip meter—Oscillator with wide frequency range used to check frequency of tuned circuits. It employs an electron tube in an oscillator circuit, and when this circuit is coupled to a tuned circuit the oscillator grid current, as shown by a milliammeter, shows a decrease or "dip" when tuned through resonance with the "unknown" circuit. This is because at resonance the unknown circuit absorbs energy from the oscillator, causing a decrease in feedback and therefore in grid current. (See Figure G-5.)

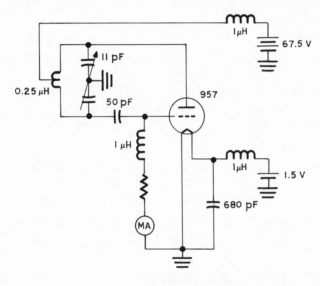

Figure G-5
Grid-Dip Meter
(125 - 160 MHz)

grid-leak detector—Radio receiver circuit which uses a single triode electron tube to both demodulate and amplify radio signals to which the circuit is tuned. A high-value resistor called a "grid leak" is connected across the grid capacitor to provide a path back to the cathode for electrons which would otherwise accumulate on the capacitor and result in "grid blocking," or cutting off tube operation. (See Figure G-6.)

grid modulation—Type of modulation circuit where the modulating signal is applied between the grid and cathode of the r.f. power amplifier. *Modulation.*

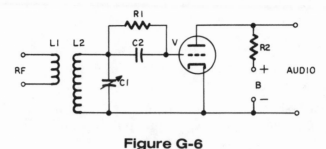

Figure G-6

Grid-Leak Detector

From *Complete Guide to Reading Schematic Diagrams,* First Edition, page 135.

grooves—In disk recording, the sound track is cut in the sides of a spiral groove that starts at the outer edge and terminates close to the label. The groove cross-section has sides sloping toward each other at 45 degrees, and the bottom of the groove is curved with a radius 0.00025 inch (monophonic), or 0.0002 inch (stereophonic). The top width is 0.0022 inch (monophonic), or 0.001 (stereophonic). *Recording.*

ground—1. Electrical connection with the ground, which remains essentially at a constant potential, assumed to be zero for reference to other potentials.   2. Electrical connection to a common point, which is the zero-voltage reference for the circuit, not necessarily at ground potential or connected to it ("chassis ground"). Also called "earth," "earth ground," "E point," "holy point," etc.

ground bus—Common ground return connection in a circuit. Where heavier currents are present, a ground bus may be a metal bar or thick wire.

ground-controlled approach (GCA)—Control of the approach of an aircraft by a ground-based human controller. *Navigation Aids.*

grounded-anode amplifier—Amplifier circuit in which the anode is connected to the common point. Also called grounded-plate or cathode-follower amplifier. *Electron-Tube Circuits.*

grounded-cathode amplifier—Amplifier circuit in which the cathode is connected to the common point. *Electron-Tube Circuits.*

grounded-grid amplifier—Amplifier circuit in which the grid is connected to the common point. *Electron-Tube Circuits.*

ground loop—In an electrical system, grounds that are "floating" and at different potentials, if a connection exists, provide an unwanted feedback or source of noise or error, which must be eliminated by correcting the design.

ground-plane antenna—Vertical antenna mounted on a mast, with grounded elements projecting horizontally and radially from its base, from which it is insulated, and which serve as an artificial "ground" to lower the angle of radiation.

ground return—Common return path in a circuit, at zero potential with reference to other potentials in the circuit.

ground rod—(See earth ground.)

ground state—Lowest energy level of an atom. *Masers.*

ground wave—Radio wave whose path remains close to the ground, as opposed to sky waves. *Electromagnetic-Wave Propagation.*

guarding—Arranging the connections to a bridge in such a way that leakage resistance does not shunt the resistance being measured, a problem when measuring high resistances.

guidance system—Equipment for altering the direction of motion of the object in which it is installed either in accordance with a predetermined program or in response to signals transmitted to it, or both.

Gunn effect—High-frequency oscillation of current flowing through a solid, such as gallium arsenide (q.v.), producing microwaves. GaAs electrons may be in a state of high mobility or of low mobility, according to the applied voltage, giving rise to a train of high-energy pulses of low-mobility with power as high as 100 W pulsed and 100 mW continuous at 2 GHz, with smaller powers at higher frequencies. *Semiconductors.*

G-Y signal—In color television the color difference signal that becomes the primary green signal by adding the Y signal.

H—1. Symbol for henry (q.v.).   2. Symbol for horizontal component of Earth's main magnetic field.   3. Chemical symbol for hydrogen (q.v.).

h—Symbol for hour

HI and HII regions—In interstellar space, HI regions contain nonionized hydrogen, HII regions contain ionized hydrogen.

half-adder—Digital circuit that can add two bits to give a sum and carry. Two half-adders make a full-adder (q.v.). *Computer Hardware.*

half-life—In radioactivity, the time required for one-half of the atomic nuclei of a radioactive sample to disintegrate. *Nuclear Physics.*

half-power point—On the response curve of an amplifier, filter, etc., that point where the cutoff frequency occurs, defined as $1/\sqrt{2}=0.707$ times its maximum amplitude. In decibels the cutoff frequency is where the amplitude is 3 dB below its maximum value. Since –3 dB represents a power ratio of 0.5, it is called the half-power. (See Figure H-1.)

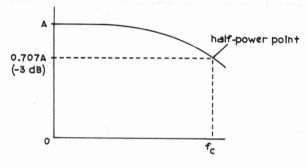

Figure H-1
Half-Power Point

half-wave dipole—Antenna consisting of two aligned conductors, slightly separated at the center for connecting a transmission line, and approximately half a wavelength long overall. *Antennas*.

half-wave rectifier—Power rectifier circuit that uses only one half, negative or positive, of the input wave. *Rectifiers and Filters*

Hall effect—The development of a transverse electric field across a solid material when it carries an electric current, and is placed in a magnetic field perpendicular to the current.

halogen—One of five nonmetallic elements comprising Group VIIa of the periodic table (see under element); fluorine, chlorine, bromine, iodine, and astatine. These elements are highly reactive and do not occur uncombined in nature.

ham—Slang name for amateur radio operator, not to be confused with a "CBer."

hand capacitance—Detuning caused by bringing the hand close to an unshielded tuned circuit.

handset—The part of a modern telephone that is lifted and held to the ear and mouth when in use. *Telephone*.

handshaking—Procedure to enable two computers, modems, or other "intelligent" equipment to verify, by exchanging coded signals, that they are connected and synchronized.

hard copy—Computer output on paper as opposed to a CRT display, etc. *Computer Hardware*.

hard-drawn copper wire—Copper wire that has not been annealed after drawing. It has approximately twice the tensile strength of annealed (soft) wire.

harmonic—Frequency that is some integral multiple of the fundamental frequency. For example, if the fundamental frequency (or first harmonic) is 100 Hz, the second harmonic will be 200 Hz, the third harmonic 300 Hz, and so on.

harmonic distortion—In sine waves harmonic distortion results from the presence of harmonics (q.v.). The higher the ratio of harmonic energy to fundamental energy, the more distorted is the waveshape. On the other hand, all nonsinusoidal waves are distorted, often highly so, being composed of a sinusoidal fundamental and various sinusoidal harmonics. For example, a square wave contains the fundamental and an infinite number of odd harmonics with amplitudes, expressed as percentages of the fundamental, as follows: 3rd harmonic, 35; 5th harmonic, 20; 7th harmonic, 15; 9th harmonic, 11; ... ad infinitum. The harmonic content of a waveform is given by the distortion factor D:

$$D = \left(\frac{\text{sum of squares of amplitudes of harmonics}}{\text{square of amplitude of fundamental}} \right)^{\frac{1}{2}} \times 100\%$$

harness—A wiring harness consists of a number of wires that travel parallel routes and are laced together. In making a harness the wires are cut, laid out, and bent to shape around pins driven in a wooden board. They are then laced together with tape ("spot-tied") and placed in the enclosure among the components they interconnect. Individual wires are identified by color coding or tagging each end.

Hartley oscillator—Feedback oscillator circuit employing an inductive-capacitive tuned circuit. *Transistor Circuits*.

H-attenuator—Balanced attenuator network, also called an H-pad or balanced T-attenuator. *Attenuators*.

Hay bridge—Bridge used for measurement of large inductance. (See Figure H-2.) Value of inductance (L_x) is given by:

$$L_x = R_1R_3C_1/ \ (1+\omega^2C_1{}^2R_2{}^2)$$

and quality (Q_x) by

$$Q_x = \omega L_x/R_x = (\omega C_1R_2)^{-1}$$

H-beacon—An LF/MF nondirectional beacon used by airborne direction finders. Same as NDB. *Navigation Aids*

H-display—A radar display in which the signal appears as two dots. The left-hand dot

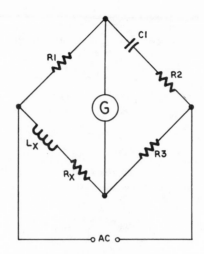

Figure H-2
Hay Bridge

From *Complete Guide to Reading Schematic Diagrams*, First Edition, page 213.

gives range and azimuth of the target, the right-hand dot, by its relative position, gives a rough indication of elevation. *Radar.*

head—1. In a tape recorder, the electromagnetic device that forms the magnetic track in the tape. In a tape player, the similar device that "reads" the track and converts it to an electrical signal. (The same head is generally used for both purposes.) Also, an erase head for obliterating the magnetic track. 2. In a tape punching device, the perforating mechanism or the reading mechanism. *Recording.*

headphone—Small loudspeaker (earphone) held over the ear by a band worn on the head. Commonly employed where surrounding noise is high. May have one earphone or two. Stereo headphones have separate channels to each earphone, with a 3-contact plug with the following connections:

Sleeve (barrel)	common
Ring	left earphone
Tip	right earphone

Positive voltage on tip or ring should move diaphragm toward listener's ear.

hearing aid—System comprising a microphone, amplifier, earphone, and power supply, to increase the loudness of sounds in the ear of the wearer. Modern microelectronic techniques have allowed hearing aids to be made that can be worn inside the ear or built into eyeglass frames.

heater—Heating element for an indirectly heated cathode. *Electron Tubes.*

heat sink—Protective structure, generally a finned metal block, on which a device that must dissipate heat is mounted.

Heaviside layer—The ionosphere (q.v.) Also called the Kennelly-Heaviside layer. *Electromagnetic-Wave Propagation.*

hectometric waves—Medium-frequency waves with wavelengths of 100 to 1000 meters (300 kHz to 3 MHz).

height control—In a television receiver, the control that adjusts the amplitude of the vertical sawtooth, and consequently the height of the picture.

helical antenna—Cylindrical, flat, or conical spring-shaped antenna that may radiate in the normal or axial mode according to frequency. *Antennas.*

helical potentiometer—Potentiometer in which the resistive element is mounted spirally on the inside of the cylindrical case, requiring as many as 10 turns of the control knob to move the sliding contact from one end to the other. *Resistors.*

helium—Inert gas, colorless and odorless, the second lightest element, used in instrument-carrying balloons, cryogenics (liquefies at -268.6°C), arc welding (inert atmosphere for aluminum welding), pressurizing rocket fuel tanks, and so on. Symbol, He; atomic weight, 4.0026; atomic number, 2.

henry—Unit of electrical inductance of a closed circuit in which an electromotive force of one volt is produced when the electric current that traverses the circuit varies uniformly at the rate of one ampere per second. Symbol, H. *S.I. Units.*

heptode—Electron tube with seven electrodes: anode, cathode, and five grids. Also called a pentagrid tube. Used in vacuum-tube converter stages. (See Figure H-3.)

hertz—Unit of frequency of a periodic phenomenon of which the period is one second. Symbol, Hz. *S.I. Units.*

heterodyne—To mix two frequencies to produce sum and difference frequencies. In modern receivers an oscillator generates a frequency that is higher than the incoming radio frequency by the amount of the intermediate frequency. The difference is held constant regardless of what station is tuned. The new signal then passes to the i.f. amplifier, and all other frequencies are filtered out. The i.f. for AM radio is 455 kHz, for FM 10.7 MHz, and for TV receivers ≈44 MHz. Also called superheterodyne.

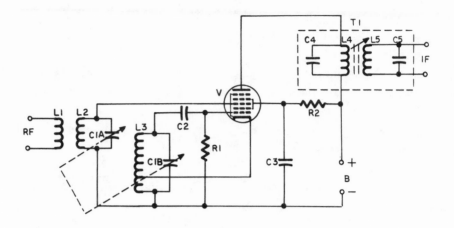

Figure H-3

Heptode (in a converter circuit)

From *Complete Guide to Electronic Test Equipment and Troubleshooting Techniques*, page 37. ©1976. Reprinted with permission of Parker Publishing Company, Inc.

heterosphere—Atmospheric region above 100 kilometers (62 miles) where the composition varies with altitude. (See atmosphere.)

hexadecimal—Number system with a base of 16, used in computer representation of numbers, as follows

Decimal	Hexadecimal	Binary
0	0	0000
1	1	0001
2	2	0010
3	3	0011
4	4	0100
5	5	0101
6	6	0110
7	7	0111
8	8	1000
9	9	1001
10	A	1010
11	B	1011
12	C	1100
13	D	1101
14	E	1110
15	F	1111

hex inverter—Microelectronic package containing six inverters.

HF—Abbreviation for high frequency (q.v.).

hi-fi—(See high fidelity sound system.)

high fidelity sound system—Sound system capable of a high degree of realism in the electronic reproduction of sound. *Electroacoustics, Recording.*

high frequency—Frequency range from 3 to 30 megahertz, or decametric waves (10 to 100 meters). (See also radio frequency.)

high-pass filter—Filter that transmits frequencies above a cutoff frequency. *Active Filters, Passive Filters.*

high-resistance joint—Poor connection, such as a "cold-solder joint," which results in a voltage drop at the joint.

high voltage probe—Probe for use with a voltmeter, oscilloscope, etc., for reducing a high voltage to a voltage within the range of the instrument, so that the high voltage may be measured. The probe contains a voltage divider with a ratio of 100/1 and is also designed to give maximum safety to the user both by its length and by the provision of a disk-shaped high-tension leakage barrier around its barrel.

hold control—In a television receiver, the vertical and horizontal hold controls adjust the frequency of the corresponding deflection circuits to enable them to lock on to the sync pulses.

hole—Quasi-positive charge that moves freely through certain types of solids. Each hole is the absence of an electron from an atom in the solid (much as a bubble is an absence of liquid in an otherwise full container). Since the electron is negative, its absence from an otherwise neutral atom confers a positive charge on the hole. *Semiconductors.*

Hollerith code—(See IBM card.)

holography—Method of forming three-dimensional images by using coherent light (as from a laser; see coherence). The images, usually formed on a photographic plate, bear no resemblance to the original object and must be reconstructed by illuminating them with coherent light.

homosphere—Atmospheric region extending to about 85 kilometers (53 miles), where the composition is essentially independent of height. (See atmosphere.)

honeycomb winding—(See lattice winding.)

hookup wire—Wire used for making connections in electronic equipment.

horizontal blanking pulse—In the composite television signal, the pedestal that carried the sync pulse and color burst, and cuts off the electron beam in the picture tube between lines to prevent retrace lines appearing on the screen. *Broadcasting.*

horizontal frequency—(See line frequency.)

horizontal hold control—(See hold control.)

horizontal polarization—Where the electric field of an electromagnetic wave lies in a plane parallel to the Earth's surface. It is standard for television in the U.S. since horizontal receiving antennas pick up less noise. (See also polarization.) *Antennas, Electromagnetic-Wave Propagation.*

horizontal sweep—In a cathode-ray tube, the horizontal deflection of the electron beam.

horn antenna—Microwave antenna made by flaring the end of the waveguide. *Antennas.*

horn speaker—Loudspeaker with a horn instead of a cone. *Electroacoustics.*

horsepower—Electric horsepower is equal to 746 watts. *S.I. Units.*

horseshoe magnet—U-shaped magnet in which the two poles are close together.

hot-electron diode—Diode with rectifying metal-semiconductor contacts in which current flows by majority carriers. When forward biased, the semiconductor injects electrons into the metal. These have greater velocities, so are termed "hot." Also called hot-carrier or Schottky barrier diodes. *Semiconductors.*

hot-wire ammeter—Ammeter movement in which the reading is proportionate to the expansion of a wire A-B as it is heated by a current. As the wire expands, the spring C is able to contract, causing the meter pointer to indicate accordingly. Useful for high-frequency currents, but rather sluggish in action. (See Figure H-4.)

H-pad—(See H-attenuator.)

h-parameter—(See hybrid parameter.)

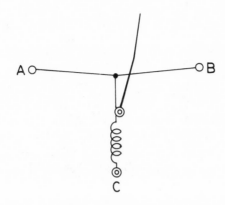

Figure H-4
Hot-Wire Ammeter (movement)

From *Complete Guide to Electronic Test Equipment and Troubleshooting Techniques*, page 37. ©1976. Reprinted with permission of Parker Publishing Company, Inc.

H-plane—Plane in which the magnetic field of an antenna lies. *Antennas.*

hum—Low frequency interference originating in the 60-hertz power-line frequency, the result of inadequate filtering, a.c. modulation of some r.-f. component ("tunable hum"), or mechanical vibration (loose transformer windings, etc.).

hum bar—Dark horizontal band across television picture due to hum.

hybrid circuit—One or more integrated circuits combined with one or more discrete devices, or a device containing more than one type of integrated circuit. *Microelectronics.*

hybrid junction—Waveguide circuit junction characterized by the fact that there is no direct coupling between the H arm and the E arm, yet power flows by virtue of reflections from the other two arms. Also called a hybrid-T or magic-T.

hybrid parameters—Parameters used in data sheets for small-signal applications of transistors. *Transistors.*

hydrophone—Type of microphone for use under water.

hyperbola—Conic section produced by the intersection of a circular cone and a plane that cuts both nappes.

hysteresis—Retardation of the effect when the forces acting on a body are changed, especially the magnetizing force acting on an iron core in an inductor. The magnetic field H causes some or all of the atomic magnets in the material to align with it, increasing the total magnetic flux density B. The aligning process, however, lags behind the application of the force. As shown in the graph, if H is strong enough all the atomic magnets become aligned in the same direction, so that B increases from 0 to A, but cannot increase further. This is called saturation. When the magnetizing force is diminished, the curve on the graph does not exactly retrace itself in reverse order, because the change in flux density B lags behind the change in field strength H. Even when H has decreased to zero, B still has a positive value (B), called remanence or retentivity, which has a high value for permanent magnets. B does not become zero (C) until H has gone somewhat negative. The value of H for which B is zero is called the coercive force, which therefore is the value of the demagnetizing force required to reduce the flux to zero. As H continues to go negative it causes the flux to build up in the opposite direction until it reaches saturation again at D. The graph shows a complete cycle, returning (via F) to the first saturation point (A). This graph is called a hysteresis loop. Broad loops are desirable for permanent magnets, narrow ones for transformer cores. *Transformers.* (See Figure H-5.)

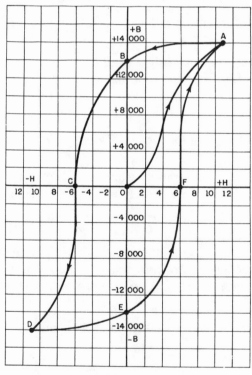

Figure H-5
Hysteresis Loop

Hz—Symbol for hertz (q.v.).

I—1. Symbol for quantity of current. 2. Symbol for iodine (q.v.). 3. Roman numeral for 1.

i—1. Symbol for instantaneous current. 2. Symbol for $\sqrt{-1}$ (an imaginary number).

IATA—Abbreviation for International Air Transport Association, an association of scheduled airlines.

IBM—Abbreviation for International Business Machines, Inc.

IBM card—Punched card measuring $3.25 \times 7.375 \times 0.0067$ inches, divided into 12 rows and 80 columns, which give 960 hole positions (Hollerith code); the card most commonly used for punched-card memory and input/output-systems. *Computer Hardware*. (See Figure I-1.)

IC—Abbreviation for integrated circuit (q.v.).

I-channel—One of two orthogonal components of the chrominance signal. The I-channel phase is 57 degrees clockwise from the reference burst, its bandwidth is 1.3

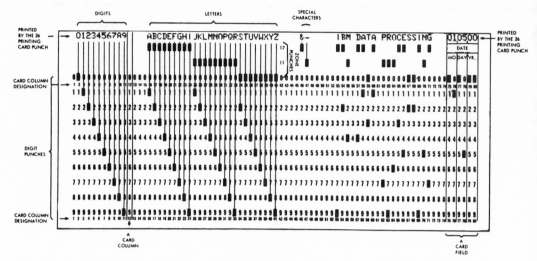

Figure I-1
IBM Card

MHz, and its equation is

$$E_I' = -0.27(E_B' - E_Y') + 0.74(E_R' - E_Y')$$

where E_Y = gamma-corrected voltage of the monochrome portion of the color
signal;

E_B', E_R' = gamma-corrected voltages corresponding to the blue and red
signals.

iconoscope—Obsolete type of television camera tube in which a screen consisting of a mosaic of minute globules of silver coated with oxides of silver and cesium acquires a positive charge where light falls on it, the level of the charge varying between the light and dark areas of the picture image focussed on it. This image is scanned by an electron beam, which neutralizes the positive charges. The sudden change of potential as each is neutralized is sensed by a metallic coating on a sheet of mica separating it from the mosaic, and becomes a series of electrical pulses. *Electron Tubes*.

I-display—Radar display in which the signal appears as a circle with radius proportional to range. The brightest part of the circle indicates the direction of the target. *Radar*.

idler—1. On a turntable, a rubber-tired wheel that is intermediate between the motor and the turntable rim.   2. On a tape recorder, a rubber wheel that transfers the motor drive to the capstan.

IEC—Abbreviation for International Electrotechnical Commission, Geneva, Switzerland.

IEEE—Abbreviation for Institute of Electrical and Electronic Engineers.

i.f.—Abbreviation for intermediate frequency (q.v.).

IGFET—Abbreviation for insulated-gate field-effect transistor (q.v.).

ignition coil—Transformer that boosts 12-volt current pulses from the battery to some 20 kilovolts for application to the spark plugs of an automobile. The primary winding consists of 200 or so turns of copper wire, the secondary of more than a mile of fine wire.

ignitron—Arc discharge tube in which the cathode is a pool of mercury. A control electrode, called an igniter, provides the initial ionization by starting an arc to the mercury pool. A heavy current can then flow to the anode. The ignitron is used in industry as a heavy duty rectifier for converting a.c. to d.c.

I²L—Abbreviation for integrated injection logic. *Microelectronics.*

illuminance—The amount of light spread over the area receiving the light, expressed in lumens per unit area. One lumen per square meter is one lux. Unobstructed daylight from a bright overcast sky is approximately 10,000 lux.

ILS—Abbreviation for Instrument Low-Approach System (q.v.).

image dissector—Type of television camera tube in which the photocathode emits electrons in proportion to the illumination from an optical image incident upon it. These electrons are accelerated along parallel paths toward the anode, which has a small aperture. The resulting electron image is deflected horizontally and vertically so that electrons pass through the aperture only from the point in the image that is coincident with it at each instant. These are then processed by an electron multiplier. *Electron Tubes.*

image frequency—In a superheterodyne receiver with an i.f. of 455 kHz, the local oscillator will be tuned for a frequency of 1365 kHz to receive a station operating on 910 kHz. The difference between 1365 kHz and 910 kHz gives the i.f. of 455 kHz. However, if another signal is present in the antenna, with a frequency of 1829 kHz, a substantial amount of it will reach the mixer or converter input because it is a harmonic of 910 kHz. As the difference between 1820 kHz and the local oscillator signal is also 455 kHz it will be amplified in the i.f. amplifier along with the desired signal. It is called an image frequency because it is as much above the local oscillator frequency as the desired signal is below it.

image iconoscope—A super-emitron camera tube (British term). (See image orthicon.) *Electron Tubes.*

image intensifier—Type of electron multiplier (q.v.).

image orthicon—Television camera tube most widely used for live commercial television because of its high sensitivity and characteristics that resemble those of the human eye. *Electron Tubes.*

image rejection—Ability of a superheterodyne receiver to reject image frequencies (q.v.).

impact avalanche and transit time diode—Silicon p-n junction diode, reverse biased to its avalanche threshold, used as a microwave oscillator. *Semiconductors.*

IMPATT—Abbreviation for impact avalanche and transit time (q.v.).

impedance—Measure of the total opposition that a circuit or a part of a circuit presents to electric current. Impedance includes both resistance and reactance (qq.v.). The resistance component arises from collisions between the charge carriers of the current and the internal structure of the conductor. The reactance component is additional opposition to the movement of charge carriers arising from the changing magnetic and electric fields when the current is alternating; in the case of direct current impedance consists of resistance only.

Although resistance and reactance are both measured in ohms, they cannot be combined indiscriminately. Resistance absorbs energy; reactance does not. Voltage and current are in phase in resistance, but differ by 90 degrees in reactance. In inductive reactance the voltage leads the current; in capacitive reactance the current leads the voltage. Because of this the impedance Z in a circuit consisting of resistance and inductance only is calculated as if it were the hypotenuse of a right triangle, in which the resistance R and inductive reactance X_L were the other two sides:

$$Z = \sqrt{R^2 + X_L^2} \qquad (1)$$

In a circuit consisting of resistance and capacitance only the formula is:

$$X = \sqrt{R^2 + X_C^2} \qquad (2)$$

In a circuit consisting of resistance, inductance, and capacitance, Z is obtained from:

$$Z = \sqrt{R^2 + (X_L - X_C^2)} \qquad (3)$$

Since reactance is proportional to applied frequency, as in:

$$X_L = 2\pi fL \qquad (4)$$

or:

$$X_C = \frac{1}{2\pi fC} \tag{5}$$

where: L = the inductance in henries
C = the capacitance in farads
f = the frequency in hertz

it follows that impedance also varies with frequency.

Since impedance is made up of resistance and reactance, Ohm's Law is applicable to it just as to resistance and reactance alone. The formulas are:

$$I = \frac{E}{Z} \tag{6}$$

$$E = IZ \tag{7}$$

$$Z = \frac{E}{I} \tag{8}$$

where: E = e.m.f. in volts
I = current in amperes
Z = impedance in ohms

As mentioned above, the voltage and current in a reactance differ in phase by 90 degrees. However, since there is always some resistance present, the phase angle θ in any circuit will be less than this, and is given by:

$$\theta = \frac{X}{R} \tag{9}$$

where $X = X_L - X_C$. In a circuit in which X_L predominates θ will therefore be positive (voltage leads current). If X_C predominates θ will be negative (voltage lags current).

As already noted, resistance absorbs power but reactance does not. The power P consumed by a circuit is given by:

$$P = I^2 R \tag{10}$$

which ignores any reactance. This is "real" power, measured in watts. If Z is substituted for R a higher power value is obtained. This is "apparent" power, measured in volt-amperes. The ratio of real power to apparent power is called the power factor, expressed as a percentage.

Characteristic impedance Z_o is a feature of transmission lines (q.v.), and is given by:

$$Z_o = \sqrt{L/C} \tag{11}$$

where L and C are the inductance and capacitance per unit length of the line, which in turn depend upon the size of the line conductors and the spacing between them. For a single coaxial line (q.v.), this is obtained from:

$$Z_o = (138/\epsilon^{1/2})\log_{10}(D/d) \tag{12}$$

where: ϵ = dielectric constant
D = inside diameter of outer conductor
d = outside diameter of inner conductor

impedance bridge—Bridge for measuring impedance or its components (inductance, capacitance, resistance, Q, and dissipation factor). In the universal impedance bridge, the function control changes the bridge circuit for each type of measurement; and the L-R-C control gives the value after it has been adjusted to balance the bridge by getting a zero reading ("null") on the detector. This value must be multiplied by the range factor. (See Figure I-2.)

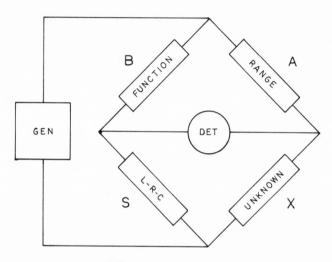

Figure I-2
Impedance Bridge (block diagram)

From *Complete Guide to Electronic Test Equipment and Troubleshooting Techniques*, page 67.

impedance matching—Arranging for the impedance of the load to be equal to the internal impedance of the source. *Transmission Lines.*

implantation, ion—Doping technique in which selected regions of a semiconductor substrate are implanted with ions of virtually any desired impurity. *Semiconductors.*

implode—To burst inwards, said of a high-vacuum envelope, of which the pieces fly inward when it shatters.

impulse noise—Irregular transients of short duration and relatively high amplitude due to relay and switch operation, lightning, and so on.

impurity—Material added to intrinsic (pure) silicon or germanium to change it to p type or n type, when it is called extrinsic (impurities added). Boron is used to create p-type semiconductor material, phosphorus and arsenic to create n-type. *Semiconductors.*

incandescence—Emission of visible light by a body at high temperature, such a filament that is heated electrically to glowing and enclosed in a vacuum or in an inert gas to keep it from burning up.

incoherent radiation—Multifrequency radiation as distinguished from the coherent radiation produced by a laser or maser (q.v.).

increment—Any small change in a variable quantity. Symbol, Δ.

indefinite integral—(See integral.)

independent variable—Mathematical variable not dependent on other variables.

index—The relation or proportion of one amount or dimension to another, as in modulation index, refractive index, etc.

index register—In a computer, a register the contents of which can be added to an instruction address to simplify and speed up a program loop. *Computer Hardware.*

indicator—Any device, such as a gauge, dial, register, or pointer, that measures or records, and visibly indicates a value, condition, etc.

indirectly heated cathode—Cathode consisting of a nickel sleeve externally coated with emitting material and heated by an internal insulated filament. (See Figure I-3.)

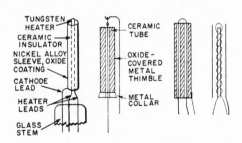

Figure I-3
Indirectly Heated Cathodes

indium—Brilliant, silvery-white metal, softer than lead (can be scratched with a fingernail), about as rare as silver. Used as a dopant in the manufacture of semiconductors (q.v.). Symbol, In; atomic weight, 114.82; atomic number, 49.

indoor antenna—Antenna located within a building, but not inside the receiver.

inductance—The electromagnetic inertia, or property of a coil, that opposes any change in a current flowing in it by storing energy in its magnetic field, This effect is greatly enhanced when the coil is wound around an iron core (see inductor). The unit of inductance is the henry (q.v.).

induction heating—Surrounding the material to be heated with a coil, which acts as the primary winding in a transformer, the work piece (which must be conductive) acting as the secondary. Alternating current (60 Hz) in the primary induces eddy currents in the "secondary," causing it to become heated.

induction motor—Alternating-current motor in which the current flowing in the stator winding induces current in the rotor winding. Interaction of the magnetic field created by the stator with the induced currents in the rotor produces a torque that causes the latter to rotate. (See also squirrel-cage induction motor.)

inductive coupling—Interstage coupling system using inductive links. (See Figure I-4.)

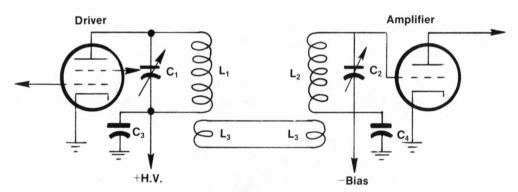

Figure I-4
Inductive Coupling

inductive feedback—Type of feedback (q.v.) using inductive coupling.

inductive reactance—Opposition to the flow of alternating current in a coil by its inductance, expressed in ohms, and given by

$$X_L - 2\pi f\ L$$

where X_L = inductive reactance, in ohms;
           π = 3.1416 . . . . .;
           f = frequency of a.c., in hertz;
           L = inductance of coil, in henries.

inductor—Electrical conductor, generally wound into a spiral or coil, used to introduce inductance, the property that opposes any change in an existing current, into a circuit. Inductors may have magnetic cores of iron, ferrite, etc., or air cores. *Coil Data.*

inertia—Property of a body that causes it to resist attempts to put it into motion or, if already in motion, to change the magnitude or direction of its velocity.

inertial guidance—Self-contained dead-reckoning navigation system that depends on the sensing of accelerations in three planes and double integration of them to obtain distance and direction traveled. *Navigation Aids.*

infinite baffle—Loudspeaker enclosure with means to prevent sound waves from the rear of the speaker finding their way to the front, where they would interfere with sound waves from the front, thus degrading low-frequency response. *Electro-acoustics.*

infinitesimal—A quantity less than any finite quantity, yet not zero.

infinity—A quantity having a greater value than any assigned number. Symbol, ∞.

infrared—Those wavelengths falling above the visible spectrum and below microwaves, classified as near (0.75-1.5 μm), middle (1.6-6.0 μm), far (6.1-40 μm) and far-far (41 μm-1 mm). *Optoelectronics.*

infrared light-emitting diode—GaAs and GaAs: Si are semiconductor materials used for IR LED's, emitting radiation with wavelengths of 905 and 940 nm respectively.

injection laser—Semiconductor laser of GaAs or GaAlAs that operates by direct-current injection into a p-n junction to stimulate the emission of photons and phonons by recombination of minority carriers. *Lasers.*

input—1. Signal or voltage applied to a circuit. 2. Terminals to which a signal or voltage is applied. 3. Data to be processed by a computer.

input admittance—Reciprocal of input impedance, expressed in siemens.

input capacitance—Capacitance at the input terminals of a circuit.

input choke—In a choke-input filter circuit, the first choke is called the input choke. *Rectifiers and Filters.*

input device—Equipment used to introduce data into a computer. *Computer Hardware*.

input impedance—The impedance at the input terminals of a circuit, transmission line, and so on, "seen" by a signal source, expressed in ohms.

insertion loss—Loss of power measured at the load end of a transmission line after the connection of some device at an intermediate point in the line.

instantaneous sampling—Type of pulse-amplitude modulation where the amplitude is determined by the instantaneous amplitude of the pulse at a single instant (i.e., center or leading edge). Also called square-topped sampling. *Modulation*.

instantaneous value—Magnitude of a varying quantity at a given instant.

instruction—Step in a computer program specifying an operation to be performed. Instructions are stored in the computer memory and fetched one by one to the instruction register in the CPU. *Computer Hardware*.

instrument—Device that performs some function, such as a measurement.

Instrument Low-Approach System (ILS)—An ICAO standard involving lateral guidance by a 108-112 MHz localizer, vertical guidance by a 330-335 MHz glide slope, and distance along the path by 75 MHz fan markers. *Navigation Aids*.

insulated-gate field-effect transistor (IGFET)—Majority-carrier semiconductor device in which the conductivity of a current path from source to drain electrodes is modulated by the field from a metal gate electrode, extending through a thin insulator into the semiconductor channel between the source and drain. *Transistors*.

insulator—Substance that completely blocks the flow of electric current, such as porcelain, glass, rubber, dry wood, dry air, some plastics, and so on. *Semiconductors, Properties of Materials*.

integers—All the positive whole numbers, their negatives, and zero.

integral—1. The indefinite integral of a function f (x), of which the derivative (q.v.) $\dfrac{df}{dx}$ is given by

$$y = \int f(x)\ dx$$

2. The definite integral of a function $f(x)$ is the difference between two values of $f(x)$ for two distinct values of the variable, and is given by

$$y = \int_{b}^{a} f(x)\,dx$$

(See Figure I-5.)

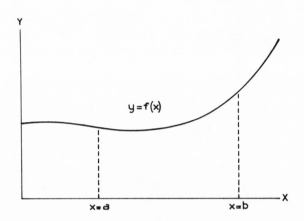

Figure I-5
Integral
The definite integral $x = \int_{b}^{a} f(x)dx$ is the area under the curve between $x = a$ and $x = b$.

integrated circuit (IC)—The physical realization of two or more circuit elements inseparably associated on or within a substance to form an electrical network. *Microelectronics.*

integration—Mathematical process inverse to that of differentiation (q.v.) for finding the function of a given derivative; for instance, given:

$$\frac{dy}{dx} = f(x)$$

find y. This means that y is some function whose derivative is $f(x)$. The answer will be given by:

$$y = \int f(x)dx$$

(See also integral.)

integrator—Circuit with an output that is an integral of the input signal. Such circuits are employed in television receivers (to integrate the vertical sync pulses) and analog computers. In the latter case an operational amplifier is used. *Operational Amplifiers.*

intensitometer—(See dosimeter.)

intensity—Term used to denote the brightness of an oscilloscope display, adjusted by the intensity control.

intercarrier noise suppression—(See squelch circuit.)

intercarrier sound system—Standard U.S. television transmission in which the aural center frequency is spaced 4.5 MHz higher than the visual carrier in the composite television signal. *Broadcasting.*

intercommunication system—Communication system within a building, vehicle, and so on, in which the stations do not require a switchboard. Commonly called "intercom."

interconnection—Connection between systems or networks.

interdigital transducer—Pattern of alternating metal fingers formed on a piezoelectric substrate to convert an electric signal to an acoustic wave, and vice versa. Used for microacoustic lines to take advantage of the much smaller wavelength of acoustic signals. (See Figure I-6.)

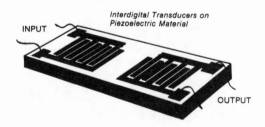

Figure I-6
Interdigital Transducers

From *Technician's Guide to Microelectronics*, page 161. ©1978. Reprinted with permission of Parker Publishing Company, Inc.

interelectrode capacitance—In an electron tube, the capacitance that exists between the tube elements, which at radio frequencies shunt the signal circuits, reducing signal strength, or providing undesirable feedback. These effects are mitigated by the use of screen-grid tubes. *Electron Tubes.*

interface—Common boundary, surface, or connection between two systems or devices.

interference—1. Wave motion produced by the combination of two or more wave trains moving in the same direction. *Electroacoustics.* 2. Extraneous signal of any kind, usually noise or another radio signal, that contributes to the degradation or reduced intelligibility of the desired radio signal. *Radio Noise and Interference.*

interferometer—1. Acoustical interferometer; device for measuring velocity or absorption of sound waves in a gas or liquid by analyzing the standing-wave pattern produced by interference between the direct and reflected waves. 2. Radio interferometer; apparatus with two or more antennas receiving radio waves from outer space and connected to one receiver in such a way that they interfere. When analyzed, this interference will give information from which the diameter of the source can be calculated.

interlacing—In television, combining two separate scanning fields to form a complete picture. The lines of the second field fall in the empty spaces between the lines of the first field. Two successive fields constitute a frame, so the field frequency is 60 Hz and the frame frequency 30 Hz, in countries where the powerline frequency is 60 Hz; and 50 Hz and 25 Hz in those where it is 50 Hz. *Broadcasting.*

interleaving—In a transformer, dividing the primary or secondary coil into two sections and placing the other winding between the two sections. This is done to reduce leakage inductance (q.v.). *Transformers*

interlocking device—Means of preventing operation of equipment unless another condition is met first. For instance, inhibition of power application until the door or access panel is safely closed.

intermediate frequency (i.f.)—New frequency to which the incoming radio signal is changed in a superheterodyne receiver. The new signal is produced by mixing the r.f. signal with one produced by a local oscillator in a mixer or converter stage, to obtain a signal equal to their difference. In most receivers the frequency of this signal is 455 kHz. The i.f. amplifier has high selectivity and gain at this frequency. (See also image rejection.)

intermediate-frequency amplifier—I.f. amplifiers usually consist of two or more stages (one stage in simple tube receiver), coupled by double-tuned i.f. transformers. (See Figure I-7.)

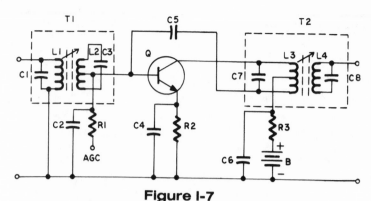

Figure I-7

Intermediate-Frequency Amplifier

From *Complete Guide to Reading Schematic Diagrams*, First Edition, page 48.

intermediate-frequency transformer—An i.f. transformer generally consists of two universal-wound coils with powdered iron cores and fixed capacitors, mounted in a metal-shield container. The capacitors are omitted for higher frequencies where distributed capacitance is adequate, and tuning is done by varying the inductance (permeability tuning). Standard frequencies are

Broadcasting (540-1600 kHz)	455 and 260 kHz
Broadcast (vehicle)	262.5 kHz
FM Broadcast	10.7 MHz
TV Receiver (video carrier)	45.75 MHz

intermittent—1. An intermittent defect is one that comes and goes.   2. Intermittent direct current is d.c. interrupted at regular intervals.

intermodulation—1. Sum and difference frequencies produced by nonlinear characteristics of the equipment.   2. Type of distortion resulting from undesired sum and difference frequencies.   3. Noise resulting from cross-talk from other channels.

International Morse Code—Standard form of the Morse Code used by all countries. *International Morse Code.*

International Morse Code

A	.—		.		.—.—.—
B	—...		, (comma)		—.—.—..
C	—.—.		;		—.—.—.
D	—..		:		—.......
E	.		?		..——..
F	..—.		' (apostrophe)		.————.

G	——·	-	—····—
H	····	/	—··—·
I	··	Ä	·—·—
J	·———	Á or Ā	·——·—
K	—·—	E	··—··
L	·—··	CH	————
M	——	Ñ	——·——
N	—·	Ö	———·
O	———	Ü	··——
P	·——·	(or)	·——·—
Q	——·—	"	·—··—·
R	·—·	—	··——··
S	···	=	—···—
T	—	SOS	···———···
U	··—	Attention	—·—·—
V	···—	CQ	—·—·——·—
W	·——	DE	—·· ·
X	—··—	Go ahead	—·—
Y	—·——	Wait	·—···
Z	——··	Break	—···—·—
1	·————	Understand	···—·
2	··———	Error	········
3	···——	OK	·—·
4	····—	End of message	·—·—·
5	·····	End of work	···—·—
6	—····		
7	——···		
8	———··		
9	————·		
0	—————		

International System of Units—English name for Le Système International d'Unités, the updated metric system, adopted in 1960 as the system of choice for general worldwide use by the General Conference on Weights and Measures. Nearly all countries now use the metric system; the U.S. is the principal hold-out, where the vast inertia of the medieval English weights and measures system, as well as popular psychology, are resulting in a very slow acceptance. *S.I. Units.*

International Telecommunications Union—The international authority located in Geneva, Switzerland, that organizes cooperation between the member nations in all matters pertaining to radio communications, by means of the Radio Regulations it issues.

interpreter—Program that translates a higher-level language into machine language one line at a time; as opposed to a compiler or an assembler, which translates the entire higher-level program into machine language at one time. *Computer Software.*

interrupt—In a computer, a process by which the controller can interrupt the program it is running, perform another operation (e.g., accept an input), and then return to the original program with no loss of information.

interrupted continuous wave (ICW)—Continuous-wave frequency that is interrrupted at an audio rate.

interrupter—Electromechanical device consisting of a solenoid energized by an electric current so that it attracts a spring-loaded armature that opens a switch, interrrupting the current. The armature is then released, the switch closes, and the cycle is repeated continuously as long as the current is applied. Used in buzzers, doorbells, choppers, etc.

interstitial atom—Atom that has forced its way into the space between two other atoms in a crystalline solid. This may be an impurity atom or one of the normal atoms. The imperfection, called a crystal defect, may alter the characteristics of the crystal. *Semiconductors.*

intrinsic-barrier diode—(See pin diode.)

intrinsic semiconductor—Chemically pure semiconductor in which the numbers of conduction electrons and holes are equal. *Semiconductors.*

invar—Alloy of iron that expands very little when heated. It contains 36 percent nickel and small amounts of manganese, silicon, and carbon. Used in many electronic components, jet engines, clockwork, and surveying equipment.

inverse feedback—(See feedback.)

inverse peak voltage—Peak voltage between the plate and cathode of a rectifier tube during the time the tube is not conducting.

inverse square law—The energy transport per unit area perpendicular to the direction of wave propagation is the intensity, which for a spherical wave falls off inversely as the square of the distance from the source. *Electromagnetic-Wave Propagation.*

inversion—When a layer of warmer air exists above a colder layer an inversion (of temperature) exists. Radio signals in the VHF region are bent downward by the upper layer (tropospheric bending) and cover abnormal distances. *Electromagnetic-Wave Propagation.*

inversion layer—Surface layer of a semiconductor material that has been inverted from its original conductivity type to the opposite conductivity type. *Microelectronics.*

inverter—1. Logic device with one input and one output, in which the output voltage level is opposite to the input level. Sometimes called a NOT gate. *Computer Hardware.* 2. D.c. to a.c. converter. *Rectifiers and Filters.* 3. Cathode-coupled amplifier with one input grounded. *Electron-Tube Circuits.*

i/o device—Input or output device used with a computer, such as a keyboard, card reader, printer, crt display, etc.

iodine—Nonmetallic, nearly black, crystalline solid that becomes a deep violet gas at room temperature without going through an intermediate liquid state (sublimation). The radioactive isotope iodine-131 is used as an x-ray tracer. It has a half-life of eight days. Symbol, I; atomic weight, 126.9044; atomic number, 53.

ion—Atom or group of atoms that bears a positive or negative charge. Positively charged ions are called cations; negatively charged ions, anions. Ions are formed by the addition of electrons to, or the removal of electrons from neutral atoms, or molecules, or other ions; by combination of ions with other particles; or by rupture of a covalent bond between two atoms in such a way that both of the electrons of the bond remain attached to one of the atoms.

ion chamber—Radiation monitor that collects the charge released by the passage of an ionizing particle through a gas. In the chamber the charges on the cathode and anode leak away in proportion to the ion current between them, so the change in voltage after exposure to radiation is a measure of the amount of radiation. *Nuclear Physics.*

ionic bond—Type of linkage between atoms with opposite charges, such as sodium (+) and chlorine (−) in sodium chloride. Also called electrostatic bond.

ion implantation—Doping technique in which precise quantities of a single dopant are driven into a semiconductor to a depth of about 1 μm by an ion accelerator. *Microelectronics.*

ionization—Any process by which electrically neutral atoms or molecules are converted to ions. In general, it occurs whenever sufficiently energetic charged particles or radiant energy travel through gases, liquids, or solids.

ionosphere—Region of the Earth's atmosphere in which a significant portion of the atoms and molecules have become electrically charged. The ionosphere begins at about 55 km (34 miles) above the Earth and extends to the outermost limits of the atmosphere, well beyond 400 km (250 miles). It has very important effects on radio communication. *Electromagnetic-Wave Propagation.*

ion rocket—Rocket with propulsion derived from positively charged ions accelerated electrostatically. Such a propulsion system has a very low thrust but a very long duration and great saving in weight. Assuming the vehicle was parked in a low-altitude orbit so that it did not require high thrust to overcome Earth's gravitation, it would be capable of deep-space missions.

ion spot—Burnt spot on cathode-ray tube phosphor screen due to ion bombardment when ion trap is misadjusted. Usually avoided in modern equipment by angling the electron gun so that ions are aimed away from the screen. The beam electrons, being much lighter, are easily kept on the proper path.

IR—Abbreviation for infrared (q.v.).

iris—Simplest type of resonant iris used in a waveguide, consisting of an inductive diaphragm and capacitive screw located in the same plane across the waveguide. The capacitive screw is tuned to resonance with the inductive diaphragm to obtain a low value of loaded Q. *Waveguides and Resonators.*

iron—Pure iron is a soft, ductile, gray-white metal of high tensile strength, which readily corrodes in moist air to form common iron rust. As commonly available, iron nearly always contains small amounts of carbon, which modify its properties. As the cheapest metal it is the most used and is particularly important in electronics for its magnetic properties (q.v.). Symbol, Fe; atomic weight, 55.847; atomic number, 26.

iron-constantan thermocouple—Thermocouple (q.v.) element with an iron wire and a constantan (60% copper, 40% nickel) wire, with a range of application from −200°C to +1100°C.

iron-core coil/transformer—Coil/transformer wound around an iron core to increase its inductance. At audio frequencies the iron core consists of laminations of silicon steel insulated from each other by varnish or shellac. At radio frequencies the core consists of powdered iron mixed in a binder which insulates the particles from each other. *Magnetic Amplifiers, Transformers, Rectifiers and Filters.*

iron-vane movement—Meter movement for alternating current with two thin, soft iron plates ("vanes"), one fixed, the other pivoted, positioned within a stationary coil. When a.c. flows in the coil, the two vanes become magnetized with the same polarity and repel each other. (See Figure I-8.)

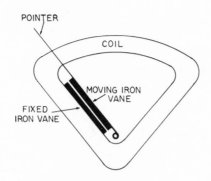

Figure I-8
Iron-Vane Movement

From *Complete Guide to Electronic Test Equipment and Troubleshooting Techiques*, page 35.

irradiation—Exposure to radiation, especially in regard to health hazards. For the nonradiation worker this is a maximum yearly dose to critical organs of 7 millijoules per kilogram (500 millirems) of high-energy gamma rays. (The natural background dose from all sources averages 90 to 200 millirems per year.) *Nuclear Physics*.

I-signal—(See I-channel.)

isobar, nuclear—Species of atom or nucleus that has the same mass number as compared with another species. (See isotones and isotopes.) *Nuclear Physics*.

isochronous cyclotron—Cyclotron in which the strength of the magnetic field increases with the acceleration of the particle to compensate for its relativistic change of mass. *Nuclear Physics*.

isolation—In silicon integrated circuits, the problem of insulation which is achieved by thermal oxidation, etching, epitaxial growth of intrinsic layers, isolation diffusions, or reverse-biased pn junctions. *Microelectronics*.

isolation amplifier—Amplifier placed between other stages to minimize their influence on each other. Also called a buffer amplifier.

isolation transformer—Transformer with a one-to-one turns ratio, connected between the a.c. power input to a piece of equipment and the a.c. line, to minimize shock hazard.

isotone—Species of atom or nucleus that has the same number of neutrons as compared with another species. *Nuclear Physics*.

isotopes—Two or more atomic species of the same element having different atomic masses, but chemically indistinguishable. *Nuclear Physics*.

isotropic antenna—An "ideal" antenna that radiates or receives equally in all directions. Also called a unipole. *Antennas*.

ITU—Abbreviation for International Telecommunications Union (q.v.).

J—Symbol for joule (q.v.).

j—Symbol for $\sqrt{-1}$ (see also i).

jack—Female receptacle for a plug. Usually mounted on a panel, but an extension jack is used on an extension cable. Jacks accept banana plugs, tip plugs, or phone plugs. (See Figure J-1.)

jamming—1. Transmitting a strong signal that overrides or obscures the signal being jammed.   2. Dropping strips of aluminum foil from aircraft during raids to produce a very large number of false echoes in the enemy's radar. Also called "window," "chaff," or "dueppel" (G). *Radar.*

"J" antenna—Vertical antenna for VHF consisting of a half-wave vertical radiator and a quarter-wave matching section. Since nearly all antennas for this band are today horizontally polarized it is becoming obsolete. *Antennas.* (See Figure J-2.)

J-display—Radar display in which the time base is circular and signals appear as radial blips. *Radar.*

JFET—Abbreviation for junction field-effect transistor (q.v.).

Figure J-1

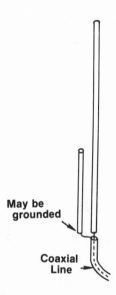

May be
grounded

Coaxial
Line

Figure J-2
"J" Antenna

jitter—Instability in a cathode-ray tube display usually due to false triggering from an extraneous signal, especially at very low signal levels.

J-K flip-flop—Flip-flop having two inputs designated J and K. At the application of a clock pulse, a 1 on the J input will set the flip-flop to a 1 or "on" state; a 1 on the K input will reset it to the 0 or "off" state. Simultaneous 1's on both inputs will cause it to change state regardless of the state it was in. *Computer Hardware*.

Johnson noise—Resistor noise caused by thermal agitation of the resistor material. It varies with the resistance value and the temperature. *Resistors.*

joint—Connection, usually one made by soldering.

joule—Unit of energy, work, quantity of heat, being the work done or energy expended when a force of one newton moves the point of application a distance of one meter in the direction of that force. *S.I. Units.*

Joule's law—The amount of heat produced in a wire by an electric current is proportional to the resistance of the wire and the square of the current, as expressed by the formula:

$$P = I^2R$$

where P = the electric power lost to heat, in watts;
I = the current, in amperes;
R = the resistance, in ohms.

jump—Instruction in a computer program to go to some other address for the next instruction instead of the next in normal sequence. For instance, if the answer to the question "Is X less than 100?" is "Yes," the computer is to execute the next instruction in the program; but if the answer is "No," it is to jump back to a previous instruction and repeat the operation. *Computer Software.*

jumper—Temporary connection between two points in a circuit.

junction—The boundary between p and n regions in a semiconductor crystal. *Semiconductors.*

junction diode—Standard semiconductor diode. *Semiconductors.*

junction field-effect transistor—Majority-carrier device in which the resistance of a current path from source to drain is modulated by the voltage applied to a gate electrode. The gate is a reverse-biased pn junction where the depletion layer extends into the conducting channel and effectively removes carriers from it. *Transistors.*

junction transistor—Standard transistor with two junctions. *Transistors.*

K—1. Symbol for potassium (q.v.).   2. Symbol for a constant (math.).   3. Roman numeral 250.   4. Symbol for cathode.   5. Symbol for Kelvin. *S.I. Units.*

k—1. Symbol for prefix kilo-. *S.I. Units.*   2. Symbol for carat.   3. Symbol for knot (nautical mile per hour) (q.v.).

KA—Abbreviation for kiloampere (1000 amperes).

kaon—Unstable particle produced in nuclear reactions with bombarding energy greater than several gigaelectron-volts. Also called a strange particle. *Nuclear Physics.*

K-band—Microwave-frequency band extending from 10.90 to 36.00 gigahertz. *Frequency Data.*

kc—Abbreviation for kilocycle (now kilohertz).

K-display—Deflection-modulated radar display similar to an A-display (q.v.), used with a lobe-switching antenna. Spread voltage splits signals from two lobes. When pips are of equal size, antenna is on target. *Radar.*

keeper—Iron bar or plate placed so as to connect the poles of permanent magnet when not in use, to protect it from demagnetization.

kelvin—Unit used to express an interval or difference of temperature. The word degree or symbol ° is not used, a temperature being stated as so many kelvins (symbol K). The kelvin is the fraction 1/273.15 of the thermodynamic temperature of the triple point of water. A kelvin is exactly equal to one degree Celsius, and a Celsius temperature (C) is related to a kelvin temperature (K) as follows:

$$K = C + 273.15$$

S.I. Units.

Kelvin bridge—Resistance-measuring bridge adapted for handling very low values without inaccuracies from stray resistances and contact resistance. The resistances a and b bear the same proportions to each other as A and B, but being much higher in value than R_s and R_x carry only a very small current, so that voltage dropped across the sliding contact of R_s, for instance, is negligible. (See Figure K-1.)

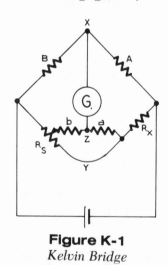

Figure K-1
Kelvin Bridge

From *Complete Guide to Electronic Test Equipment and Troubleshooting Techniques,* page 57.

Kelvin clips—Spring clips and leads for connecting a Kelvin or similar bridge to a low-value resistance. The jaws of the clips are insulated from each other, one being the current connection, the other the potential connection. Since the important potential drops due to stray resistances are all in the current path, measurements are made with the potential path where the current is almost non-existent. (See Figure K-2.)

Kerr cell—Device that employs the Kerr effect to interrupt a beam of light up to 10^{10} times per second. In the Kerr effect an electric field applied to a transparent substance varies its index of refraction. *Lasers.*

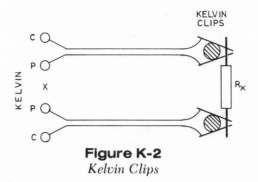

Figure K-2
Kelvin Clips

From *Complete Guide to Electronic Test Equipment and Troubleshooting Techniques,* page 60.

keV—Symbol for kiloelectron-volt.

key—1. Telegraphic hand-operated switch for sending signal in code. Also called a Morse key or telegraph key. 2. Guiding pin or slot to ensure the correct mating of connectors, etc.

keyboard—Device like a typewriter for making entries to a computer, calculator, etc.

key chirp—Change in transmitter frequency during keying due to oscillator loading or line-voltage changes that affect the oscillator, when the power amplifier is energized.

key click—Click heard when transmitter is keyed due to too-short rise and decay times. A filter may be necessary to lengthen these times. (See Figure K-3.)

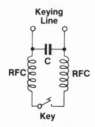

Figure K-3
Key Click Filter
Values of RFC and C must be found by experiment, since they vary with transmitter characteristics. RFC ranges from 2.5 to 80 millihenries; C from 0.05 to 0.5 μF

key punch—Device with a keyboard that punches corresponding holes in an IBM card or paper tape. *Computer Hardware.*

keystone distortion—When the picture on a television screen is wider at the top than the bottom, or vice versa, due to a defect in the yoke. Also called trapezoidal distortion.

kHz—Symbol for kilohertz (q.v.).

kilo-—Prefix meaning $\times 10^3$, or 1000 times the unit to which it is attached. Symbol, k. *S.I. Units.*

kiloampere—1000 amperes.

kilogram—Unit of mass, being the weight of a cylinder of platinum-iridium alloy kept at Sèvres, France, by the International Bureau of Weights and Measures. The U.S. has a duplicate in the custody of the National Bureau of Standards. It was intended to be exactly equal to the mass of 1000 cubic centimeters of water and is very nearly so. A kilogram is equal to 2.204 622 6 pounds avoirdupois. *S.I. Units.*

kilohertz—1000 hertz.

kilohm—1000 ohms.

kilometer—1000 meters.

kilometric waves—Low-frequency waves with a frequency range of from 30 to 300 kilohertz (10-1 km).

kilovolt—1000 volts.

kilowatt—1000 watts, or about 1.34 horsepower.

kilowatt-hour—1000 watts used steadily for one hour. The unit used for calculating consumption of electrical energy.

kinescope—A television picture tube.

Kirchhoff's circuit rules—1. The sum of the currents into a specific junction in a circuit equals the sum of the currents out of the same junction.   2. Around each loop in an electric circuit the sum of the e.m.f.'s is equal to the sum of the potential drops across each of the resistances in the same loop.

klystron—Velocity-modulated microwave electron tube. *Electron Tubes.*

knife switch—Switch consisting of a hinged blade that is sandwiched between two contacts when closed. (See Figure K-4.)

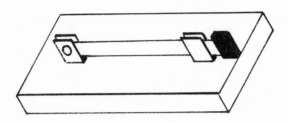

Figure K-4
Knife Switch

knot—One nautical mile per hour. A nautical mile is 6080 feet, 800 feet longer than a statute mile. (It is incorrect to say "knots per hour.")

Kovar—Iron-nickel-cobalt alloy used in glass-to-metal seals.

kraft paper—Paper used as an insulating layer in capacitors and transformers.

L—1. Symbol for inductance (q.v.). 2. Symbol for lambert (q.v.). 3. Roman numeral for 50. 4. Radiance, in watts per square meter per steradian.

La—Symbol for the element lanthanum.

labyrinth—(See acoustic labyrinth.)

ladder attenuator—Attenuator consisting essentially of a number of cascaded sections. *Attenuators.*

lag—When two waves of the same frequency start their cycles at slightly different times, the one starting later may be said to lag the other. The interval between them is given in degrees, each degree being 1/360 of a cycle. (See also lead.) (See Figure L-1.)

lag circuit—Keying filter to eliminate key clicks (q.v.).

lambda—Greek letter Λ (cap.) or λ (small). Λ is used to designate permeance. λ is used to designate wavelength and attenuation constant.

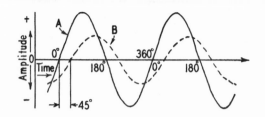

Figure L-1
Lag
Waveform B lags A by 45 degrees

lambert—Unit of luminance (brightness), defined as the brightness of a perfectly diffusing surface that radiates or reflects one lumen per square centimeter. In S.I.units, one lambert is equal to π^{-1} lm sr^{-1} m^{-2}, or $\pi \times 10^{-4}$ candela per square meter. *S.I. Units.*

laminations—Thin strips or plates of iron, coated with an insulator such as varnish or shellac, that are stacked together to make a transformer or choke core, to reduce loss of power from eddy currents induced in the core. *Transformers.*

lamp—Device for producing light. In electronics, usually an indicator lamp or pilot light. Lamps may have bayonet, screw, or flanged metal bases for installing in corresponding holders, or leads (for very small lamps).

lamp cord—Power supply cord for lamps, small appliances, etc., usually consisting of two 18-gauge stranded wires with plastic insulation, and often with a plastic male two-prong connector molded on one end; rated for current up to 10 amperes at 125 volts.

landing beam—Usually refers to Instrument Low-Approach System. *Navigation Aids.*

language—Computer programming code, such as "machine language," "assembly language," and so on. *Computer Software.*

L-antenna—Antenna with a horizontal radiator and a vertical lead-in connected at one end. *Antennas.*

Laplace transform—Mathematical procedure for solving transient problems, most of which fall into a class for which Laplace transforms exist and for which tables are available.

large-scale integration (LSI)—Monolithic integrated circuit with very large number of gates, etc., such as a microprocessor chip, calculator chip, or electronic watch chip. *Microelectronics.*

LASCR—Abbreviation for light-activated silicon controlled rectifier (q.v.).

LASCS—Abbreviation for light-activated silicon controlled switch (q.v.)

laser—Acronym formed from the initial letters of "light amplification by stimulated emission of radiation." The laser is a quantum electronics device in which a suitable active material is excited by an external stimulus. The active material is in an optical resonant cavity, and its output includes a narrow linewidth amplitude stabilized oscillation greatly amplified compared to the input. *Lasers.*

LASERS

Like a man on a ladder, an atom can be at different levels, called energy states. With an input of energy it can step up to a higher level. It can also step down by emitting energy. But the levels are fixed. The man on the ladder can stand only on the rungs that are there; he cannot stand between them. So with the atom: it has "allowed" states; anything between is "forbidden."

An atom that has stepped up to a higher level by absorbing energy does not stay there very long. In fact, it gets back to its former level in about 10^{-8} second by emitting a photon at frequency f according to the formula:

$$E_2 - E_1 = hf \tag{1}$$

where $E_2 - E_1$ equals the energy emitted in falling back from the higher state E_2 to the lower E_1, and h is Planck's constant. This is called a spontaneous transition.

Atoms can also be deliberately subjected to radiation, in which case the changes in level are termed stimulated transitions. If the radiation is intense enough the number of atoms in the higher energy state will exceed those in the lower, and what is called a "population inversion" will exist.

If the formula (1) above is rearranged so that:

$$f = \frac{E_2 - E_1}{h}$$

it will be obvious that for a given frequency a photon will have a certain energy and, conversely, a photon with a given energy must have a certain frequency. In the population of excited atoms (all of the same kind), photons of a certain energy are being emitted as the atoms fall back to the lower energy level, although the downward transitions for the moment are fewer than those in the opposite direction. These photons also have a specific frequency corresponding to their energy. If a light wave of this frequency is now made to pass through the mass of atoms, the photons in the light wave will have exactly the same characteristics as those being emitted. Every time a photon in the light wave strikes an upper-level atom, it is stimulated to make a downward transition, emitting an identical photon, so now two photons exist instead of one, both moving in the same direction. These two speed on, striking two more atoms with the same result, so now there are four photons, all traveling in one direction. The new emission thus amplifies the wave that stimulated it, and if multiplied sufficiently gives rise to a superradiant beam of light having a relatively narrow bandwidth.

This process is carried further in the laser (acronym for light amplification by stimulated emission of radiation), which is an optical resonating cavity. The earliest ones (still very much in use) were solid-state, made of synthetic ruby (aluminum oxide) doped with triply ionized chromium. This material is formed into a rod. The ends are cut off square, highly polished and silvered, so they become two mirrors, facing and parallel (see Figure L-2). One mirror, however, is only 95 percent reflecting, so 5 percent of the light striking it can pass through. The rod is surrounded by a xenon flash lamp in the form of a coiled glass tube. The radiation from this has a wavelength of about 550 nanometers, which raises the chromium ions to a higher energy state. This is called optical pumping. The photons emitted from the chromium ions travel along in the ruby rod, strike one of the end mirrors, and are reflected back to the other mirror. The mirrors are spaced so that they form a cavity resonant at the photon frequency (432.1 terahertz). As the photons flash back and forth along the rod, they stimulate more and more emission and the power of the beam builds up until it bursts through the mirror that is only 95 percent silvered. This beam consists of light all of a single frequency, deep red in color, and with all its components in phase, or *coherent*. The beam is very narrow and its waves so parallel that its divergence over ordinary distances is negligible; therefore its full power is concentrated in a very small area.

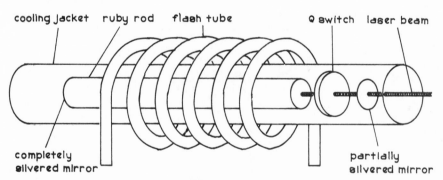

Figure L-2
Ruby Laser

Solid-State Lasers

The description just given of the operation of the ruby laser pretty much describes that of all solid-state lasers. The most widely used material is not ruby, however, but glass, YAG (yttrium aluminum garnet), and some other materials, all doped with neodymium ions. Their laser wavelengths are all in the near infrared.

Gas Lasers

Atoms in a gas can be stimulated to make transitions between different energy levels in the same way as solids. In the helium-neon laser the neon gas is made to glow, as it does in a neon tube, and because the photons emitted by it have almost the same energy as helium photons they cause laser action as described above for the ruby laser. Gas lasers can be made to operate between more than one pair of levels, resulting in several possible frequencies for the resultant laser beam. In the helium-neon laser the most commonly used wavelength is 632.8 nanometers, which is a reddish-orange color. By using argon instead of helium, beams of 488 nanometers (blue) and 514 nanometers (green) can be obtained.

These noble-gas lasers have beam powers ranging from 0.05 watt to several watts and are used for a great many different purposes. They possess a higher degree of coherence and a much narrower bandwidth than any other kind of laser, so that the beam provides a nearly perfect straight line, making it ideal for surveying and similar uses.

A much more powerful type of gas laser uses a tube containing carbon dioxide mixed with helium and nitrogen (sometimes steam as well). By applying a high voltage to opposite ends of the tube the gas mixture is made to glow. The excited atoms emit electrons which collide repeatedly with the carbon dioxide molecules, raising them to a higher level of vibrational-rotational energy. Laser action then takes place with the emission of a beam with a wavelength of 10.6 micrometers (in the infrared region). Some of these lasers have power outputs of over 10 kilowatts which can be used in much the same way as a welding torch, but with far greater precision. An even more powerful beam ($>$30 kilowatts) can be obtained from a gas laser using carbon monoxide and nitrogen, in which the burning monoxide is expanded through jet nozzles. These lasers are called gas dynamic lasers.

Liquid Lasers

The simplest type of liquid laser is a transparent cavity of glass or quartz containing neodymium oxide or chloride dissolved in a suitable solvent such as selenium oxychloride, and operated like a solid-state laser. Using a liquid has the advantage of not being susceptible to damage from heat at high power levels.

A special type of liquid laser employs organic dyes such as rhodamine 6G and methylumbelliferone. When these are excited by another laser they can emit beams with wavelengths ranging from ultraviolet to yellow, and peak powers up to 10^8 watts in pulsed operation. Continuous-wave power is very much less.

Semiconductor Lasers

When a pn junction is forward biased the lowering of the potential barrier allows the minority carriers to recombine. In some semiconductors the energy is released mostly in photons, the best-known example being gallium arsenide, which is used in light-emitting diodes (LED's). When a large electric current is passed through a junction of differently doped gallium arsenide, laser light emerges from the junction region.

Power output is low, only a few milliwatts, but the modest cost and small size of these devices make them suitable for use in surveying instruments in short-range communication equipment.

Q-Switching

In Figure L-2 a device called a Q switch is shown. This blocks the partially silvered mirror, preventing laser action but, as optical pumping continues, practically all the atoms are raised to the higher energy level. If the blockage is now removed at the proper instant the stored energy is released in a "giant" pulse which, while it lasts only 10^{-9} to 10^{-10} seconds, may exceed 10^{12} watts in large solid-state lasers. It can be used to punch holes so fast that the surrounding material is unaffected. The Q switch may be an optical shutter of liquid or solid that is normally opaque but can be made transparent by the application of an electrical pulse or light pulse from another laser.

latch—Set of flip-flops that serves as an interface between two subsystems by accepting a value from the first subsystem and holding it until the second subsystem is ready for it. Also called a buffer. *Computer Hardware*.

latching relay—Relay that locks in either the on or off condition until reset manually or by a signal.

lattice—Simplified way of representing the orderly arrangement of atoms in a crystalline solid. *Semiconductors*.

lattice winding—Type of crisscross coil winding designed to reduce distributed capacitance. Also called honeycomb winding.

lavalier microphone—Microphone designed to be suspended from a band or cord hung around the neck, so as to leave the user's hands free and permit movement within the range of the cable.

layers, ionosphere—(See ionosphere.)

layer winding—Coil winding in which the turns are closely wound with no space between.

lazy-H antenna—Four-element array resembling a letter H lying on its side, consisting of two half-wave dipoles one above the other, popular at one time with amateurs, but superseded by half-wave antennas with parasytic elements, which have greater gain and directivity. *Antennas*.

L-band—Microwave band with frequencies 0.390 through 1.550 gigahertz.

LCD—Abbreviation for liquid crystal display (q.v.).

LC product—The product of multiplying together the inductance (in henries) and the capacitance (in farads) in a circuit.

L-display—Radar display in which the signals from two antenna lobes are placed back to back. When the blips are of equal size the antenna is on target.

lead—1. When two waves of the same frequency start their cycles at slightly different times, the one starting first may be said to lead the other. The interval between them is given in degrees, each degree being 1/360 of a cycle. (See also lag.)   2. Insulated flexible wires used for test purposes.

lead—Soft, silvery-white or grayish metal, very malleable, ductile and dense, and a poor conductor of electricity; used in the manufacture of storage batteries, solder, coverings for electric cables, shielding around nuclear reactors, x-ray equipment. Symbol, Pb; atomic weight, 207.19; atomic number, 82.

lead-acid cell—Cell in a lead-acid storage battery, the essential parts of which are lead and lead dioxide plates immersed in dilute sulfuric acid. The lead-acid cell, or "accumulator," is rechargeable, so is classified as a secondary cell. (See Figure L-3.)

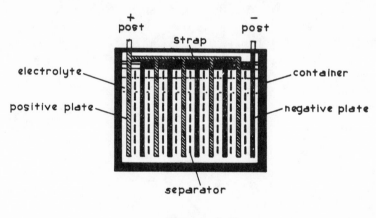

Figure L-3
Lead-Acid Cell

lead-in—Wire connecting antenna to receiver. Wire connecting a transmitter to an antenna is a transmission line, which can also be used as a lead-in. *Transmission lines.*

lead-in groove—Groove on the outer edge of a phonograph in which the stylus is set down and which leads it into the sound track.

lead-out groove—Spiral groove on phonograph disk that stylus enters at end of recording.

leakage—In a transformer, leakage flux is flux that "cuts" one winding but not the other. This flux causes self-induction termed leakage inductance, which has reactance called leakage reactance. Leakage reactance increases with frequency; at 60 hertz, in a well-designed transformer, it should not cause a loss of secondary voltage at full load of more than 10 percent. *Transformers.*

leakage current—In a capacitor, the current flowing between two or more electrodes by any path other than the interelectrode space. *Capacitors.*

leap second—Second inserted as required between the end of the 60th second of the last minute of the last day of a month and the beginning of the next minute in order to keep Coordinated Universal Time (UTC)—formerly called Greenwich Mean Time (GMT)—in step with UT1 (solar time corrected for periodic variations in the Earth's rotation). If the leap second is positive (+1 s) the last minute will have 61 seconds; if negative (-1 s), it will have 59 seconds. *Frequency Data.*

least significant bit (LSB)—In a binary number, the right-most bit, which says how many 2^0's are in the number. *Computer Hardware.*

Lecher wires—Two parallel wires forming a transmission line and equipped with a sliding contact that short-circuits them. When loosely coupled to the output of a transmitter, the distance x between two points (current loops) at which a dip in output is indicated as the shorting bar is moved along the wires, is equal to half the wavelength. The output indication may be provided by a flashlight bulb soldered to a one-turn loop of wire.

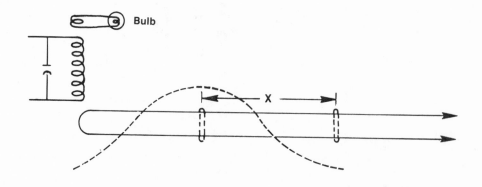

Figure L-4
Lecher Wires

Leclanché cell—Glass jar containing an ammonium chloride ("sal ammoniac") solution. The anode is a zinc rod, coated with mercury to prevent corrosion, the cathode a carbon rod within a porous pot filled with a mixture of manganese dioxide and carbon powder. (The manganese dioxide combines with hydrogen, which otherwise would collect on the cathode and stop the operation.) The flashlight battery works on the same principal, except the electrolyte is in the form of a paste—hence the name "dry cell."

LED—Abbreviation for light-emitting diode (q.v.).

lens—1. In optics, a piece of glass or other transparent substance used to form an image of an object by focussing rays of light from the object. (See Figure L-5.)   2. An electronic device that focuses a beam of electrons, as in a cathode-ray tube or electron microscope.

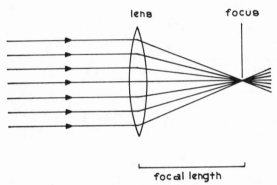

Figure L-5
Convex Lens

Lenz's law—An induced electric current flows in a direction such that the current opposes the change that produced it.

lepton—Any member of the class of lightest subatomic particles that include the electron family (electron and electron neutrino) and the muon family (negative muon and muon neutrino). Each has an antiparticle that together comprise the antileptons. *Nuclear Physics.*

level—Usually a d.c. voltage value with reference to zero or other base value; but may also be used of other values or of a ratio expressed in decibels.

Leyden jar—Device for storing static electricity consisting of an insulating jar with its inner and outer surfaces coated with metal foil. The outer coating is grounded and a suitable connection is made with the inner coating through a central brass rod which projects through the mouth of the jar. The Leyden jar, invented in 1745, was a prototype of the capacitor.

LF—Abbreviation for low frequency (q.v.).

lie detector—(See polygraph.)

life support systems—Devices enabling man to live and work in hostile environments such as outer space and under water.

light—That part of the electromagnetic spectrum detected by the human eye, with wavelengths between 370 and 750 nanometers. *Optoelectronics.*

light, speed of (symbol, c)—Fundamental physical constant according to current theories of relativity and electromagnetism. In a vacuum the speed of light and all other electromagnetic radiation is $2.997\,925 \times 10^8$ meters per second (approximately 186,000 miles per second). For general purposes the speed of light is assumed to be 3×10^8 meters per second. It is less in matter, the actual speed varying with the medium.

light-activated silicon controlled rectifier (LASCR)—Pnpn device consisting of a silicon controlled rectifier (SCR) installed in a housing having a transparent window or collection lens. Operation of the LASCR is similar to that of a conventional SCR except an optical signal replaces the gate electrical signal. *Semiconductors.*

light-activated silicon controlled switch (LASCS)—Pnpn device with both anode and cathode gate terminals, so it is bidirectional. Its operation is the same as that of a silicon controlled switch, but it is installed in a housing having a transparent window or lens so that it can be controlled by an optical signal instead of by electrical gating signals. *Semiconductors.*

light-dimmer control—Circuit containing a diac, a triac, and an RC phase-control network. Varying the resistance varies the RC constant, changing the point in the a.c. cycle when the triac is triggered, and consequently the power applied to the load (in this case, the lamp). *Semiconductors.* (See Figure L-6.)

light-emitting diode (LED)—Electroluminescent diode, usually of gallium arsenide (GaAs), in which radiative recombination predominates over other equilibrium byproducts. *Semiconductors.*

lighthouse tube—Type of electron tube designed for reduced transit time and interelectrode capacitance, used in VHF transmitters, and capable of operation at frequencies up to 3 gigahertz. Also called a megatron. *Electron Tubes.* (See Figure L-7.)

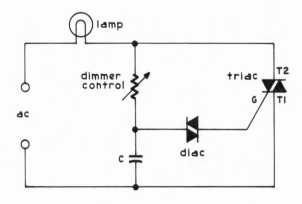

Figure L-6
Light Dimmer Circuit

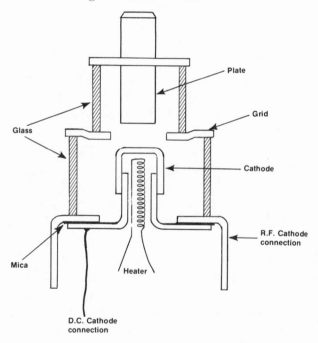

Figure L-7
Lighthouse Tube

light meter—(See exposure meter.)

lightning—Visible discharge of atmospheric electricity that occurs when a region of the atmosphere acquires an electrical charge sufficient to overcome the resistance of the air. For cloud-to-ground discharges the potential is on the order of hundreds of millions of volts. The current in a lightning flash is typically some 20,000 amperes, but may occasionally be ten times as much. An average flash consists of three or four strokes in each direction lasting about a tenth of a second.

lightning protection—Protection from destruction caused by lightning is obtained by providing a low-resistance path to ground so the flash does not have to pass through non-conducting materials. The lightning rod is one such device, and lightning arrestors (spark gaps between overhead wire and ground that can be bridged by lightning) and attenna switches (to ground the antenna) are among others.

light pen—Light-sensitive device by which a person can select some portion of a computer display for action. The computer scans the display, lighting each display element in sequence. When the display element in the field of view of the light pen is lit, the light pen signals the computer that this element has been selected.

light pipe—Optical fiber, or bundle of fibers, which conducts light in much the same way a waveguide conducts microwave radiation.

light ray—Straight line along which any part of a light wave is regarded as traveling from its source to any given point.

light-year—The distance traveled in a vacuum by light in one year, approximately 9.4605×10^{12} kilometers (5.878×10^{12} miles). About 3.262 light years equal one parsec (q.v.).

limited space-charge accumulation diode—Transferred-electron device similar to a Gunn diode, but in which the high-energy domain has been broadened by changing the doping so that it is as wide as the length of the diode, thus overcoming the frequency limitation imposed by the minimum transit time obtainable in a Gunn diode. *Semiconductors.*

limiter—Circuit for limiting the amplitude of a signal to a predetermined level. A basic circuit consists of a voltage-reference (Zener) diode shunted across the signal path. Assuming a diode breakdown voltage of 3 volts, it will pass everything up to that level, but anything higher will be shunted to ground. This circuit is also called a shunt clipper. (See Figure L-8.)

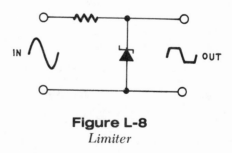

Figure L-8
Limiter

limit switch—Microswitch positioned so that it is actuated by a moving part, in order to shut off or reverse the power to the motor driving the part when it reaches the limit set for its travel.

line—1. Wire or wires connecting stations in a telephone or telegraph system.   2. Wire conducting electric power from one place to another.   3. 1/12 of an inch.   4. The path of a moving point, having length but no breadth.   5. One of 525 scanning lines in a television transmission.

linear accelerator—Traveling-wave linear accelerator that moves electrons along a straight path by means of a radio-frequency wave. *Nuclear Physics.*

linear amplifier—Amplifier that faithfully reproduces the waveform pf of the input signal, as opposed to a digital amplifier, which is operated as a switch. Usually operated Class B in a transmitter, Class A in voltage amplifiers. (See Figure L-9.)

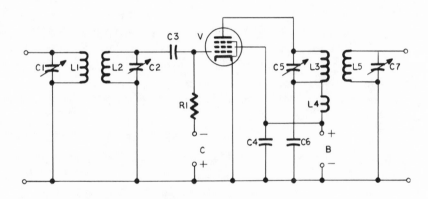

Figure L-9

Class B Linear RF Power Amplifier

From Complete Guide to Reading Schematic Diagrams, First Edition, page 71. ©1972. Reprinted with permission of Parker Publishing Company, Inc.

linear circuit—Circuit whose output is an amplified version or a predetermined variation of its input. *Microelectronics.*

linear control—Control whose effect is linear, or directly proportionate, to its movement.

linear distortion—(See distortion.)

line driver—Buffer amplifier designed to match impedances and boost low-level signals along long lines.

line drop—Potential drop along a power line due to resistance, reactance, or leakage.

line filter—Filter to prevent noise on the power line from entering equipment, or radio-frequency, etc., signals from being coupled to the power line. *Passive Filters.*

line frequency—1. Power line frequency in the U.S. and Canada is 60 hertz; Europe uses 50 hertz; Japan uses both. 2. Horizontal scanning frequency in television broadcasting is 15 734.264 ± 0.044 hertz, or 2/455 times the chrominance subcarrier frequency of 3.579545 MHz.

line of force—Path followed by an electric charge free to move in an electric field. In a magnetic field the field lines are called lines of force only in the sense that a small magnet is forced to align itself in the direction of those field lines, there being no such thing as an isolated magnetic "charge" (unit pole).

line-of-sight propagation—For radio waves, line of sight propagation is further than for visible waves because of bending in the more refractive levels of the atmosphere near the ground. With an average value of the refractive index, the radio line of sight maximum distance in miles over smooth terrain is given by $(2h_t)^{1/2} + (2h_r)^{1/2}$, where $h_t =$ the height of the transmitting antenna and $h_r =$ the height of the receiving antenna, both in feet above mean sea level. *Electromagnetic-Wave Propagation.*

line printer—Print-wheel printer equipped with 120 rotary print wheels side by side in a row, each of which has 48 characters of type around its rim, including letters, numerals, punctuation, etc. At the time of printing, all 120 print wheels are correctly positioned to represent the data to be printed, which is then printed as a complete line. Lines are printed at a rate of 150 per minute. *Computer Hardware.*

line regulation—Measure of change in output voltage with variation of power-line voltage of a regulated power supply. Generally expressed as a percentage over a specified range of power-line voltage, usually from 105 to 125 volts.

line-voltage regulator—1. A variable transformer, such as a Variac, Powerstat, etc., used to adjust the power-line voltage manually (see autotransformer). 2. An electronic automatic voltage regulator or stabilizer using silicon controlled rectifiers or magnetic amplifiers to provide a predetermined output regardless of input fluctuations.

link coupling—Means of coupling two coils that cannot be coupled to each other, using link coils connected by a short transmission line. The coils usually have a small number of turns compared with the resonant-circuit coils. (See Figure L-10.)

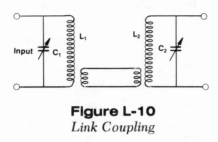

Figure L-10
Link Coupling

liquid-crystal display (LCD)—Segmented display using the ability of a twisted nematic liquid crystal to polarize light, so that the normally transparent segments become opaque when an electric field is applied. *Microelectronics.*

liquid laser—Laser in which the lasing medium is a liquid or dissolved in a liquid. Media include rare-earth ions and organic dyes. *Lasers.*

Lissajous figures—Curves that can be made to appear on the screen of an oscilloscope, produced by the application of two sinusoidal signals of the same frequency, or of harmonically related frequencies, to the X and Y inputs (with the sweep disabled). Used mostly to indicate phase shift or distortion in an audio amplifier. (See Figure L-11.)

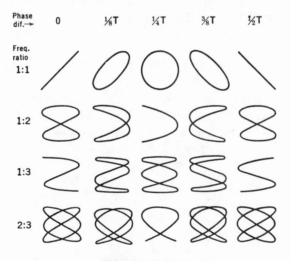

Figure L-11
Lissajous Figures

lithium—Soft white metal with a silver luster, the lightest of the solid elements, softer than lead, it floats on water. It must, however, be kept in nonsolvents such as kerosene and naphtha because it reacts with the oxygen in moist air. Used in heat-transfer applications (specific heat 0.83), scintillators, photocells, and storage batteries. Symbol, Li; atomic weight, 6.941; atomic number, 3.

Litz wire—Type of wire consisting of many fine individually insulated strands, used in winding high-frequency coils to minimize eddy-current losses generated in the winding by stray flux from the core cutting the winding.

load—All active devices (transistors, electron tubes, etc.) work into a load of one type or another. Ideally, the load should be capable of absorbing the maximum power output of the device. This occurs when the impedance of the load is equal to the internal impedance of the device.

loaded Q—The Q of an impedance under working conditions in a circuit.

load impedance—The impedance of the load as seen by the generator or source.

loading—Addition of inductance to an antenna or, at periodic intervals, to a transmission line. Loading an antenna is done by inserting a coil in series with it, to increase its electrical length. This is often done with auto radio antennas, which are usually too short to be resonant at the radio frequency for which they are used. Loading coils are provided in telephone lines to counteract the effects of capacitance.

load line—Sloping line drawn on a family of plate-voltage, plate-current curves to enable the current flow through the tube to be determined for any value of plate and grid voltage. The d.c. load line HK represents the effect of the load resistor alone: the a.c. load line AB represents the effect of the load resistor shunted by the grid resistor. The point where they intersect E is called the quiescent point. JK is the operating range. A similar graph can be drawn for a transistor. (See Figure L-12.)

load regulation—Measure of the ability of a power supply to maintain a constant output despite changing loads.

load resistor—Resistor in series with the plate of an electron tube or the collector of a transistor, and the power supply, across which the output voltage is developed. For a voltage amplifier, the value of this resistor should be as high as possible, consistent with the characteristics of the active device. *Electron-Tube Circuits, Transistor Circuits.*

lobe—In the radiation pattern of a directional antenna one of the directions in which transmisson or reception is stronger. (See radiation pattern.)

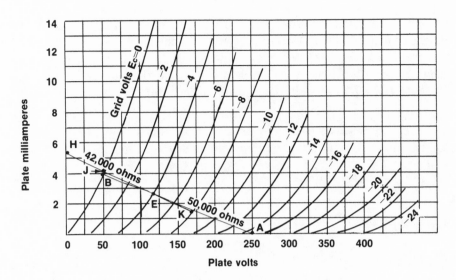

Figure L-12
Load Lines for a 6J5 Triode Electron Tube

local oscillator—In a superheterodyne receiver, the high-frequency oscillator whose signal is mixed with the incoming signal to produce the intermediate frequency (q.v.).

locking in—Phenomenon commonly observed when an oscillator is coupled to a source of a.c. voltage of approximately the frequency at which the oscillator is operating; the more stable of the two frequencies assumes control over the other. Also called "plug-in."

lock-on—When a radar system changes over from searching for to tracking a target automatically. *Radar.*

log—Station operating record that shows (1) date and time of each transmission, (2) all calls and transmissions made, (3) input power to last stage of transmitter, (4) frequency band used, and (5) time of ending each communication.

logarithm—The exponent or power to which a number must be raised to yield a given number. In the expression $100 = 10^2$, 2 is the logarithm of 100 to the base 10; i.e., the number 10 must be raised to the power 2 to equal 100. Logarithms to the base 10 are common logarithms, those most commonly used. Another base used frequently in science and engineering is e (=2.71828...). Logarithms with this base are called natural or Napierian. To distinguish them, common logarithms are denoted "log," Napierian "ln." A logarithm is written in the form X.YYYY ..., where X is the characteristic and YYYY ... the mantissa. The common logarithm 2.4567 is $10^{2.4567}$, or $(10^2 \times 10^{0.4567})$. 10^2= 100, of course. Looking up 0.4567 in log tables gives 2.8622 ...; therefore $10^{2.4567}$=100× 2.8622=286.22. On the other hand, log $\overline{2}$.4567 (the bar over the 2 means –2) is $(10^{-2} \times 10^{0.4567})$ = (0.01 × 2.8622) = 0.028622. *Mathematical Tables.*

logarithmic scale—Scale calibrated so that the distances from the lower end of the scale to each value shown on it are proportional to their logarithms, as in a slide rule.

logic circuit—Digital circuit in which discrete voltage or current levels are used to represent the binary bits 0 and 1 to carry out one of the logic functions. *Microelectronics, Computer Hardware.*

logic, symbolic—As applied in logic networks in computers, the three basic Boolean operations ∧, ∨,'(corresponding to AND, OR, and NOT), are provided by AND gates, OR gates, and inverters. Also called switching algebra, it is a two-state logic, expressed by the binary digits 0, 1. *Computer Hardware.*

log-periodic antenna—Antenna designed so that its characteristics are periodic with the logarithm of the frequency. Such an antenna is practically frequency-independent. *Antennas.*

long-play (lp) record—Microgroove record played at 33 1/3 r.p.m., usually 10 or 12 inches in diameter. *Recording.*

long-wire antenna—Antenna more than a half-wave long. Also called a harmonic antenna. *Antennas.*

loop—1. In a computer program, an iterated set of instructions repeated until some condition is satisfied. *Computer Software.* 2. Circuit providing feedback, especially a closed loop as in a feedback control system.

Loran—Acronym for long-range navigation, used by both aircraft and ships, in which a master station broadcasts an uninterrupted series of pulses, which each slave station (minimum of two), 200-300 miles away, repeats, but with a fixed time difference. The time difference between the master and one slave locates the receiving station on one curve on a Loran chart; the time difference between the master and another slave locates it on a second curve. Where they intersect is the position of the receiving station. Loran-C operates at 100 kilohertz; the higher frequency Loran-A is being phased out. *Navigation Aids.*

lossy—Term meaning "dissipating" or "attenuating."

loudness control—Combined volume and tone control on a hi-fi amplifier that increases the amplifier's low-frequency response at low output levels to compensate for the decreased sensitivity of the human ear to low frequencies at low sound volume. *Electroacoustics.*

loudspeaker—Device for converting electrical audio-frequency signals into sound waves. The electrical signal is applied to a voice coil, which is positioned in a strong magnetic field, so that the variations in the signal cause it to move accordingly. The voice coil is attached to a conical diaphragm that conveys the voice coil's vibrations to the surrounding air. *Electroacoustics, Recording.* (See Figure L-13.)

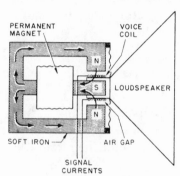

Figure L-13
Loudspeaker

low-frequency (LF)—Frequency range from 30 to 300 kilohertz (10 kilometers to 1 kilometer), also called kilometric waves. *Frequency Data.*

low-pass filter—Filter circuit that passes all frequencies below the cutoff frequency and blocks frequencies above it. *Active Filters, Passive Filters.*

LSA diode—Abbreviation for limited space-charge accumulation diode (q.v.).

LSB—Abbreviation for least significant bit (q.v.).

LSI—Abbreviation for large-scale integration (q.v.).

lug—1. Flat U-shaped metal tag, usually insulated, on the end of a test lead that is used to connect the wire to a screw-type terminal or binding post. 2. Soldering lug, similar to 1., but not insulated, used to connect a circuit hook-up wire to the chassis or a terminal block, etc.

lumen (lm)—Unit of luminous flux, being the luminous flux emitted in a unit solid angle by an isotropic point source having a luminous intensity of one candela (q.v.). *S.I. Units.*

luminance—Intensity of light emitted, reflected or transmitted, per unit area of a surface, as projected on a plane at right angles to the line of vision. Also called brightness. The older unit, the lambert, is equal to $\pi \times 10^{-4}$ candela/meter^2.

luminescence—Emission of light by certain materials, without heating as in incandescence. Luminescence occurs when the material is irradiated by energetic radiation or particles, which raise the atoms of the material to an excited energy level. Since this is an unstable level, the atoms soon fall back to their unexcited level, with the emission of light or heat or both.

luminous intensity—Quantity of visible light. (See lumen.) *Optoelectronics.*

lumped constants—Values of capacitance and inductance in a circuit are said to be lumped when they are present in discrete components or locations as opposed to being distributed generally throughout the circuit.

M—1. Symbol for mutual inductance (in henries). 2. Symbol for element in electrolysis. 3. Symbol for prefix mega- (q.v.). 4. Roman numeral equal to 1000.

m—Symbol for prefix milli- (q.v.).

MA—Symbol for megampere (one million amperes).

mA—Symbol for milliampere (one thousandth of an ampere).

machine language—Binary digit patterns comprising instructions or data which directly control the operations of a computer. *Computer Software*.

Mach number—Ratio of the speed of a body to the speed of sound in undisturbed air. In dry air, at standard temperature and pressure, the velocity of sound is approximately 741 m.p.h., so an aircraft traveling at Mach 1 is traveling at 741 m.p.h.; at Mach 2 its speed would be 1482 m.p.h.; and so on.

macroinstruction—Source program statement that generates a sequence of machine language instructions. *Computer Software*.

magic T—See hybrid junction.

magnesium—Silvery-white metal, the lightest structural metal, used extensively in the aerospace industry, generally alloyed with aluminum, zinc, or manganese to improve its structural strength, resistance to corrosion, and workability. Symbol, Mg; atomic weight, 24.312; atomic number, 12.

magnet—A device that generates a magnetic field. (See electromagnet, permanent magnet, magnetism.)

magnetic amplifier—Saturable-core reactor or inductor with a control winding, in which d.c. flows, and a power winding, in which a.c. flows, both on the same magnetic core. The amount of a.c. power applied to a load is determined by the level of the d.c. voltage, which causes the core to saturate early or late in each a.c. cycle. *Magnetic Amplifiers.*

MAGNETIC AMPLIFIERS

Figure M-1 shows the schematic symbol for a transductor, saturable-core inductor or saturable-core reactor, all terms used for the basic circuit of a magnetic amplifier. It consists of two windings on the same magnetic core. The winding with five scallops is the control winding; the one with three scallops is the power winding. The saturable properties symbol superimposed on them indicates that they are magnetically coupled by a saturable core.

Figure M-1
Schematic Symbol for Magnetic Amplifier

In the absence of a control signal the core does not saturate. The flux induced in the core rises toward the saturation level with each peak of the a.c. current in the power winding, but does not reach it. The power winding therefore presents a continuous high impedance to the power supply so that essentially all the power-supply potential is dropped across the power winding and none across a load in series with it, as in Figure M-2a.

If a current is now made to flow in the control winding, it will induce additional flux in the core. When the currents in the control winding and the power winding are flowing in directions that cause the flux due to each winding to add, the total flux drives the core into saturation. When this happens the power winding becomes a very low impedance, and practially all of the supply voltage appears across the load, as in Figures M-2b and M-2c.

The exact moment when the core saturates will depend upon the current level in the

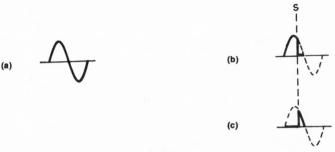

Figure M-2

A.-c. Voltage Across Power Winding:

(a) with no control signal;

(b) with control signal that causes core to saturate at S;

(c) resultant a.-c. voltage across load.

control winding. The flux induced by the power winding rises on each half cycle until saturation occurs, and this will be early or late in the half cycle according to the level of flux induced by the control winding. This enables the duration of the power applied to the load during each half cycle to be set as desired, as illustrated in Figures M-2b and M-2c.

However, power can flow in the load only on every other half cycle, since the core cannot saturate when the fluxes subtract. Consequently, full-wave power utilization requires that *two* saturable-core inductors be provided with their power windings connected oppositely, as shown in Figure M-3.

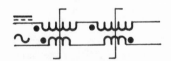

Figure M-3

Full-Wave Magnetic Amplifier

This also enables the saturated core to be reset to non-saturation in time for the next half cycle that saturates it. As the other core saturates, the current pulse in its power winding induces a current pulse in its control winding. Since both control windings are in series, the pulse also appears in the other control winding, providing the necessary coercive force to reduce the induction to zero.

There are, of course, many different configurations for magnetic amplifiers, depending upon the method of control and type of output desired. Although most power-control circuits are now designed using semiconductors such as triacs, magnetic amplifiers continue to be used where extreme environmental or overload conditions would be intolerable for solid-state devices. Another valuable characteristic is that they allow the summing of a number of input signals while keeping them electrically isolated, when this is necessary. Figure M-4 shows a transductor with two control circuits that accepts two inputs and saturates according to their addition while keeping them isolated from each other. Magnetic amplifiers, therefore, continue to be used in many types of instrumentation.

magnetic damping—Reduction of energy in a mechanical system by using the resistance to motion in a magnetic field of a conductor through which a current flows.

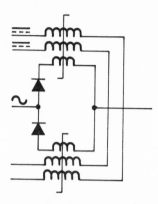

Figure M-4
Magnetic Amplifier with Two Control Circuits

magnetic deflection—In a television picture tube, the method of deflecting the electron beam horizontally and vertically by the magnetic fields produced by sawtooth currents in the corresponding coils of the deflection yoke. *Electron Tubes.*

magnetic storm—Disturbance of the Earth's upper atmosphere caused by plasma ejected from a solar flare pressing on the magnetosphere (q.v.) with subsequent deformation of the geomagnetic field. The storm may last from a few hours to a couple of days and is accompanied by aurora displays and radio interference. *Radio Noise and Interference.*

magnetic tape recording—Preserving electrical signals as magnetic patterns in finely powdered iron oxide in a coating on one side of a thin plastic tape. *Recording, Computer Hardware.*

magnetism—An electron, whether flowing independently in an electric current, or in orbital motion in an atom, generates a magnetic field around itself. In most atoms the fields of individual electrons conflict, but in a few (the iron and rare-earth groups) enough coincide that, if oriented parallel to each other, combine to form a powerful external field. Similarly, in a conductor, when a large number of electrons move in the same direction, their individual magnetic fields add together to make a strong external field, which may be further enhanced by forming the conductor into a coil around an iron core. Conversely, an external magnetic field will generate a current in a conductor by interaction with the magnetic fields of the conduction electrons. *Coil Data, Nuclear Physics, Transformers.*

magneto—Alternator used to provide the ignition current for an internal-combustion engine, in which a permanent magnet replaces the field coil of the usual alternator or generator, so that a battery and coil are not required.

magnetohydrodynamic (MHD) generator—A plasma (gas at high temperature, and therefore ionized and conductive) flows through a duct. As it does so it passes through a strong magnetic field, which deflects negative and positive ions in opposite directions to charge two electrodes accordingly. If an external circuit is connected between them, a current will flow from the negatively charged electrode to the positively charged electrode. (See Figure M-5.)

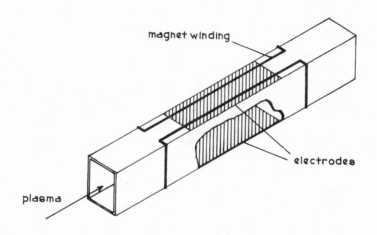

Figure M-5
Magnetohydrodynamic Generator

magnetohydrodynamics—Study of the behavior of electrically conducting fluids (liquids or gases) in the presence of electric and magnetic fields.

magneton—Unit of magnetic moment, equal to the product of a magnet's pole strength and the distance between its poles. The Bohr magneton (μ_B) = 9.293×10^{-24} joules per tesla. The nuclear magneton (μ_n) = 5.051×10^{-27} joules per tesla.

magnetosphere—Region in the atmosphere characterized by phenomena caused by the Earth's magnetic field and high atmospheric conductivity produced by ionization. Due to the pressure of the solar wind this region is shaped like a comet with a "tail" extending several astronomical units (1 AU = 1.5×10^8 km) from Earth on the side opposite to the Sun, while its "head" reaches some 6.5×10^4 km from Earth toward the

Sun. The lower boundary of the magnetosphere is several hundred kilometers above the Earth's surface.

magnetostriction—Slight changes in the dimensions of a ferromagnetic material when placed in a magnetic field, resulting from realignment of its magnetic domains to conform to the field. This property is made use of in a certain type of ultrasonic transducer.

magnetron—Electron-tube diode consisting of a cylindrical anode structure with radial resonant cavities surrounding the cathode. An external magnet provides a magnetic field in the interelectrode space that causes electrons traveling from the cathode to the anode to take a circular path, generating extremely high-frequency oscillation. Used in radar and in microwave ovens. *Electron Tubes.*

magnified sweep—In an oscilloscope, a sweep whose time per division has been decreased by amplification of the sweep waveform rather than by changing the time constants used to generate it.

majority carrier—Carrier of the same polarity as the host semiconductor material (holes in p-type material, electrons in n-type material). *Semiconductors.*

make-before-break—Action of a type of switch in which one contact closes before another opens. (See Figure M-6.)

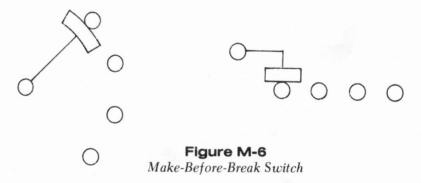

Figure M-6
Make-Before-Break Switch

male contact—In a connector, a contact that mates by insertion into a hollow contact in another connector.

manganese—Gray-white, hard, brittle metal, essential for making steel and used also to improve corrosion resistance in aluminum and magnesium. Manganese dioxide (MnO_2) is used as a depolarizer in electric dry cells. Symbol, Mn; atomic weight, 54.938; atomic number, 25.

manganin—Alloy of copper, manganese and nickel, used to make wire for wire-wound resistors. *Resistors.*

man-made noise—Radio interference due to electric motors, neon signs, power lines, and ignition systems. *Radio Noise and Interference.*

mantissa—(See logarithm.)

maritime radio navigation bands—Radio navigation services for benefit of ships. *Frequency Data.*

marker beacon—Radio transmitter that gives a signal to assist an aircraft to determine its location.

maser—Acronym for microwave amplification by stimulated emission of radiation. *Masers.*

Masers

Like a laser (q.v.), a maser oscillator requires a source of atoms or molecules raised to a higher energy level, and a resonant cavity to store their radiant energy long enough to enable it to multiply by stimulating emission from more atoms or molecules. However, the frequencies involved are in the microwave region instead of the infrared or visible light region, so the acronym maser is used, meaning microwave (or molecular) amplification by stimulated emission of radiation.

One form of maser, called a three-level or cavity maser, uses a chromium-doped ruby crystal, the same material as in the original solid-state laser (see Figure L-2). This material has several energy levels, and radiation is applied to it to raise atoms from level 1 (the lower level) to level 3. If sufficient radiative energy is applied, there will be as many atoms at level 3 as at level 1. There will actually be more at the intermediate level 2, thus providing the population inversion necessary for maser action.

There is a snag, however. The three-level maser has to be cooled to a very low temperature before this population inversion can be obtained. High performance is possible only at the temperature of liquid helium (4.2 K), although considerably reduced to performance with liquid nitrogen (77 K) and dry ice (195 K) has been obtained.

However, the solid-state maser has an advantage to compensate for this, and that is that it is tunable to different frequencies. This is because the ions in the crystal can be made to change their energy levels by the application of a suitable magnetic field. Further improvement is obtained by use of a slow-wave structure that makes the wave pass through the crystal at a lower speed, thus giving more time for interaction between the wave and the excited atoms. This type of maser is called a traveling-wave maser, by analogy with the traveling-wave electron tube (q.v.). Various types of slow-wave structure are shown in Figure M-7.

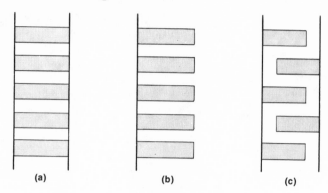

(a)   (b)   (c)

Figure M-7
*Examples of Slow-Wave Structures Used in Maser
Traveling-Wave Devices:
(a) ladder; (b) comb; (c) interdigital*

Three-level masers are excellent low-noise amplifiers of the weak signals from distant sources in space, so they are used for radio astronomy and communication with satellites. Maser amplifiers were the first to pick up radio waves from Venus, from which we learned that, far from being a celestial paradise, the planet was as hot as hell. Frequencies operated with masers range from 300 megahertz to 96.3 gigahertz, using temperatures down to 1.2 K.

Gas masers originally used ammonia. A beam of ammonia molecules was made to travel along the axis of a cylindrical cage. Alternate rods of this cage were charged positive and negative, so a nonuniform field was created that concentrated molecules at the higher energy level in the center of the beam. The beam then passed through a small hole into a cavity. The hole admitted only the excited molecules, which then released their energy in the cavity. This maser was not tunable, so it was superseded by cavity masers for radio and radar use.

However, the frequency of the ammonia maser ($^{15}NH_3$) is extremely stable at 22 789 421 701 ±1 hertz. The hydrogen-beam maser is even more stable at 1 420 405 751.800 ±0.028 hertz, so that it will probably be used as a time standard for atomic time (A_1 time) in place of the presently used cesium atomic clock. The hydrogen clock would not be more than one second off in 100,000 years.

Another type of gas maser is the rubidium gas-cell maser, in which the gas is pumped optically. Although less accurate than the hydrogen and ammonia masers, it is still stable enough to be used as a secondary frequency standard.

mask—1. Implement used to shield selected portions of a wafer or substrate during a deposition process. *Microelectronics.* 2. Shadow mask in a picture tube. *Electron Tubes.*

mass—Measure of the inertia of a body, or its resistance to change in position or velocity. Unlike weight, it is not dependent on gravitational attraction and is the same everywhere. The kilogram is the unit of mass. *S.I. Units.*

master oscillator—In a transmitter, the oscillator generating the carrier frequency.

matching attenuator—Network designed to provide a known reduction in the amplitude of a signal while matching the source and load impedances between which it is connected. Also called a pad. *Attenuators.*

mathematical table—Array of tabulated values of a mathematical function, as in logarithm tables, trigonometric tables, and so on. *Mathematical Tables.* (See Figures M-8—M-15.)

matrix—1. Rectangular array of numbers in rows and columns of a determinant, a procedure used to solve simultaneous equations in the analysis of electric circuits.   2. In computers, a term synonymous with translator (q.v.).   3. Network used in a color television transmitter to convert the primary color signals into color difference signals, and in a color receiver to perform the reverse.   4. In an FM stereo transmitter, a network to obtain the sum and difference signals from the left and right inputs.

maximum usable frequency (MUF)—Highest frequency at which an ionized layer reflects a vertically incident wave, termed f_c. At oblique incidence, $\text{MUF} = f_c \sec \phi$, where ϕ is the angle of incidence, ignoring the curvature of the Earth and the ionosphere. *Electromagnetic-Wave Propagation.*

maxwell—Unit of magnetic flux now replaced by the weber (q.v.).

Maxwell bridge—Bridge for measuring series inductance. (See Figure M-16.)

Maxwell's equations—Four fundamental equations of the physics of electromagnetic fields:

$$(1)\ \text{div } E = 0$$
$$(2)\ \text{div } H = 0$$
$$(3)\ \text{curl } E = -\frac{1}{c}\frac{\partial H}{\partial t}$$
$$(4)\ \text{curl } H = \frac{1}{c}\frac{\partial E}{\partial t}$$

where: E = the electric field strength
  H = the magnetic field strength
  c = the velocity of light
  t = the time coordinate

The meanings of the symbols are as follows: E and H represent electric and magnetic field strengths which, because they vary in time and from place to place, are functions of the space coordinates x, y, and z (not shown) and of the time coordinate, t. The velocity of light, c, enters the equations as a conversion factor between electrostatic and electromagnetic units; div (abbreviation for divergence) and curl (abbreviation for rotation) are mathematical operations whose physical meaning appears below. The symbol ∂ indicates partial differentiation with respect to time, t. Divergence is essentially a measure of source strength, and curl is essentially a measure of vorticity. In words, equation (1) says that in the absence of charges electric lines of force can be neither created nor destroyed. If one conceives of an electric field, in Maxwell's idiom, as a fluid, div E = 0 says that as much fluid flows out of each tiny volume of space in a given time as flows in. Equation (2) makes the same assertion for magnetic lines that equation (1) makes for electric lines, but is more general: there *are* no magnetic charges. Equation (3) is a statement of Faraday's law of electromagnetic induction. It says that the limiting value of electromotive force per unit area is proportional to the rate of change of H at a limit point P of the area; loosely stated, a changing magnetic field creates an electric field at right angles to the magnetic field change. Equation (4) says that a changing electric field produces a magnetic field. The time rate of change of E, $\partial E/\partial t$, is Maxwell's displacement current, which led him to predict the existence of electromagnetic waves, discovered later by Hertz.

mayday—International distress call equivalent to SOS, used in radio-telephone transmission. It is the phonetic spelling of *m'aidez*, French for "help me."

mc—Abbreviation for megacycle, now obsolete and replaced by megahertz (q.v.).

m-derived filter—Basic filter with an additional reactive element added in series or shunt to give a sharper cutoff. *Active Filters, Passive Filters.*

M-display—Radar display similar to the A-display, but with a superimposed marker that can be moved horizontally by means of a dial. When the marker is made to coincide with the target blip, the range can be read from the dial.

mean—In mathematics, a quantity with a value intermediate between the values of two or more other quantities; especially the average (*arithmetic mean*) obtained by adding the quantities and dividing the total by the number of quantities added. The *geometric mean* is obtained by multiplying two numbers together and taking the square root of their product.

medical electronics—Branch of electronic technology concerned with diagnosis, therapy, and other applications in the field of medicine.

medium frequency—Frequency range from 300 to 3000 kilohertz, or hectometric waves (1000 to 100 meters). *Frequency Data.*

Common Logarithms

N	0	1	2	3	4	5	6	7	8	9	N
10	0000	0043	0086	0128	0170	0212	0253	0294	0334	0374	10
11	0414	0453	0492	0531	0569	0607	0645	0682	0719	0755	11
12	0792	0828	0864	0899	0934	0969	1004	1038	1072	1106	12
13	1139	1173	1206	1239	1271	1303	1335	1367	1399	1430	13
14	1461	1492	1523	1553	1584	1614	1644	1673	1703	1732	14
15	1761	1790	1818	1847	1875	1903	1931	1959	1987	2014	15
16	2041	2068	2095	2122	2148	2175	2201	2227	2253	2279	16
17	2304	2330	2355	2380	2405	2430	2455	2480	2504	2529	17
18	2553	2577	2601	2625	2648	2672	2695	2718	2742	2765	18
19	2788	2810	2833	2856	2878	2900	2923	2945	2967	2989	19
20	3010	3032	3054	3075	3096	3118	3139	3160	3181	3201	20
21	3222	3243	3263	3284	3304	3324	3345	3365	3385	3404	21
22	3424	3444	3464	3483	3502	3522	3541	3560	3579	3598	22
23	3617	3636	3655	3674	3692	3711	3729	3747	3766	3784	23
24	3802	3820	3838	3856	3874	3892	3909	3927	3945	3962	24
25	3979	3997	4014	4031	4048	4065	4082	4099	4116	4133	25
26	4150	4166	4183	4200	4216	4232	4249	4265	4281	4298	26
27	4314	4330	4346	4362	4378	4393	4409	4425	4440	4456	27
28	4472	4487	4502	4518	4533	4548	4564	4579	4594	4609	28
29	4624	4639	4654	4669	4683	4698	4713	4728	4742	4757	29
30	4771	4786	4800	4814	4829	4843	4857	4871	4886	4900	30
31	4914	4928	4942	4955	4969	4983	4997	5011	5024	5038	31
32	5051	5065	5079	5092	5105	5119	5132	5145	5159	5172	32
33	5185	5198	5211	5224	5237	5250	5263	5276	5289	5302	33
34	5315	5328	5340	5353	5366	5378	5391	5403	5416	5428	34
35	5441	5453	5465	5478	5490	5502	5514	5527	5539	5551	35
36	5563	5575	5587	5599	5611	5623	5635	5647	5658	5670	36
37	5682	5694	5705	5717	5729	5740	5752	5763	5775	5786	37
38	5798	5809	5821	5832	5843	5855	5866	5877	5888	5899	38
39	5911	5922	5933	5944	5955	5966	5977	5988	5999	6010	39
40	6021	6031	6042	6053	6064	6075	6085	6096	6107	6117	40
41	6128	6138	6149	6160	6170	6180	6191	6201	6212	6222	41
42	6232	6243	6253	6263	6274	6284	6294	6304	6314	6325	42
43	6335	6345	6355	6365	6375	6385	6395	6405	6415	6425	43
44	6435	6444	6454	6464	6474	6484	6493	6503	6513	6522	44
45	6532	6542	6551	6561	6571	6580	6590	6599	6609	6618	45
46	6628	6637	6646	6656	6665	6675	6684	6693	6702	6712	46
47	6721	6730	6739	6749	6758	6767	6776	6785	6794	6803	47
48	6812	6821	6830	6839	6848	6857	6866	6875	6884	6893	48
49	6902	6911	6920	6928	6937	6946	6955	6964	6972	6981	49
50	6990	6998	7007	7016	7024	7033	7042	7050	7059	7067	50
51	7076	7084	7093	7101	7110	7118	7126	7135	7143	7152	51
52	7160	7168	7177	7185	7193	7202	7210	7218	7226	7235	52
53	7243	7251	7259	7267	7275	7284	7292	7300	7308	7316	53
54	7324	7332	7340	7348	7356	7364	7372	7380	7388	7396	54
N	0	1	2	3	4	5	6	7	8	9	N

Figure M-8

Common Logarithms (Continued)

N	0	1	2	3	4	5	6	7	8	9	N
55	7404	7412	7419	7427	7435	7443	7451	7459	7466	7474	55
56	7482	7490	7497	7505	7513	7520	7528	7536	7543	7551	56
57	7559	7566	7574	7582	7589	7597	7604	7612	7619	7627	57
58	7634	7642	7649	7657	7664	7672	7679	7686	7694	7701	58
59	7709	7716	7723	7731	7738	7745	7752	7760	7767	7774	59
60	7782	7789	7796	7803	7810	7818	7825	7832	7839	7846	60
61	7853	7860	7868	7875	7882	7889	7896	7903	7910	7917	61
62	7924	7931	7938	7945	7952	7959	7966	7973	7980	7987	62
63	7993	8000	8007	8014	8021	8028	8035	8041	8048	8055	63
64	8062	8069	8075	8082	8089	8096	8102	8109	8116	8122	64
65	8129	8136	8142	8149	8156	8162	8169	8176	8182	8189	65
66	8195	8202	8209	8215	8222	8228	8235	8241	8248	8254	66
67	8261	8267	8274	8280	8287	8293	8299	8306	8312	8319	67
68	8325	8331	8338	8344	8351	8357	8363	8370	8376	8382	68
69	8388	8395	8401	8407	8414	8420	8426	8432	8439	8445	69
70	8451	8457	8463	8470	8476	8482	8488	8494	8500	8506	70
71	8513	8519	8525	8531	8537	8543	8549	8555	8561	8567	71
72	8573	8579	8585	8591	8597	8603	8609	8615	8621	8627	72
73	8633	8639	8645	8651	8657	8663	8669	8675	8681	8686	73
74	8692	8698	8704	8710	8716	8722	8727	8733	8739	8745	74
75	8751	8756	8762	8768	8774	8779	8785	8791	8797	8802	75
76	8808	8814	8820	8825	8831	8837	8842	8848	8854	8859	76
77	8865	8871	8876	8882	8887	8893	8899	8904	8910	8915	77
78	8921	8927	8932	8938	8943	8949	8954	8960	8965	8971	78
79	8976	8982	8987	8993	8998	9004	9009	9015	9020	9025	79
80	9031	9036	9042	9047	9053	9058	9063	9069	9074	9079	80
81	9085	9090	9096	9101	9106	9112	9117	9122	9128	9133	81
82	9138	9143	9149	9154	9159	9165	9170	9175	9180	9186	82
83	9191	9196	9201	9206	9212	9217	9222	9227	9232	9238	83
84	9243	9248	9253	9258	9263	9269	9274	9279	9284	9289	84
85	9294	9299	9304	9309	9315	9320	9325	9330	9335	9340	85
86	9345	9350	9355	9360	9365	9370	9375	9380	9385	9390	86
87	9395	9400	9405	9410	9415	9420	9425	9430	9435	9440	87
88	9445	9450	9455	9460	9465	9469	9474	9479	9484	9489	88
89	9494	9499	9504	9509	9513	9518	9523	9528	9533	9538	89
90	9542	9547	9552	9557	9562	9566	9571	9576	9581	9586	90
91	9590	9595	9600	9605	9609	9614	9619	9624	9628	9633	91
92	9638	9643	9647	9652	9657	9661	9666	9671	9675	9680	92
93	9685	9689	9694	9699	9703	9708	9713	9717	9722	9727	93
94	9731	9736	9741	9745	9750	9754	9759	9763	9768	9773	94
95	9777	9782	9786	9791	9795	9800	9805	9809	9814	9818	95
96	9823	9827	9832	9836	9841	9845	9850	9854	9859	9863	96
97	9868	9872	9877	9881	9886	9890	9894	9899	9903	9908	97
98	9912	9917	9921	9926	9930	9934	9939	9943	9948	9952	98
99	9956	9961	9965	9969	9974	9978	9983	9987	9991	9996	99
N	0	1	2	3	4	5	6	7	8	9	N

Figure M-9

Table of Sines, Cosines, and Tangents

Angle	Radians	Sine	Cosine	Tangent	Angle	Radians	Sine	Cosine	Tangent
0°	.0000	.0000	1.0000	.0000	45°	.7854	.7071	.7071	1.0000
1	.0175	.0175	.9998	.0175	46	.8029	.7193	.6947	1.0355
2	.0349	.0349	.9994	.0349	47	.8203	.7314	.6820	1.0724
3	.0524	.0523	.9986	.0524	48	.8378	.7431	.6691	1.1106
4	.0698	.0698	.9976	.0699	49	.8552	.7547	.6561	1.1504
5	.0873	.0872	.9962	.0875	50	.8727	.7660	.6428	1.1918
6	.1047	.1045	.9945	.1051	51	.8901	.7771	.6293	1.2349
7	.1222	.1219	.9925	.1228	52	.9076	.7880	.6157	1.2799
8	.1396	.1392	.9903	.1405	53	.9250	.7986	.6018	1.3270
9	.1571	.1564	.9877	.1584	54	.9425	.8090	.5878	1.3764
10	.1745	.1736	.9848	.1763	55	.9599	.8192	.5736	1.4281
11	.1920	.1908	.9816	.1944	56	.9774	.8290	.5592	1.4826
12	.2094	.2079	.9781	.2126	57	.9948	.8387	.5446	1.5399
13	.2269	.2250	.9744	.2309	58	1.0123	.8480	.5299	1.6003
14	.2443	.2419	.9703	.2493	59	1.0297	.8572	.5150	1.6643
15	.2618	.2588	.9659	.2679	60	1.0472	.8660	.5000	1.7321
16	.2793	.2756	.9613	.2867	61	1.0647	.8746	.4848	1.8040
17	.2967	.2924	.9563	.3057	62	1.0821	.8829	.4695	1.8807
18	.3142	.3090	.9511	.3249	63	1.0996	.8910	.4540	1.9626
19	.3316	.3256	.9455	.3443	64	1.1170	.8988	.4384	2.0503
20	.3491	.3420	.9397	.3640	65	1.1345	.9063	.4226	2.1445
21	.3665	.3584	.9336	.3839	66	1.1519	.9135	.4067	2.2460
22	.3840	.3746	.9272	.4040	67	1.1694	.9205	.3907	2.3559
23	.4014	.3907	.9205	.4245	68	1.1868	.9272	.3746	2.4751
24	.4189	.4067	.9135	.4452	69	1.2043	.9336	.3584	2.6051
25	.4363	.4226	.9063	.4663	70	1.2217	.9397	.3420	2.7475
26	.4538	.4384	.8988	.4877	71	1.2392	.9455	.3256	2.9042
27	.4712	.4540	.8910	.5095	72	1.2566	.9511	.3090	3.0777
28	.4887	.4695	.8829	.5317	73	1.2741	.9563	.2924	3.2709
29	.5061	.4848	.8746	.5543	74	1.2915	.9613	.2756	3.4874
30	.5236	.5000	.8660	.5774	75	1.3090	.9659	.2588	3.7321
31	.5411	.5150	.8572	.6009	76	1.3265	.9703	.2419	4.0108
32	.5585	.5299	.8480	.6249	77	1.3439	.9744	.2250	4.3315
33	.5760	.5446	.8387	.6494	78	1.3614	.9781	.2079	4.7046
34	.5934	.5592	.8290	.6745	79	1.3788	.9816	.1908	5.1446
35	.6109	.5736	.8192	.7002	80	1.3963	.9848	.1736	5.6713
36	.6283	.5878	.8090	.7265	81	1.4137	.9877	.1564	6.3138
37	.6458	.6018	.7986	.7536	82	1.4312	.9903	.1392	7.1154
38	.6632	.6157	.7880	.7813	83	1.4486	.9925	.1219	8.1443
39	.6807	.6293	.7771	.8098	84	1.4661	.9945	.1045	9.5144
40	.6981	.6428	.7660	.8391	85	1.4835	.9962	.0872	11.43
41	.7156	.6561	.7547	.8693	86	1.5010	.9976	.0698	14.30
42	.7330	.6691	.7431	.9004	87	1.5184	.9986	.0523	19.08
43	.7505	.6820	.7314	.9325	88	1.5359	.9994	.0349	28.64
44	.7679	.6947	.7193	.9657	89	1.5533	.9998	.0175	57.29

Figure M-10

Number	Number²	√Number	√10 × Number	Number³
1	1	1.000000	3.162278	1
2	4	1.414214	4.472136	8
3	9	1.732051	5.477226	27
4	16	2.000000	6.324555	64
5	25	2.236068	7.071068	125
6	36	2.449490	7.745967	216
7	49	2.645751	8.366600	343
8	64	2.828427	8.944272	512
9	81	3.000000	9.486833	729
10	100	3.162278	10.00000	1,000
11	121	3.316625	10.48809	1,331
12	144	3.464102	10.95445	1,728
13	169	3.605551	11.40175	2,197
14	196	3.741657	11.83216	2,744
15	225	3.872983	12.24745	3,375
16	256	4.000000	12.64911	4,096
17	289	4.123106	13.03840	4,913
18	324	4.242641	13.41641	5,832
19	361	4.358899	13.78405	6,859
20	400	4.472136	14.14214	8,000
21	441	4.582576	14.49138	9,261
22	484	4.690416	14.83240	10,648
23	529	4.795832	15.16575	12,167
24	576	4.898979	15.49193	13,824
25	625	5.000000	15.81139	15,625
26	676	5.099020	16.12452	17,576
27	729	5.196152	16.43168	19,683
28	784	5.291503	16.73320	21,952
29	841	5.385165	17.02939	24,389
30	900	5.477226	17.32051	27,000
31	961	5.567764	17.60682	29,791
32	1,024	5.656854	17.88854	32,768
33	1,089	5.744563	18.16590	35,937
34	1,156	5.830952	18.43909	39,304
35	1,225	5.916080	18.70829	42,875
36	1,296	6.000000	18.97367	46,656
37	1,369	6.082763	19.23538	50,653
38	1,444	6.164414	19.49359	54,872
39	1,521	6.244998	19.74842	59,319
40	1,600	6.324555	20.00000	64,000
41	1,681	6.403124	20.24846	68,921
42	1,764	6.480741	20.49390	74,088
43	1,849	6.557439	20.73644	79,507
44	1,936	6.633250	20.97618	85,184

Figure M-11
Table of Number Functions

John D. Lenk, *Practical Semiconductor Data Book for Electronic Engineers and Technicians*, ©1970. Reprinted by permission of Prentice-Hall, Inc.

Number	Number^2	$\sqrt{Number}$	$\sqrt{10 \times Number}$	Number^3
45	2,025	6.708204	21.21320	91,125
46	2,116	6.782330	21.44761	97,336
47	2,209	6.855655	21.67948	103,823
48	2,304	6.928203	21.90890	110,592
49	2,401	7.000000	22.13594	117,649
50	2,500	7.071680	22.36068	125,000
51	2,601	7.141428	22.58318	132,651
52	2,704	7.211103	22.80351	140,608
53	2,809	7.280110	23.02173	148,877
54	2,916	7.348469	23.23790	157,464
55	3,025	7.416198	23.45208	166,375
56	3,136	7.483315	23.66432	175,616
57	3,249	7.549834	23.87467	185,193
58	3,364	7.615773	24.06319	194,112
59	3,481	7.681146	24.28992	205,379
60	3,600	7.745967	24.49490	216,000
61	3,721	7.810250	24.69818	226,981
62	3,844	7.874008	24.89980	238,047
63	3,969	7.937254	25.09980	250,047
64	4,096	8.000000	25.29822	262,144
65	4,225	8.062258	25.49510	274,625
66	4,356	8.124038	25.69047	287,496
67	4,489	8.185353	25.88436	300,763
68	4,624	8.246211	26.07681	314,432
69	4,761	8.306624	26.26785	328,509
70	4,900	8.366600	26.45751	343,000
71	5,041	8.426150	26.64583	357,911
72	5,184	8.485281	26.83282	373,248
73	5,329	8.544004	27.01851	389,017
74	5,476	8.602325	27.20294	405,224
75	5,625	8.660254	27.38613	421,875
76	5,776	8.717798	27.56810	438,976
77	5,929	8.774964	27.74887	456,533
78	6,084	8.831761	27.92848	474,552
79	6,241	8.888194	28.10694	493,039
80	6,400	8.944272	28.28427	512,000
81	6,561	9.000000	28.46050	531,441
82	6,724	9.055385	28.63564	551,368
83	6,889	9.110434	28.80972	571,787
84	7,056	9.165151	28.98275	592,704
85	7,225	9.219544	29.15476	614,125
86	7,396	9.273618	29.32576	636,056
87	7,569	9.327379	29.49576	658,503
88	7,744	9.380832	29.66479	681,472
89	7,921	9.433981	29.83287	704,969
90	8,100	9.486833	30.00000	729,000

Figure M-12
Table of Number Functions (cont'd.)

John D. Lenk, *Practical Semiconductor Data Book for Electronic Engineers and Technicians*, ©1970. Reprinted by permission of Prentice-Hall, Inc.

Number	Number^2	$\sqrt{Number}$	$\sqrt{10 \times Number}$	Number^3
91	8,281	9.539392	30.16621	753,571
92	8,464	9.591663	30.33150	778,688
93	8,649	9.643651	30.49590	804,357
94	8,836	9.695360	30.65942	830,584
95	9,025	9.746794	30.82207	857,375
96	9,216	9.797959	30.98387	884,736
97	9,409	9.848858	31.14482	912,673
98	9,604	9.899495	31.30495	941,192
99	9,801	9.949874	31.46427	970,299
100	10,000	10.00000	31.62278	1,000,000

Number	$\sqrt[3]{Number}$	$\sqrt[3]{10 \times Number}$	$\sqrt[3]{100 \times Number}$
1	1.000000	2.154435	4.641589
2	1.259921	2.714418	5.848035
3	1.442250	3.107233	6.694330
4	1.587401	3.419952	7.368063
5	1.709976	3.684031	7.937005
6	1.817121	3.914868	8.434327
7	1.912931	4.121285	8.879040
8	2.000000	4.308869	9.283178
9	2.080084	4.481405	9.654894
10	2.154435	4.641589	10.00000
11	2.223980	4.791420	10.32280
12	2.289428	4.932424	10.62659
13	2.351335	5.065797	10.91393
14	2.410142	5.192494	11.18689
15	2.466212	5.313293	11.44714
16	2.519842	5.428835	11.69607
17	2.571282	5.539658	11.93483
18	2.620741	5.646216	12.16440
19	2.668402	5.748897	12.38562
20	2.714418	5.848035	12.59921
21	2.758924	5.943922	12.80579
22	2.802039	6.036811	13.00591
23	2.843867	6.126926	15.20006
24	2.884499	6.214465	13.38866
25	2.924018	6.299605	13.57209
26	2.962496	6.382504	13.75069
27	3.000000	6.463304	13.92477
28	3.036589	6.542133	14.09460
29	3.072317	6.619106	14.26043
30	3.107233	6.694330	14.42250
31	3.141381	6.767899	14.58100
32	3.174802	6.839904	14.73613
33	3.207534	6.910423	14.88806
34	3.239612	6.979532	15.03695
35	3.271066	7.047299	15.18294

Figure M-13

Table of Number Functions (cont'd.)

John D. Lenk, *Practical Semiconductor Data Book for Electronic Engineers and Technicians,* ©1970. Reprinted by permission of Prentice-Hall, Inc.

Number	$\sqrt[3]{Number}$	$\sqrt[3]{10 \times Number}$	$\sqrt[3]{100 \times Number}$
36	3.301927	7.113787	15.32619
37	3.332222	7.179054	15.46680
38	3.361975	7.243156	15.60491
39	3.391211	7.306144	15.74061
40	3.419952	7.368063	15.87401
41	3.448217	7.428959	16.00521
42	3.476027	7.488872	16.13429
43	3.503398	7.547842	16.26133
44	3.530348	7.605905	16.38643
45	3.556893	7.663094	16.50964
46	3.583048	7.719443	16.63103
47	3.608826	7.774980	16.75069
48	3.634241	7.829735	16.86865
49	3.659306	7.883735	16.98499
50	3.684031	7.937005	17.09976
51	3.708430	7.989570	17.21301
52	3.732511	8.041452	17.32478
53	3.756286	8.092672	17.43513
54	3.779763	8.143253	17.54411
55	3.802952	8.193213	17.65174
56	3.825862	8.242571	17.75808
57	3.848501	8.291344	17.86316
58	3.870877	8.339551	17.96702
59	3.892996	8.387207	18.06969
60	3.914868	8.434327	18.17121
61	3.936497	8.480926	18.27160
62	3.957892	8.527019	18.37091
63	3.979057	8.572619	18.46915
64	4.000000	8.617739	18.56636
65	4.020726	8.662391	18.66256
66	4.041240	8.706588	18.75777
67	4.061548	8.750340	18.85204
68	4.081655	8.793659	18.94536
69	4.101566	8.836556	19.03778
70	4.121285	8.879040	19.12931
71	4.140818	8.921121	19.21997
72	4.160168	8.962809	19.30979
73	4.179339	9.004113	19.39877
74	4.198336	9.045042	19.48695
75	4.217163	9.085603	19.57434
76	4.235824	9.125805	19.66095
77	4.254321	9.165656	19.74681
78	4.272659	9.205164	19.83192
79	4.290840	9.244335	19.91632
80	4.308869	9.283178	20.00000

Figure M-14
Table of Number Functions (cont'd.)

John D. Lenk, *Practical Semiconductor Data Book for Electronic Engineers and Technicians,* ©1970. Reprinted by permission of Prentice-Hall, Inc.

LOGARITHMS TO BASE 10

let $N = 10^n$

$\log N = n$ *see Table 24-2*

$\log 10 = 1$ $\log 2 = 0.3010$

$\log 100 = 2$ $\log 20 = 1.3010$

$\log 1 = 0$ $\log 200 = 2.3010$

$\log 0.1 = -1$ $\log 0.2 = -1.3010$

$\log 0.01 = -2$ $\log 0.02 = -2.3010$

$\log a + \log b = 10^a \cdot 10^b = \log(a + b)$

$\log a - \log b = \dfrac{10^a}{10^b} = \log(a - b)$

$\log(-a) = -\log(+a)$

Number	$\sqrt[3]{Number}$	$\sqrt[3]{10 \times Number}$	$\sqrt[3]{100 \times Number}$
81	4.326749	9.321698	20.08299
82	4.344481	9.359902	20.16530
83	4.362071	9.397796	20.24694
84	4.379519	9.435388	20.32793
85	4.396830	9.472682	20.40828
86	4.414005	9.509685	20.48800
87	4.431048	9.546403	20.56710
88	4.447960	9.582840	20.64560
89	4.464745	9.619002	20.72351
90	4.481405	9.654894	20.80084
91	4.497941	9.690521	20.87759
92	4.514357	9.725888	20.95379
93	4.530655	9.761000	21.02944
94	4.546836	9.795861	21.10454
95	4.562903	9.830476	21.17912
96	4.578857	9.864848	21.25317
97	4.594701	9.898983	21.32671
98	4.610436	9.932884	21.39975
99	4.626065	9.966555	21.47229
100	4.641589	10.00000	21.54435

Figure M-15
Table of Number Functions (cont'd.)

John D. Lenk, *Practical Semiconductor Data Book for Electronic Engineers and Technicians,* ©1970. Reprinted by permission of Prentice-Hall, Inc.

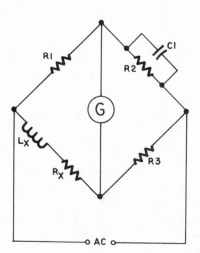

Figure M-16
Maxwell Bridge

medium-scale integration (MSI)—Integrated circuits such as registers, decade counters, and the like, that do not have the density of large-scale integrated circuits, but still have a considerable number of circuits.

mega—Prefix of S.I. Units meaning one million times (10^6); symbol M.

megabit—One million bits (binary digits).

megaelectron-volt (MeV)—One million electron-volts, equal to 1.6×10^{-13} joule.

megahertz (MHz)—One million hertz.

megametric waves—Extremely low-frequency waves with a frequency range of from 30 to 300 hertz (10-1 Mm).

megampere (MA)—One million amperes.

megatron—(See lighthouse tube.)

megavolt (MV)—One million volts.

megawatt (MW)—One million watts.

megger—(See megohmmeter.)

megohm (MΩ)—One million ohms.

megohmmeter—Instrument for measuring the extremely high resistances of insulation, earth-ground return paths, and the like. With the input terminals shorted together the same current flows through each coil, but as their fields oppose each other the pointer does not move. A resistance connected to the input reduces the current through the deflecting coil so the pointer now responds to the difference. The actual voltage of the hand generator (≈ 1000) does not matter. Except for the addition of the control coil the movement is essentially that of a permanent-magnet moving-coil meter movement (q.v.). (See Figure M-17.)

memory—In a computer, the section that stores data and instructions in binary code. Memories are both internal registers in the central processing unit and external systems such as magnetic tape, punched cards, and so on. *Computer Hardware*.

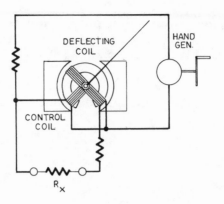

Figure M-17
Megohmmeter

mercury—Only elemental metal that is liquid at ordinary temperatures (its melting point is –39°C), mercury is heavy, silvery white, slowly tarnished in moist air, freezes into a soft metal like tin or lead, and alloys with most metals, forming amalgams. Much used for thermometers, barometers, sealed electric switches and relays, and mercury-vapor lamps. Symbol, Hg; atomic weight, 200.59; atomic number, 80.

mercury-arc rectifier—Arc discharge tube in which the ionized gas atmosphere is mercury vapor. *Electron Tubes.*

mercury cell—Alkaline dry primary cell with a zinc anode, cathode of mercuric oxide mixed with graphite, and electrolyte of potassium hydroxide contained in an absorbent material. Capable of far higher current output than other alkaline cells, its use is limited by its high cost to applications where cheaper cells are unsuitable, such as hearing aids (small size and long life) and photo flash units (high current pulses).

mercury switch—Switch consisting of a glass capsule in which is a globule of mercury. Tilting the capsule causes the globule to roll so as to make or break a connection between two electrodes fused through the glass. Commonly used in thermostats.

mercury-vapor lamp—Lighting device in which mercury vapor is ionized by the applied voltage so that it gives off a bluish light.

mesa transistor—Transistor formed by double diffusion of impurities into a silicon substrate. After masking and etching, the base and emitter regions are in the form of a residual "island" or "mesa" on the substrate, which is the collector. *Semiconductors, Transistors.*

meson—Subatomic particle with a mass between that of a lepton and that of a baryon. *Nuclear Physics.*

metal—Substance with high electrical and thermal conductivity, metallic luster, malleability, and ductility.

metal detector—Electronic device that detects the presence of a metal object by the change in resonance of a search coil brought near to it. Used to locate buried metal objects such as pipes, coins, etc.

metallization—Deposition of a metal film on a substrate by evaporative deposition, cathode sputtering, vapor plating, or electroplating. *Microelectronics.*

metal-oxide semiconductor field-effect transistor (MOSFET)—Insulated-gate field-effect transistor (IGFET) in which the gate is insulated from the source-to-drain channel by silicon dioxide. *Transistors, Microelectronics.*

meter—1. Unit of length defined as 1 650 763.73 wavelengths in a vacuum of the orange-red line of the spectrum of krypton-86. Also spelled metre. *S.I. Units.* 2. Instrument for measuring; e.g., voltmeter, ammeter, ohmmeter, wattmeter, etc.

meter-kilogram-second-ampere (MKSA) system—Another name for the metric system. Also called the Giorgi system. *S.I. Units.*

meter resistance—Internal d.c. resistance of a meter movement. This should never be measured with an ohmmeter. Measurement is performed by adjusting R1 for a full scale reading with S open. Then R2 is adjusted for a half-scale reading with S closed. R2 is then disconnected and its resistance measured with an ohmmeter. The value obtained is equal to the meter resistance. A safe value for R1 is equal to twice the battery voltage divided by the full-scale current of the meter in amperes. (See Figure M-18.)

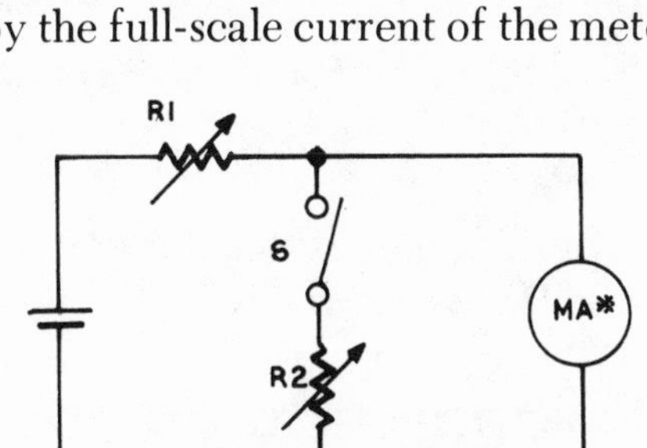

*any type of dc
movement

Figure M-18
Meter Resistance Measurement

metric system—The set of physical units used internationally by scientists and for general purposes in nearly all nations, the United States being the only notable exception. *S.I. Units.*

metric waves—Very high-frequency waves with a frequency range of from 30 to 300 megahertz (10 – 1 m).

MeV—Symbol for megaelectron-volt (q.v.).

MF—Abbreviation for medium frequency (q.v.).

mH—Symbol for millihenry (q.v.).

mho—Former unit of conductance, now replaced by the siemens (q.v.).

MHz—Symbol for megahertz (q.v.).

mica capacitor—Fixed capacitor with mica dielectric. *Capacitors.*

micas—Family of common silicate minerals characterized by platy structure easily separated into sheets. Some mica, containing little iron, is used as a dielectric or insulator. Also called isinglass.

micro—Prefix of S.I. Units meaning one millionth of (10^{-6}); symbol, μ.

microammeter—Meter for measuring currents in microamperes.

microampere (μA)—One millionth of an ampere.

microcircuit—Another name for integrated circuit (q.v.).

microcomputer—Single LSI chip which contains a central processing unit memory and input/output circuitry. *Computer Hardware.*

microelectronics—Branch of electronics associated with extremely small electronic parts, assemblies, or systems. *Microelectronics.*

MICROELECTRONICS

Microelectronics is the field of electronics that includes the design, fabrication, and use of monolithic integrated circuits. These are formed of elements that are prepared within and upon a semiconductor substrate. In almost all cases this substrate consists of a thin slice that has been cut from a large single crystal of silicon. The fabrication of microelectronic devices is, therefore, completely different from the assembly of conventional circuits from discrete components connected by wires. It bears to the latter about the same relationship that printing does to handwriting.

Fabrication

Silicon is used because it is a semiconductor and because its surface is easily oxidized to provide an insulating layer of silicon dioxide (SiO_2), a substance resembling glass. This layer in turn may be photoengraved to provide windows through which impurities may be introduced into selected areas of the silicon substrate, and connections made to an overlying pattern of conductors. Figure M-19 illustrates the steps in the process of forming a metal-oxide-semiconductor field transistor (MOSFET) on a silicon substrate.

a. Silicon substrate is exposed to a current of dry oxygen in an oven heated to a temperature between 1000° and 1300° C to form a layer of SiO_2 from 0.7 to 1.8 μm thick.

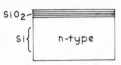

b. A layer of photoresist from 0.4 to 0.7 μm is applied evenly over the SiO_2, and hardened by baking in a low-temperature oven (<120° C).

c. A photomask with circuit details is positioned over the photoresist and exposed to ultraviolet light. Where the light passes through the mask it makes the photoresist insoluble.

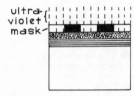

d. The soluble (unexposed) portions of the photoresist are dissolved.

e. Immersion in a saturated aqueous solution of ammonium fluoride, bifluoride, or Etchall cream removes the SiO_2 not protected by photoresist. The exposed photoresist is then removed also.

f. In an oven with a temperature about 1200° C and containing an atmosphere of gaseous boron, boron is deposited on the surface of the silicon.

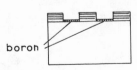

Figure M-19
MOSFET Fabrication

g. In a second oven with a higher temperature and containing an atmosphere of oxygen, the boron is "driven in" to form two p-type regions about 7 μm deep. The oxygen grows a new SiO_2 layer over these regions.

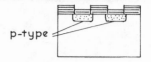

h. A new coating of photoresist is applied, masked, and exposed to ultraviolet light.

i. The soluble (unexposed) portion of the photoresist is dissolved.

j. The SiO_2 not protected by photoresist is removed, as in step e, and then the exposed photoresist.

k. A new SiO_2 layer only 0.1 μm thick is formed on the exposed silicon, as in step a.

l. A new coating of photoresist is applied, masked, and exposed to ultraviolet light.

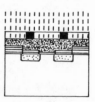

Figure M-19
(continued)
MOSFET Fabrication

m. Contact windows are etched in the SiO_2 as in steps i and j.

n. The device is placed in a vacuum chamber at a pressure of about 1 mPa and a temperature between 200° and 400°C. A small amount of aluminum is vaporized in the chamber (by radio-frequency, electron beam, or laser heating) and condenses on the surface of the device.

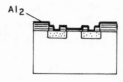

o. A new coating of photoresist is applied, masked, exposed to ultraviolet light, etc., as in steps h and i, and the unwanted parts of the aluminum layer are removed by etching with ferric chloride or sodium hydroxide.

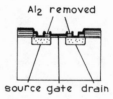

Figure M-19
(continued)
MOSFET Fabrication

The MOSFET is preferred over the bipolar transistor for applications where high switching speed is not essential, because it requires fewer steps in fabrication, many steps requiring less rigid control; also the device is smaller (so more of them can be packed on a silicon chip); hence, production costs are reduced. MOS LSI (large-scale integration) is used for a great many purposes, including microcomputers, microprocessors, calculators, watches, microwave-oven controls, and so on.

A cross-section of a bipolar transistor formed on a silicon substrate is shown in Figure M-20. It is more complex than the MOSFET and includes a "buried layer" to reduce collector saturation resistance.

The substrate itself is single-crystal silicon in which the atoms are arranged in a perfectly regular pattern throughout. Polycrystalline silicon is melted in an inert gas, and at the same time donor or acceptor impurities are added as required. Then a seed crystal of single-crystal silicon is lowered on to the surface of the molten silicon, which is very close to its melting point (1415°C). The silicon at the point of contact is cooled below its melting point and recrystallizes around the seed, adopting the same crystalline structure. This is called epitaxial growth. The seed is gradually raised as the silicon crystallizes around it, resulting in a cylinder of single-crystal silicon with a diameter of about 50 millimeters. (See Figure M-21). This is sliced into wafers, which after grinding and polishing have a mirror-like surface and a thickness of some 200 micrometers. Each of these wafers will finally make a large number of individual chips but they are not separated until after the oxidation, etching, diffusion, and metallization processes shown in Figure M-19, in which all the chips are formed simultaneously on each wafer.

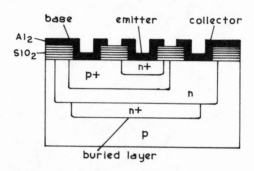

Figure M-20
Bipolar Transistor Formed on Silicon Substrate

Figure M-21
Cylinder of single-crystal silicon is sliced into wafers with a diameter of 50 mm. and a thickness of 200 μm. after grinding, lapping and polishing.

After fabrication and testing, the wafer is divided into the individual chips, each of which is packaged in one of the capsules illustrated in Figure M-22. Leads have to be connected between the metallized areas on the chip and the inner ends of the external terminal pins or leads of the package. These leads are of fine gold or aluminum wire and are attached by ball bonding or ultrasonic bonding (qq.v.).

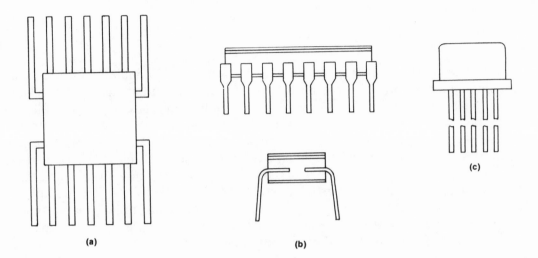

Figure M-22

Types of Packaging for IC's:

(a) *flatpack, usually a plastic case with metal pins;*

(b) *dual-in-line (DIP), plastic case with two rows of metal pins;*

(c) *to metal can with wire leads:*

From *Technician's Guide to Microelectronics*, pages 41, 47 and 49.

In the course of the evolution of monolithic integrated circuits, film integrated circuits were developed and are still used for some purposes. They are divided into thin-film and thick-film types. The thin-film IC uses glass or ceramic as a substrate, upon which resistors and capacitors with their interconnections are formed in somewhat the same manner as for monolithic IC's. However, transistors cannot be formed on a standard thin-film substrate, so the process is limited to passive components. A thick-film IC is fabricated using a silk-screen process and special inks. In both types precise resistor values can be obtained by removing a portion of the film with a laser beam or small grinder.

Because active devices have to be added, many of these IC's consist of a thin or thick-film ceramic chip with silicon-chip active devices mounted on it. They are then called hybrid or multichip IC's. The silicon chip is attached by making raised contacts on it, which match the conductor pattern on the ceramic chip. It is then mounted face down ("flip chip"). Another technique uses beam leads, relatively thick metal fingers projecting beyond the edge of the chip, which are welded to the film conductor pattern.

Film integrated circuits are very much larger than monolithic IC's, and their component density is far lower.

Circuit Functions

The applications of microelectronic circuits can be divided into:

> Digital (logic and memory);
> Analog (linear);
> Microwave.

Digital or logic circuits comprise the major category of monolithic circuits. The capability of fabricating a very large number of transistors simultaneously on a silicon wafer, which is then subdivided into many small chips, has made possible the production of large-scale integrated circuits (LSI) containing the masses of switching devices required by computers and other data processing systems. The first electronic computer, ENIAC, contained 18,000 electron tubes and 1500 relays. These were all hand-wired, taking a large number of assemblers (mainly housewives) some three years, at a cost to the taxpayers of about $50,000 in 1946 dollars. The completed machine occupied the whole of a very large room and weighed tons. Today, an entire computer can be fabricated on a silicon chip 4 millimeters square, and computers for use at home sold for less than $500 in 1980 dollars!

Not all of the transistors on the chip are used as active devices. Because of their small size, and because it simplifies fabrication, many of them are connected so that they can substitute for resistors or diodes. In Figure M-23, the MOS circuit (a) shows how this is done. The diodes and resistors shown in the bipolar circuits are also really modified transistors, but it has been customary to show them as diodes and resistors, since those are their functions.

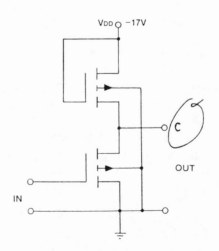

Figure M-23
Principle Logic Circuits Used in IC's
(a) MOS inverter (upper device is loaded resistor)
From *Technician's Guide to Microelectronics*, page 87.

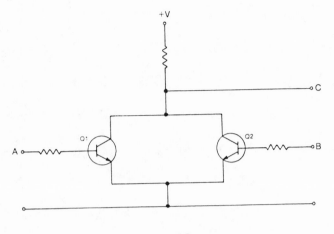

(b) RTL NOR gate

Ibid, page 93.

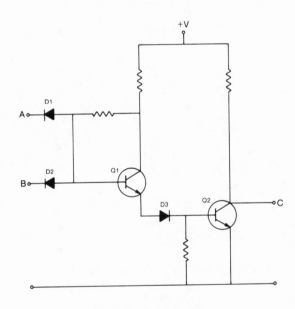

(c) DTL NAND gate

Ibid, page 97.

Whether MOS or bipolar, the transistors in these circuits act as switches. Each is connected in a circuit between the high side of the power supply (V_{dd} or V_{cc}) and the low side (ground). A load resistor is also connected between the transistor and V_{dd} or V_{cc}. With no signal applied to its gate, or base, the transistor switch is "open." In this state the entire power supply voltage appears at the output C, since no current is flowing through the load resistor to produce any voltage drop. When a signal is

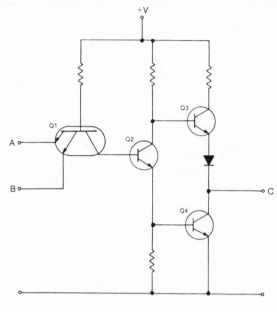

(d) TTL NAND gate

Ibid, page 98.

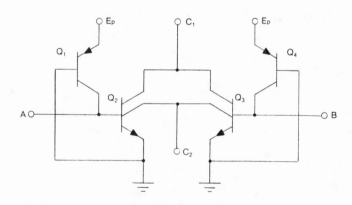

(e) I²L NOR gate

Ibid, page 101.

applied to the gate, or base, so that the transistor saturates, the switch is "closed." Now
C is practically shorted to ground, so it is at zero or a very low voltage. All the power
supply voltage is being dropped across the load resistor. This voltage is generally from
–5 to –30 volts for p-channel enhancement-type MOS devices, –5 volts for bipolar. The
polarity is reversed, of course, for n-channel enhancement-type MOS devices. For
designs with conventional bipolar integrated circuits, the circuit action is usually
described in terms of positive logic, because C swings between zero and a positive

voltage. The operation of p-channel enhancement-type MOS devices may sometimes be described in terms of negative logic, however. The rule is:

In positive logic, 1 ("true") is the most positive level;
0 ("false") is the most negative voltage level.

In negative logic, 1 ("true") is the most negative voltage level;
0 ("false") is the most positive level.

Figure M-23 shows basic circuits for inverter, NOR, and NAND gates in both MOS and bipolar circuits. Since each manufacturer has his own version, there are minor differences between the circuits illustrated in different books, but the principle of operation is the same. MOS circuits are preferred for general use, because they are simpler and less costly to fabricate and, being smaller, more of them can be packed on a chip. However, the parasitic capacitance of the gate of a MOSFET makes it slower, since it must be charged to the threshold voltage before the device can turn on. This is an important factor in some circuits, making the use of bipolar devices mandatory.

On the other hand, the gate capacitance of a MOS device is used advantageously in some circuit schemes to store information temporarily pending the arrival of a clock pulse. A further benefit conferred by enhancement MOSFET's is that they only consume power when turned on, so they require smaller power supplies and present less of a heat-dissipation problem. (For further information on the uses of logic gates see *Computer Hardware.*)

Memory devices are numerous and varied, ranging from bubble memories to punched cards. Microelectronic memories, of course, are those formed on a chip, and are therefore broadly categorized as transistor memories and bubble memories.

Transistor memories are arrays of transistors which may be operated as transistors or diodes. The simplest type is the read-only memory (ROM) illustrated in Figure M-24. In this memory, MOSFET's are arranged in columns and rows. All the sources are connected to ground and the drains, through load transistors T_{L1}, etc., to V_{dd}.

In manufacturing this matrix some transistors are fabricated without gates. Signals applied to rows A through N will have no effect on these transistors, but will cause those with gates to switch. To extract the data stored in row A, a negative voltage is applied to row-select line A. Those transistors with gates then turned on, and voltages on column-select lines 2, 3 ... M go to zero. Column-select line 1 remains at $-V_{dd}$, because its transistor in row A cannot switch. Using positive logic, this gives an output of 011 ... 1, the data stored in row A.

The pattern of active and inactive transistors therefore makes up a permanent memory that cannot be changed. The data can be read out over and over again and is not lost if power is interrupted, but there is no way to write data in.

ROM's require only one transistor to store each data bit, which makes it possible to store a lot of instructions in a small area. Random-access memories (RAM's), into which data can also be written, are somewhat more complex and come in two types, only one of which has a comparable packing density. This is the dynamic RAM, which makes use of the parasytic capacitance of the MOSFET to store a data bit as a charge.

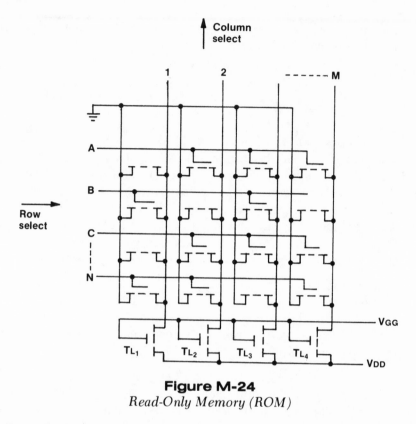

Figure M-24
Read-Only Memory (ROM)

Figure M-25 shows the basic circuit for a one-transistor cell RAM. A control instruction applied to the gate makes the MOSFET conduct, so that a signal on the data-bit line can charge the parasytic capacitance to write in a 1. If the data bit is 0 there will be no voltage on the data bit line, so no charge is stored. To read the data the transistor is turned on again, so that the charge, or absence of charge, appears as a voltage or lack of same on the data-bit line.

There is a problem, however, with dynamic cells. The charge will not remain indefinitely. It leaks away rapidly and is gone in a matter of milliseconds. To keep the data stored requires that each cell be read repeatedly at regular intervals by a separate feedback circuit that recharges the parasytic capacitance in each cell storing a 1. This is called refreshing and is a standard feature of all dynamic RAM's.

RAM's of this type are fabricated on silicon chips less than 50,000 square mils in size, with storage capacities up to 16,384 bits. However, newer designs are coming with a storage capacity of 65,536 bits. The refresh circuitry is also on the chip.

The other type of RAM is called a static RAM. It does not need refreshing, but is still volatile (i.e., it loses its data if the power is cut off). A memory cell in a static RAM may have as many as eight transistors, so it is comparatively large. It is also slower in operation. It is used mostly for the internal registers of the computer's central processing unit (CPU). These are small data-storage devices used for holding the data while it is being manipulated. (See *Computer Hardware.*) The basic circuit of a static

RAM cell comprises a flip-flop (q.v.) and gates operated by control signals that connect the flip-flop to the data-bit line as required. The output state of the flip-flop will be high (1) or low (0) according to what state the data bit to be stored set it in. A typical circuit is shown in Figure M-26.

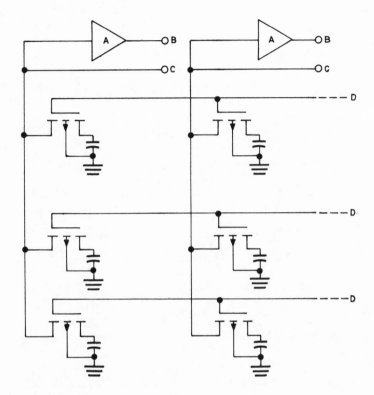

Figure M-25

Portion of Dynamic RAM

A and B: Refresh and sense amplifier and line

C: Data-bit line

D: Row-select line

(Collector capacitances are between metallization and substrate with SiO_2 dielectric.)

A more advanced memory system, expected eventually to have the capability of storing a million bits of data on a single chip, is the bubble memory. It consists of a substrate of nonmagnetic garnet, on which is a single-crystal thin film of magnetic garnet or some other ferrite or magnetic material. In this film, magnetic domains form under the influence of a stationary external magnetic field applied perpendicularly by a pair of permanent magnets. The strength of the field is such as to shrink these domains into small cylindrical magnetic dipoles called bubbles.

A pattern of permalloy tracks on the surface of the film is magnetized by a rotating magnetic field induced by two coils wrapped around the chip at right angles to each other in which flow alternating currents with frequencies of 100 or 150 kilohertz. The manner in which the rotating field magnetizes the permalloy structure pulls the bubbles along. The tracks are in the form of loops, around which the bubbles travel.

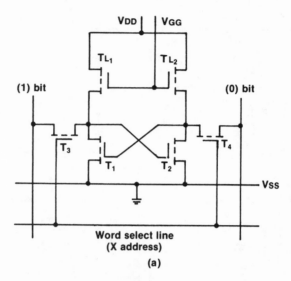

Figure M-26

Static RAM, Using a Flip-Flop

From *Technician's Guide to Microelectronics*, page 81.

A block of data entering the input of the device generates a bubble for each 1 bit, no bubble for each 0 bit. The bubbles are transferred into a major loop (see Figure M-27) and are lined up so that each bubble or space is opposite the end of one of an array of permalloy tracks perpendicular to the major loop. These tracks are called minor loops. The entire data block is then sent, all at once, into the minor loops. Typically, there are 256 minor loops, so the data block consists of 256 bits. Each minor loop can hold 1025 bits, so this number of data blocks can be entered, for a total storage capability of 262,400 bits.

The bubbles keep circulating synchronously in their loops, so that each data block arrives in turn opposite the read major loop. Control circuitry keeps track of the data blocks, so when the desired one appears new duplicate bubbles are produced in the read track by a transfer/replicate structure, without destroying the original ones. The new bubbles are then converted to electrical signals by a detector, before being eliminated.

Analog, or linear circuits are mainly amplifiers. In these, the magnitude of the signal is the important characteristic. Unlike logic circuits that carry out binary arithmetic functions by switching between two voltage levels, the purpose of a linear amplifier is to produce an output that is a faithful reproduction of the input signal, apart from being amplified.

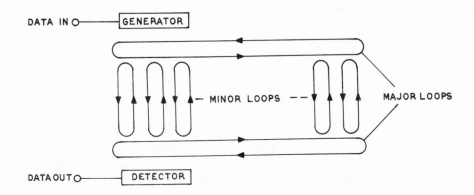

DATA IN o——[GENERATOR]

— MINOR LOOPS — —   MAJOR LOOPS

DATA OUT o——[DETECTOR]

Figure M-27
Principle of Bubble Memory (see text)

Conventional amplifiers employ a large variety of circuit elements, most of which are difficult to integrate. Monolithic circuits, on the other hand, use many more active devices, because they take up less area on the chip than large resistors or capacitors, and are therefore less expensive. Transistors with fixed bias are preferred as substitutes for resistors, and capacitors are not used at all if this can be avoided. If a capacitor must be used it is generally a discrete device connected externally.

Most amplifiers are therefore direct-coupled, using the minimum number of components. However, direct coupling is prone to d.c. drift, and with no coupling capacitor to isolate the stages the drift in one stage is amplified by the following stages, so that the final d.c. ouput level may be considerably altered.

To avoid this, nearly all linear IC's employ differential amplifiers. As shown in Figure M-28, a differential amplifier stage consists of two identical amplifiers, each with its own input. If identical signals are applied to the inputs, there will be identical outputs. Since the output signals are the same there will be no instantaneous voltage difference between them. Consequently any signal that is common to both amplifiers is balanced out. This includes d.c. drift or any other unwanted interference that affects both amplifiers equally. A signal applied to one input only, however, appears across the output, since the amplifier responds to the difference between its inputs. The resistors shown in the schematic are transistors with fixed bias in reality. The emitter "resistor" has a fairly high value so as to maintain a constant current, so that as the current through one amplifier increases, the current through the other decreases, so enhancing the differential amplification of the stage. The active devices may be MOSFET's or bipolar transistors.

The largest class of linear IC's is that of the operational amplifier ("op-amp"), which is a dual-input differential amplifier followed by one or more direct-coupled gain stages (see Figure M-29). Such an amplifier will have a very high voltage-gain which, however, is usually adjusted downwards by the application of external feedback. As shown in Figure M-30, a portion of the output signal is applied to the input, with the result that the closed-loop gain is given by R_f/R_{in}. If R_f is 5 kilohms and R_{in}. 50 ohms, the closed-loop gain will be 100.

An ideal op-amp would have

(1) Infinite open-loop gain

(2) Infinite input impedance

(3) Zero output impedance

(4) Zero output for zero input

(5) Infinite bandwidth

Practical op-amps do not, of course, achieve these ideal characteristics, but approach them. A typical op-amp with a CMOS input stage has an input impedance of 1.5×10^{12} ohms, an open-loop (without feedback) gain of 110 dB (320,000 times), and a unity-gain bandwidth of 15 megahertz.

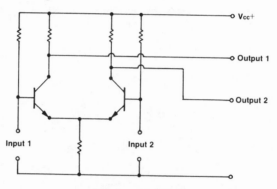

Figure M-28

Basic Differential Amplifier Stage

From *Technican's Guide to Microelectronics*, page 144.

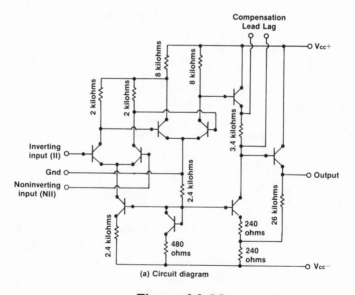

(a) Circuit diagram

Figure M-29

Operational Amplifier

Ibid, page 147.

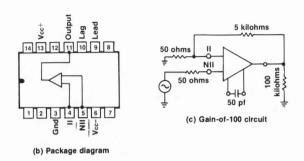

(b) Package diagram

(c) Gain-of-100 circuit

Figure M-30
How Feedback Is Used with an Op-Amp

Ibid, page 147.

The same op-amp can be used for different purposes, depending on the feedback circuit used. (For further details see *Operational Amplifiers.*)

There are also many special-purpose linear IC's available. As their numbers keep growing it is not possible to give an up-to-date list here, but they include FM stereo demultiplexers and demodulators, phase-locked loops, function generators, TV circuits, tone generators, and many more.

Microwave circuits operate with wavelengths of only a few centimeters, so ordinary circuit connections are comparable to these wavelengths, resulting in a variety of problems of signal propagation. It is necessary, therefore, to employ coaxial cable or waveguide in conventional circuits. However, the dimensions of monolithic circuits are so microscopic that open interconnections are again possible, with considerable savings in cost, size, and weight.

Transmission lines are formed by coating the underside of the substrate with a layer of deposited metal, to act as a ground plane, and depositing conductive lines of metal on the upper side. These are called microstrop (Figure M-31).

A different approach is to use a length of acoustically propagating material overlaid epitaxially on to a suitable substrate, with a transducer at each end. (See Figure M-32.) The input transducer converts the electromagnetic signal to an acoustic one that travels along the surface of the microacoustic transmission line, to be reconverted to an electromagnetic signal by the output transducer. The wave can be guided around bends by cutting grooves in the surface of the acoustically propagating material. Because the velocity of the acoustic wave is about 100,000 times slower than that of the electromagnetic wave, its wavelength, given by $\lambda = \frac{v}{f}$, is only 3 μm, where f = 1 gigahertz, compared to 0.3 m for the electromagnetic wave. The microacoustic line's attenuation increases with frequency at a faster rate than that of a microstrip line, so the latter is preferable for frequencies of 2 gigahertz and above.

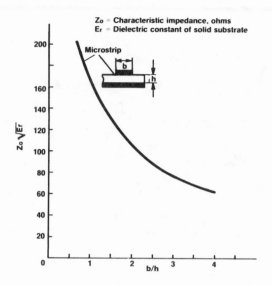

Figure M-31
Microstrip Dimensions and Calculation of Characteristic Impedance

Ibid, page 160.

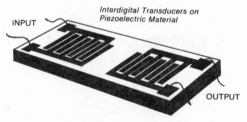

Figure M-32
Principle of Microacoustic Line

Ibid, page 161.

The small size and light weight of microeletornic devices, together with greater reliability, has stimulated a great deal of research directed toward replacing conventional heavy-weight and bulky airborne radar with solid-state equipment. One such concept consists of a sapphire substrate about 1.5 inches long and 0.5 inch wide with a dipole antenna deposited along the edge at one end (see Figure M-33). This is connected by microstrip to a ferrite circulator (duplexer). When transmitting, a Gunn-effect oscillator is pulsed and emits a one-watt signal at 10 gigahertz. This module is intended to be used with other modules in a scanned complete radar system, in which each transmits on the same frequency but with a different phase. The phase shifting is handled by a master synchronizing and phase-programming controller, so that an electronically-directed beam is propagated, doing away with the former motor-driven dish antenna and all its machinery, waveguides, and so on. These modules can be separated and installed conformally, following the contours of the vehicle.

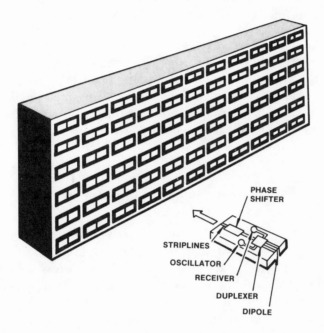

PHASE / SHIFTER

STRIPLINES

OSCILLATOR

RECEIVER

DUPLEXER

DIPOLE

Figure M-33

Principle of Phased-Array Microelectronic Radar

Ibid, page 167.

The total output power depends on the total number of modules. A single Gunn-effect diode cannot compete with a magnetron or a klystron (see under *Electron Tubes*), but a large number can. A great deal of effort is being devoted to further development of these bulk-effect devices (described in more detail under *Semiconductors*).

microfarad (μF)—One millionth of a farad.

microfiche—Photographic reproduction in reduced size, in which the picture is reduced to a mere dot, a large number of which can be printed on a single small negative fitted to a file card. Special viewers greatly enhance the image for viewing.

microgroove—The size of groove used for 33-1/3 and 45-rpm records, so called because it is approximately half the width of the groove used on the earlier 78-rpm record. *Recording.*

microhenry (μH)—One millionth of a henry.

microhm (μΩ)—One millionth of an ohm.

micrometer μm)—One millionth of a meter, or one micron.

micrometer caliper—Instrument for making precise linear measurements of dimensions such as diameters, thicknesses, and lengths of solid bodies. Shaped like a C-clamp, it has a jaw that is moved in or out by a very fine-pitch screw, the amount of movement being indicated by a scale engraved on the screw's sleeve. The scale indicates the distance between the face of the jaw and the anvil mounted on the opposite end of the C-frame.

microminiaturization—Fabrication of microelectronic devices.

microphone—Device for converting acoustic power into electric power with analogous waveforms. The greatest number of microphones is used in telephone instruments, next in hearing aids, tape recorders, public address systems, and radio. Sound waves impinge on a diaphragm, which is turn causes a corresponding change in some property of an electric circuit. Microphones are named according to the means used; e.g., carbon, condenser, dynamic, ribbon, crystal, and so on. *Telephone, Recording.*

microphonics—Production of electrical noise by mechanical disturbance. For instance, an electron tube may generate microphonic noise by vibration of its elements caused by external shock.

microprocessor—Computer central processing unit fabricated on a single silicon chip. *Microelectronics, Computer Hardware.*

microsecond (μs)—One millionth of a second.

microstrip—Microelectronic transmission line for microwave frequencies, consisting of a metal strip on a substrate, on the under side of which is a corresponding ground plane not less than three times as wide as the strip.

microvolt (μV)—One millionth of a volt.

microwatt (μW)—One millionth of a watt.

microwave—Loosely applied term for the range of radio frequencies from one gigahertz to one terahertz. *Frequency Data, Waveguides and Resonators.*

midrange—Audio-frequency range between bass and treble, usually considered as being between 800 and 8000 hertz. *Electroacoustics.*

mil—Unit of length equal to one thousandth of an inch or 2.54×10^{-5} meter, used in measuring the diameter of wire. A circular mil is equal to the area of a circle with a diameter of one mil.

milli—Prefix of S.I. Units, meaning one thousandth of $(+10^{-3})$. Symbol, m.

milliammeter—Meter for measuring currents in milliamperes.

milliampere (mA)—One thousandth of an ampere.

millihenry (mH)—One thousandth of a henry.

millimetric waves—Extremely high-frequency waves with a frequency range of from 30 to 300 gigahertz (10 - 1 mm).

milliohm (mΩ)—One thousandth of an ohm.

millisecond (ms)—One thousandth of a second.

millivolt (mV)—One thousandth of a volt.

milliwatt (mW)—One thousandth of a watt.

minicomputer—Desk-top computer that preceded microcomputers.

minority carrier—Carrier with opposite polarity to the host semiconductor material (holes in n-type material, electrons in p-type material).

mirror galvanometer—Galvanometer with a small mirror attached to the movement instead of a pointer. A ray of light from a lamp is reflected by the mirror so that it falls on a translucent scale, the illuminated spot moving according to the deflection of the movement. This has the effect of a very large pointer and a great number of divisions on the scale, providing much higher resolution in the reading.

mirror scale—Arc of reflecting material placed on a meter dial beneath the calibrated scale. By keeping the pointer and its reflected image in line, parallax error (q.v.) is avoided.

mismatch—Condition arising when two units with different impedances are coupled together, preventing maximum transfer of power from one to the other. *Transmission Lines.*

mixer—Circuit for combining two or more signals, as the external and local-oscillator signals in a superheterodyne receiver (q.v.), or audio signals from different microphones, tapes, etc.

MKSA system—Abbreviation for meter-kilogram-second-ampere system (q.v.).

mm—Symbol for millimeter (q.v.).

mnemonics—Devices for aiding the memory. For example, a mnemonic to recall the value of π, in which the number of letters in each word gives the corresponding figure in the value, goes:

    "How I wish I could recollect pi easily today ..."
      3   1   4   1   5   9   2   6   5   ...

Mnemonics are used frequently in computer programming. For instance LDA means "load accumulator" on an 8080 system and is much easier to remember than the machine operating code 00111010.

MNOS—Abbreviation for metal-nitride-oxide-semiconductor, descriptive of an insulated-gate field-effect transistor in which the gate dielectric consists of a layer of silicon nitride deposited over a very thin layer of silicon dioxide. This raises the dielectric constant to 6.8, compared to 4.0 for silicon dioxide alone, which in turn lowers the threshold voltage to a value between -1.9 and -2.9 V, which is compatible with TTL devices. *Microelectronics.*

mobile radio communication—Radio communication between stations intended to be used while in motion or during halts at unspecified points or between such stations and fixed stations. *Frequency Data.*

mobility (μ)—Limiting carrier velocity in a semiconductor resulting from collisions with imperfections in the crystal lattice and ionized impurities. *Semiconductors.*

mode—Wave pattern in waveguide, designated transverse electric (TE), or transverse magnetic (TM), according to the orientation of the field. *Waveguides and Resonators.*

modem—Acronym for modulator/demodulator. Device used to interface a computer with a telephone line. It converts the computer's binary code into a form that can be transmitted by telephone line and be impervious to noise and static on the line. At the other end a second modem converts the signal back to binary code. Used for communication between computers or computers and terminals at distant locations. *Computer Hardware.*

modulation—Process whereby certain characteristics of a carrier wave are varied or selected in accordance with a message signal. *Modulation.*

MODULATION

In a telephone, sound waves in the air vibrate a diaphragm which presses against an assembly of carbon particles and causes its electrical resistance to vary, so that an electric current flowing through the particles is altered and fluctuates in accordance with the pressure on the particles. The current has been *modulated* so that its amplitude varies directly with that of the sound waves.

This current can be transmitted along a wire to the receiver, where it flows through an electromagnet. The fluctuations of the current cause variations in the strength of the electromagnet's pull on an adjacent steel diaphragm so that it vibrates, moving the air and producing sound waves corresponding to the original waves that made the transmitter's diaphragm vibrate.

The transmission of sound waves by radio requires a similar process, but without the use of a connecting wire. The current in the wire is replaced by an electromagnetic wave in space. This *carrier wave* is a simple form of alternating current, practically a sine wave. A sine wave has three dimensions that can be modulated: its amplitude, its frequency, and its phase.

Amplitude modulation. In standard broadcasting and in the transmission of the picture portion of a television signal, amplitude modulation (AM) is used. As in telephony, the amplitude or strength of the carrier is varied in accordance with the amplitude and frequency of the information to be put on it. For example, in transmitting a tone with a frequency of 400 hertz the *amplitude* of the carrier is varied at a frequency of 400 hertz.

Frequency modulation. In frequency modulation (FM) the amplitude of the carrier remains constant. The information to be transmitted is made to vary its frequency in accordance with that of the information to be transmitted. For example, in transmitting the same musical tone of 400 hertz the frequency of the carrier is made to vary above and below its normal frequency at a *rate* of 400 hertz. The *amount* of the variation is in proportion to the amplitude of the information.

Phase modulation. In phase modulation (PM) the phase of the carrier is advanced or retarded by variations in the amplitude of the information to be transmitted, and at a *rate* according to their frequency. The transmitter power does not vary but the carrier frequency does.

FM and PM are two forms of *angle modulation.* The reasons for this name will become apparent in the more detailed explanation given later.

Sidebands. Modulation sets up new groups of radio frequencies above and below the carrier frequency, which are called sidebands. Consequently a modulated signal occupies a band of frequencies rather than the single frequency of the carrier alone. The carrier and its sidebands are called a channel. AM broadcasting stations are limited to modulating frequencies that must not exceed 5000 hertz. If a carrier of 1000 kilohertz is modulated with a 5000-hertz signal, an upper side frequency of 1005 kilohertz and a lower side frequency of 995 kilohertz will be produced. The maximum width of the channel is therefore 10 kilohertz for an AM station. Since the modulating signal seldom consists of a single tone, but rather of a wide range of audio frequencies, there will in fact be many more side frequencies than two in the 10-kilohertz channel bandwidth.

In AM the upper and lower sidebands carry the same information; therefore either one is superfluous. Either can be filtered out at the transmitter. The same applies to the carrier, which contains no information at all. AM transmission can be divided into:

> *Double sideband, full carrier (A3),* the standard broadcast practice, in which both sidebands and the carrier are transmitted. This requires no special circuitry in the receiver, so is economically more practical for widespread use, but a considerable wastage of power at the transmitter makes it the least efficient system.

> *Single sideband, suppressed carrier (A3J),* on the other hand, is the most efficient, since the transmitter does not radiate superfluous power. The receiver, however, has to generate and insert the missing carrier so that demodulation (recovery of the information) can take place. Single-sideband (SSB) transmission of this type is in common use for all forms of point-to-point communication.

> *Vestigial sideband (A5C)* is the method used by television stations. As in the case of AM radio broadcasting, both sidebands and carrier must be transmitted if special circuitry in the receiver is to be avoided, but the modulation bandwidth of the picure information is 4.5 megahertz. When allowance is made for the sound signal and guard bands the channel width for the whole transmission would be almost double the six megahertz allocated if both sidebands were transmitted in full. The lower sideband is therefore rolled off so that at and beyond 1.25 megahertz below the carrier frequency it is at least 20 decibels below the carrier level.

> Other types of sideband transmission used with AM carriers are *single sideband, reduced carrier (A3A), two independent sidebands, reduced carrier (A3B), single sideband, reduced carrier (A4A and A7A),* and *two independent sidebands (A9B).* Only the first two are used for telephony in the ordinary sense. The others are used for facsimile and multi-channel telegraphy, or combinations of telegraphy and telephony, etc.

In FM and PM, sidebands are set up as in AM, but they are more complex. A single tone gives a whole series of pairs of side frequencies that are harmonically related to its frequency. The number of extra sidebands depends on the modulation index, which is given by:

$$\text{Modulation index} \quad = \quad \frac{\text{Carrier frequency deviation}}{\text{Modulating frequency}}$$

In the case of FM, the larger the modulation index the more effectively this type of transmission performs in combating noise and interference, and the greater the channel bandwidth, which is approximately twice the frequency deviation plus twice the modulating signal frequency. However, sideband or carrier suppression is not possible because the energy that goes into the sidebands is taken from the carrier, the *total* power remaining the same regardless of the modulation index. In AM the sideband power is supplied by the modulation circuit in the case of plate modulation, or by changing the power input and efficiency in the case of grid modulation.

Methods of modulation. In AM the audio signal is impressed on the carrier by modulating the power supply or bias to the output power amplifier. Modulating the power supply or bias results in varying the amplification of the carrier, so that the output power fluctuates in correspondence with the information to be transmitted. Figure M-34 shows an example of plate modulation, in which the signal from the microphone after suitable amplification drives a push-pull audio power amplifier with the same power output capability as the r.f. power amplifier. The plate supply for the r.f. power amplifier has to pass through the secondary winding of the modulation transformer T3. The audio output signal from V3 and V4 is also present in this winding, and its variations in amplitude add to or subtract from the plate supply of V1 and V2, resulting in variations in the carrier output.

The degree to which the carrier is modulated depends upon the power of the audio signal and is expressed as a percentage. In 100-percent modulation the carrier is driven all the way to zero at its minimum amplitude and to double its unmodulated amplitude at the maximum value. This is illustrated in Figure M-35.

The r.f. power amplifier in plate modulation is operated Class C. In the figure two triodes are shown, as would be used in a broadcast station, although many now use beam-power tubes. In citizen's band and similiar low-power applications where the circuits are solid state, plate modulation is not used, of course, and modulation may be done at the oscillator stage, as shown in Figure M-36, where the single-stage differential amplifier Q2 and Q4 is oscillating at the r.f. frequency determined by crystal Y. The modulating signal is applied to the base of Q3, which is the constant-current emitter "resistor" common to Q2 and Q4, so its resistance is varied in accordance with the variations of the signal, resulting in variations in amplitude of the r.f. output. The principle involved is that the operating-current conditions of the differential pair are determined by the base-bias circuit and emitter resistance of Q3, which in turn depend on the modulating signal. The area in Figure M-36 within the dashed lines will usually be provided in the form of an intergrated circuit.

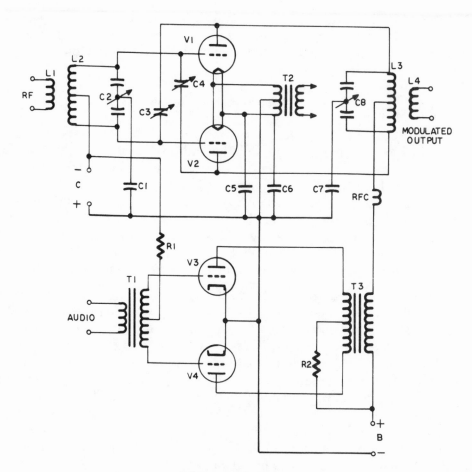

Figure M-34
Plate Modulation

From *Complete Guide to Reading Schematic Diagrams,* First Edition, page
102. ©1972. Reprinted with permission of Parker Publishing Company, Inc.

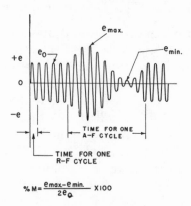

$$\% M = \frac{e_{max} - e_{min}}{2e_0} \times 100$$

Figure M-35
Percentage of Modulation

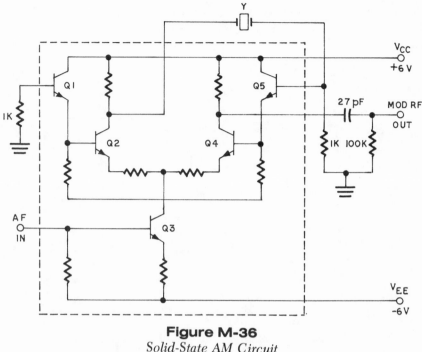

Figure M-36
Solid-State AM Circuit

Another method of amplitude modulation used with vacuum-tube circuits is grid modulation (Figure M-37). In this method the grid bias supply to the r.f. amplifier is modulated, resulting in corresponding variations in the gain of the tube, which is operated Class B.

Many methods have been used for FM and PM, including reactance tubes (circuits that act as variable capacitors or inductors), special tubes such as the Phasitron, and balanced modulators.

The balanced modulator is the heart of the Armstrong FM system used today. As shown in Figure M-38, it consists of two tubes connected in push-pull. The master oscillator signal is applied in phase to both grids and the modulating signal to the screen grids via the transformer. In the absence of modulation, the master oscillator signal on the grids causes both plate currents to vary in phase. Since these flow in opposite directions through L1 and L2 and are of equal amplitude, they cancel each other out. No signal is therefore induced in L3. However, when a modulating signal appears on the screen grids, one grid goes more positive while the other goes less positive, so the plate currents are no longer balanced and a voltage is induced in L3. This voltage lags the current by 90 degrees.

In the absence of modulation, as stated above, there is no signal in L3. In effect, the unmodulated carrier has been suppressed. Therefore, the signals that appear in L3 as a result of modulation are sidebands, upper and lower. The resultant of the two sideband voltages is amplified and recombined with the master oscillator signal in another amplifier to give an overall resultant signal swinging back and forth through a range $\pm\theta$, thereby producing angle modulation, which can be predominantly phase or frequency modulation, depending on additional circuitry. For instance, if FM is desired rather than PM, the audio signal is first processed in a correction amplifier before being applied to the balanced modulator.

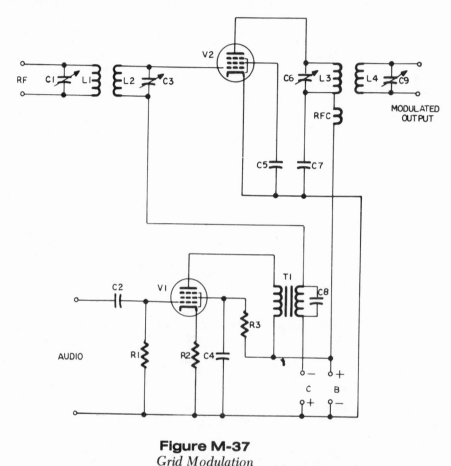

Figure M-37
Grid Modulation

From *Complete Guide to Reading Schematic Diagrams*, First Edition, page 102. ©1972. Reprinted with permission of Parker Publishing Company, Inc.

Pulse modulation. The modulation techniques described so far produce continuously varying (analog) modulation of a continuous carrier. In pulse modulation the carrier is usually a series of regularly recurrent pulses. Information is conveyed by modulating some aspect of the carrier pulses, such as amplitude, duration, time of occurrence, or shape. This is done by taking samples of the modulating waveform at regular intervals and at a rate higher than twice the highest significant signal frequency, and using them to modulate the pulses, as shown in Figure M-39. In pulse-amplitude modulation (PAM) the height of the pulse is varied in accordance with the amplitude of the modulating signal. In the example shown the pulses are all positive going. In another type of PAM the pulses also go negative when the modulating signal does. The first type is therefore called single-polarity PAM to distinguish it from double-polarity PAM, the second type. Either type can have pulse tops which follow the amplitude variation of the modulating signal during the sampling interval, as in the example, which is called "natural" or "top" sampling; or the pulse tops can be flat, called "square-topped" or "instantaneous" sampling, where the height corresponds to the instantaneous value of the modulating signal at the center or edge of the pulse.

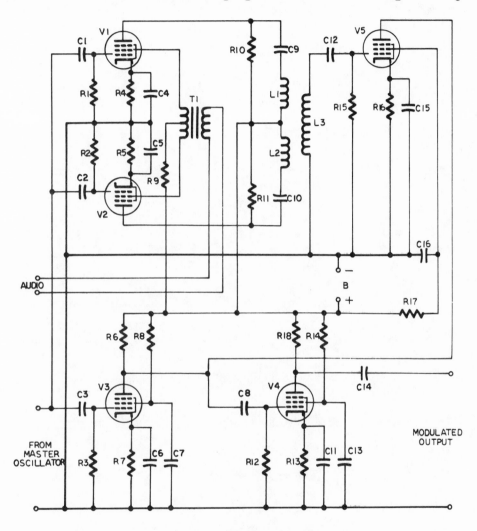

Figure M-38

Armstrong Balanced Modulator

Ibid, page 108.

In pulse-duration modulation (PDM) the width or duration of the pulse is proportional to the amplitude of the modulating signal. Like PAM, sampling may be "natural," the samples coinciding with the pulses, but as the pulses vary in width the sampling intervals are not constant. In "uniform" sampling the samples are uniformly spaced instead of varying with the pulse width. PDM is also called pulse-length or pulse-width modulation (PLM or PWM).

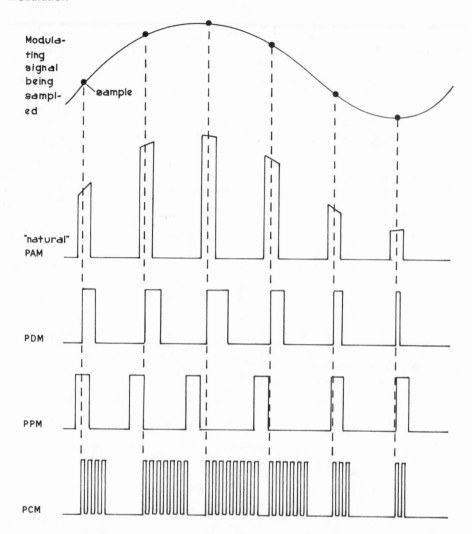

Figure M-39
Types of Pulse Modulation

Another form of pulse-time modulation is pulse-position (or phase) modulation (PPM) in which the occurrence of the pulse in time is advanced or retarded in accordance with the amplitude of either natural or uniform samples of the modulating signal. Similar to PPM is pulse-frequency modulation (PFM), except that rate of occurrence of the pulses varies in proportion to the amplitude of the modulating signal.

The pulse-modulation systems described so far are all uncoded pulse systems. In pulse-coded modulation (PCM) the modulating signal is sampled uniformly, but the sampled values are then converted to code groups of pulses in binary, quaternary, or octonary notation. Another code system called delta modulation (DM) encodes only changes in modulation signal amplitude, using positive and negative pulses to indicate the direction of change.

molecule—Smallest unit into which a pure substance can be divided and still retain the composition and chemical properties of the substance. It may consist of a single atom, as in the noble gases, or an aggregation of atoms held together by valence forces, and acting as a unit.

molybdenum—Silver-gray metal with a very high melting point (2610°C) and the ability to retain its strength and hardness at extremely high temperatures. Used as an alloy to impart these qualities to steel. Resembles tungsten and is similarly used for filament supports, heaters, etc. Symbol, Mo; atomic weight, 95.94; atomic number, 42.

monochrome—Single color, but usually meaning black-and-white television reception.

monolithic circuit—Single flat-surfaced chip of silicon on which patterns may be etched, scribed, diffused, etc., the result being an integrated circuit (q.v.). *Microelectronics.*

monostable multivibrator—Multivibrator (q.v.) with only one stable state, to which it returns as soon as the signal triggering it to its other state is removed. It is therefore analogous to a one-shot or single-shot multivibrator. (See Figure M-40.)

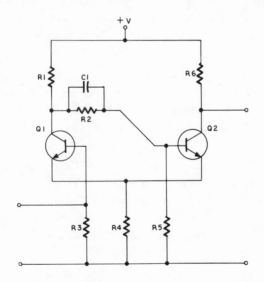

Figure M-40
Monostable Multivibrator (Schmitt trigger)

From *Complete Guide to Reading Schematic Diagrams*, First Edition, page 228. ©1972. Reprinted with permission of Parker Publishing Company, Inc.

Morse code—System of signals in which dots and dashes are combined to represent letters, numerals, and punctuation marks. See *International Morse Code.*

MOSFET—Abbreviation for metal-oxide-semiconductor field-effect transistor (q.v.).

most significant bit (MSB)—In a binary number the left-most bit. *Computer Hardware.*

motherboard—In a system employing several printed-circuit boards, a board that is the communication center, or nexus, for the others.

motor—Electromechanical device that converts electrical energy into mechanical energy.

motorboating—Oscillation at audio frequency giving rise to a "putt-putt" sound in a radio speaker. Caused by a combination decoupling and filter system defect, most often resulting in a high common impedance in the plate circuit of an audio amplifier. Check for open bypass or filter capacitors, especially electrolytics.

moving-coil meter—(See permanent-magnet moving-coil movement.)

moving-coil microphone—Microphone in which sound waves cause vibrations in a diaphragm, to which is attached a coil moving in the field of a permanent magnet, so that electrical signals are induced in the coil analogous to the sound waves. Also called a dynamic microphone.

moving-coil pickup—Phonograph pickup in which the movements of the stylus cause a coil to move in the field of a permanent magnet, thereby inducing electrical signals analogous to the sound track on the record. Also called a dynamic pickup.

moving-coil speaker—Loudspeaker in which the electrical audio signals pass through a coil located in the field of a permanent magnet. This "voice coil" is mounted on a diaphragm or cone, and the movements of the coil resulting from interaction between the audio signals and the magnetic field cause the cone to move also, setting up sound waves in the surrounding air. Also called a dynamic speaker, the most common type in general use. (See under loudspeaker for illustration.) *Electroacoustics.*

ms—Abbreviation for millisecond (q.v.).

MSB—Abbreviation for most significant bit (q.v.).

MSI—Abbreviation for medium-scale integration (q.v.).

mu—Twelfth letter of Greek alphabet (M, μ), corresponding to English M, m. (See *Greek Alphabet.*) Symbol for: 1. S.I. Unit prefix micro-; 2. Micron (unit of length); 3. Amplification factor; 4. Mobility. (See individual entries.)

MUF—Abbreviation for maximum usable frequency (q.v.).

multimeter—Instrument that measures several functions, such as d.c. voltage, a.c. voltage, current and resistance, using one meter movement or digital readout, and selecting for each function and range by means of a multiposition switch. (See Figure M-41.)

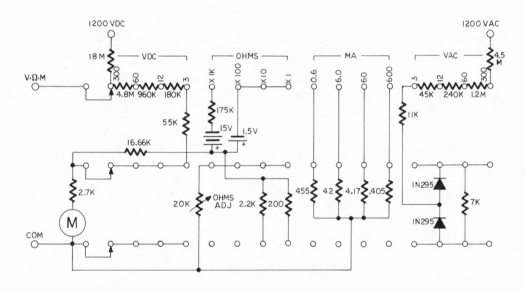

Figure M-41
Multimeter

multiplexing—Modulating one carrier with two or more information signals. This may be done by frequency division, in which each channel is allocated a different frequency within the wideband carrier; or time division, in which each channel consists of a series of pulses, interleaved on a time-sharing basis. *Modulation.*

multiplier tube—(See photomultiplier tube.)

multivibrator—Relaxation oscillator with two amplifier stages cross-connected to provide a large excess of positive feedback, causing the circuit to operate in abrupt transitions between saturation and cutoff. Multivibrators may be monostable, bistable, or astable (qq.v.).

music power—Peak power rating of an audio amplifier as opposed to continuous or r.m.s. power rating, preferred for use in hi-fi manufacturers' specifications because it sounds more impressive. Rated in watts into a specified load, using tone bursts instead of a continuous signal, it is also called dynamic power.

mutual inductance—When two coils are placed near each other (or wound on the same iron core) the flux lines of one coil will cut through the other coil, and the coils are said to be mutually coupled. The amount of coupling depends on the nearness of the coils to each other and their shape and size. This coupling is called mutual inductance and is expressed in henries. *Transformers.*

mW—Symbol for milliwatt (q.v.).

Mylar capacitor—Polyester-film dielectric capacitor with metal foil or evaporated-metal electrodes. *Capacitors.*

myriametric waves—Very low-frequency waves with a frequency range of from 3 to 30 kilohertz (100-10 km).

mux—Abbreviation for multiplex (q.v.).

MV—Symbol for megavolt (q.v.).

mV—Symbol for millivolt (q.v.).

MW—Symbol for megawatt (q.v.).

N—1. Symbol for nitrogen.   2. Symbol for north.   3. Number of turns in a winding.

n—1. Symbol for S.I. Unit prefix nano- (q.v.).   2. Symbol for an indefinite number (e.g., nth).   3. Symbol for neutron (q.v.).   4. Symbol for semiconductor type in which electrons are the majority carriers.

Na—Symbol for sodium (q.v.).

nA—Symbol for nanoampere (q.v.).

NAB—Abbreviation for National Association of Broadcasters, standardization authority for the industry. *Recording.*

nail-head bonding—(See ball bonding.)

NAND—Basic logic circuit in which the output is true (1) as long as all the inputs are not. *Computer Hardware, Microelectronics.*

nano—Prefix of S.I. Units meaning one thousandth millionth of (10^{-9}). Symbol, n.

nanoampere (nA)—One thousandth millionth (10^{-9}) of an ampere.

nanofarad (nF)—One thousandth millionth (10^{-9}) of a farad. Also equal to 0.001 microfarad or 1000 picofarads.

nanohenry (nH)—One thousandth millionth (10^{-9}) of a henry. Also equal to 0.001 microhenry.

nanosecond (ns)—One thousandth millionth (10^{-9}) of a second.

nanovolt (nV)—One thousandth millionth (10^{-9}) of a volt.

nanowatt (nW)—One thousandth millionth (10^{-9}) of a watt.

Napierian logarithms—Logarithms to the base e, in which $e = 2.71828 \ldots$ The notation ln is used for Napierian logarithms to distinguish them from common logarithms, denoted log. Also called natural logarithms.

National Electrical Manufacturer's Association (NEMA)—Organization that establishes standards in the manufacture of electronic components, parts, etc.

National Electric Code (NEC)—Code used in the U.S. to control wiring and other electrical installation in the interest of public safety.

National Television System Committee (NTSC)—Organization responsible for formulation of standards for television broadcasting in the U.S.

natural logarithms—(See Napierian logarithms.)

navigation aids—Radio and radar systems used to determine position and provide guidance to aircraft and ships. *Navigation Aids.*

NAVIGATION AIDS

Navigation is the art of "driving" a ship from one place to another, knowing at all stages of the journey the ship's position, arriving on time, and doing this so as to avoid collision with other vessels, or running out of fuel. Even trains and motorists are not immune to some of these problems, but all of them must be overcome by ships, a term which includes maritime vessels, aircraft, and space vehicles.

In essence, navigation consists of deducing the ship's present position from knowing how long it has been going in a certain direction at a known speed. In practice, this method, called *dead reckoning*, is complicated by uncertainties arising from wind velocity, ocean currents, instrument inaccuracies, and human error. It is necessary, therefore, to have means of checking the ship's position in case the real position differs from the assumed position.

In the past this was solely the responsibility of the navigator. When out of sight of land or the surface of the Earth, he could only supplement dead reckoning by obtaining position information from the sun and stars, using a sextant to measure the angle between the heavenly body and the horizon, noting the time of the observation, and performing a difficult calculation. In conditions of poor visibility this was impossible, of course, and even in the best of conditions an aircraft would be many miles away from that position by the time the necessary calculations had been done.

The development of electronic aids to navigation has consequently been of tremendous value to the navigator. The earliest of these was the loop antenna, which is a vertical antenna comprising several circular turns of wire that can be rotated in azimuth. Mounted on the top of an aircraft's fuselage or an equivalent position on a vessel's superstructure, the loop antenna is connected to a radio receiver. The latter is tuned to a radio station or beacon whose position is known, and the loop rotated for minimum signal pickup, which occurs when the loop is perpendicular to the direction in which the radio waves are traveling. The direction is read off a scale and plotted on the navigator's chart as a line drawn from the radio station. At the precise moment of the reading the ship (vessel or aircraft) must have been located somewhere on this line. To determine where, a second bearing (direction reading) is obtained by tuning another radio station or beacon and repeating the foregoing procedure to obtain a second position line. After making due allowance for the ship's movement between the two readings the intersection of the two lines gives the navigator a "fix," or position, for the ship at a particular time.

On the outbreak of hostilities with Germany in 1939, all British broadcasting stations were reset to transmit on one of three frequencies and radio beacons were shut down, so that loop antennas on hostile aircraft could not use them for navigation. Instead, friendly aircraft used a ground-based direction-finding system in which the aircraft transmitted a signal that was picked up by an *Adcock antenna* array. This is a set of four vertical antennas placed at the corners of a square, with a fifth in the center, connected to a goniometer. The goniometer consists of two coils, one of which is rotated with respect to the other, which is fixed. The effect of rotating the movable coil is the same as if the Adcock array was a large rotating loop antenna, so that a bearing can be obtained of the aircraft's signal. If two DF stations in different locations tuned the aircraft's signal, they could determine its position, and so inform it.

Neither of these methods so far described is ideal, because reflections from the ionosphere and refractions in the atmosphere distort the direction of arrival of the radio signals, so that other more accurate systems have been developed since. However, loop antennas on aircraft are still widely used, since they will operate on almost any type of station in the 200-1600 kilohertz band. Some radio beacons also exist for this service. Called *nondirectional beacons* (NDF), they operate in the 200 to 415 kilohertz band.

From this it can be seen that electronic navigation aids fall into two broad divisions: those operated by the navigator or pilot, and those operated by a ground controller.

The tendency is also to place the more complex part of the equipment on the ground so that the instrumentation in the vehicle is simplified as much as possible. The exception to this is found in military aircraft, which obviously prefer methods in which they keep the knowledge of their position to themselves.

The major radio navigation aids in use at present, in addition to the airborne direction finder (ADF), or loop antenna, are:

AEROSAT	Omega
ATCRBS	Radar (airborne)
DME	TACAN
ILS	VOR
LORAN	Consol

In addition, the new microwave landing system (MLS) is in limited production and is expected to have replaced ILS before the year 2000.

AEROSAT (Aeronautical Satellite), which is now being deployed, is the use of satellites to provide improved communication and surveillance for oceanic and air traffic control.

ATCRBS (Air Traffic Control Radar Beacon System), which is also called *secondary surveillance radar (SSR),* is a system where the aircraft carries a *transponder* (a radar transmitter-receiver which replies automatically to the ground interrogator). The ground station has a rotating directional antenna that transmits 400 times a second on 1030 megahertz. If an aircraft with a transponder is within range it responds with a coded signal that gives its identity and altitude. This information is displayed on the controller's PPI beside the primary radar "blip," which shows direction and range. The transponder frequency is 1090 megahertz.

DME (Distance Measuring Equipment) works in the opposite direction to ATCRBS. An airborne interrogator transmits to a ground transponder which replies, and the airborne transmitter-receiver displays the interval between the transmitted and received pulse in nautical miles. DME operates on any of 126 channels between 1025 and 1150 megahertz, with replies on channels with frequencies between 962 and 1024, or 1151 and 1213 megahertz.

ILS (Instrument Low-Approach System), illustrated in Figure N-1, is the standard means of guiding an aircraft during its approach to the runway. This requires keeping the aircraft lined up with the runway *(lateral guidance),* bringing it down at the proper rate so it will arrive at the runway at the proper altitude *(vertical guidance),* and advising the pilot as to how close he is to the runway *(along-course guidance).*

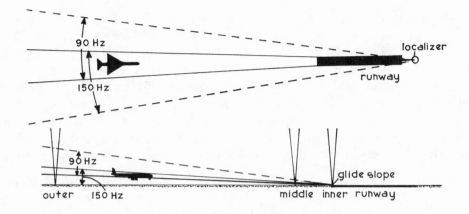

Figure N-1
ILS:

(a) lateral guidance:

(b) vertical guidance and along-course guidance (fan markers)

The pilot has an instrument on his panel called a "cross-pointer," with two pointers. One hangs down; the other is horizontal. The vertical pointer will remain centered as long as the aircraft stays lined up with the runway. If it wanders, say, to the left, the pointer will swing to the right, telling the pilot to correct the aircraft's course to the right; and vice versa, if the pointer swings to the left. The horizontal pointer operates in a similar manner, pointing upward if the aircraft is too low and downward if it is too high.

To achieve this, two transmitters are provided on the ground. One, called the *localizer*, is situated in line with the runway and at the end opposite the one where the aircraft will touch down. It transmits a thin horizontal beam divided in two, with a frequency in the band between 108 to 112 megahertz. The right-hand part of the beam, as "viewed" by the aircraft, is modulated with a 90-hertz signal; the left-hand part with a 150-hertz signal. The two parts of the beam overlap by five degrees. As long as the aircraft is in the overlap, both signals are received and the pointer remains vertical; but if it strays outside the overlap it can receive only one signal, hence the deviation of the pointer.

The operation of the horizontal pointer is very similar. The second transmitter, called the *glide slope*, is situated beside the point on the runway where the aircraft will touch down. It transmits a thin vertical beam divided in two, with a frequency in the band between 329 and 335 megahertz. The upper part of the beam is modulated with a 90-hertz signal which causes the pointer to point downward, and the lower part of the beam is modulated with a 150-hertz signal which causes the pointer to point upward. The overlapping portion of the beam slopes upward at three degrees, corresponding to the normal angle of descent of an aircraft approaching the runway.

Along-course guidance is provided by *fan markers*. These transmit signals in the form of dots and dashes. An outer marker is situated about five miles out from the touchdown point, where the aircraft would be about 1400 feet up. The middle marker is about 3500 feet out, where the aircraft should be at about 200 feet. The inner marker is close to the end of the runway, where the aircraft comes "over the hedge," and begins to hold off or flare out. These markers are all at 75 megahertz, and as their name implies, they give a fan-shaped radiation pattern at right angles to the approach path, so the aircraft receives the transmission for only a brief period. The pilot hears the marker in his headset and is also notified by means of a flashing light. The information enables him to check his actual altitude and position against the glide-slope indication.

LORAN (Long Range Navigation) is what is called a *hyperbolic system*. Its principle is easy to understand by the "two-gun" analogy. If an observer is positioned at an equal distance from two guns that are fired simultaneously he will hear only one report. If he is closer to one gun he will hear two reports, the nearer gun first. Since sound travels at 1100 feet (335 meters) per second, an interval of one second between the reports would tell the observer he was 1100 feet closer to the nearer gun than to the farther one.

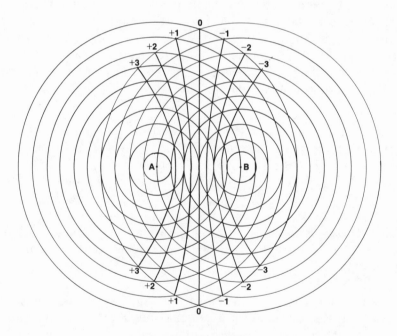

Figure N-2a
Why Constant-Time-Difference Hyperbolas Are Used in Loran

If the positions of the guns are plotted on a chart and circles drawn around each with radii of multiples of 1100 feet, the circles will intersect, and curves may be drawn through the intersection points where the intervals between hearing the guns are the same. As shown in Figure N-2a, these curves are hyperbolas. At any point on a hyperbola the interval will be the same, regardless of the distance or direction of the guns.

In the Loran system the two guns are replaced by two radio stations 300 miles apart, called a master and a slave (see Figure N-2b). A chart is used that has a family of hyperbolas for these stations. On the same chart a second family of hyperbolas for another pair of stations is printed also, so by finding where the hyperbolas of constant time difference for each pair intersect, the aircraft's (or ship's) position may be determined. Loran stations used a frequency of 100 kilohertz.

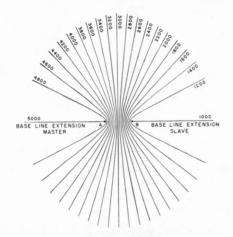

Figure N-2b
System of Lines-of-Position from Actual Loran Station Pair

Other hyperbolic systems are Omega, Decca, and Consol. Omega is very low frequency, Decca is low frequency, and Consol is in the 300-kilohertz region. Omega is a worldwide system which complements the U.S. Loran system, while Decca is widely used by coastal shipping in Europe. Consol was invented by Germany during World War II, when it was called Sonne. It is still in limited use.

Airborne radar provides a new technique for aircraft equipped with a navigation computer. The geographical coordinates of a check point are previously entered in the computer, which then proceeds to calculate the range and bearing of the check point at some future convenient time from information acquired of the aircraft's heading, speed, etc. At the chosen time the aircraft's radar obtains the actual range and bearing of the check point. This information is then compared by the computer with the calculated data, and the data are corrected as necessary.

TACAN (Tactical Air Navigation) consists of a ground radio beacon operating in the 960 through 1215-megahertz band in conjunction with airborne equipment providing a visual presentation of the bearing and distance of the beacon from the aircraft.

VOR (VHF Omnidirectional Range) requries two ground transmitters. One sends out a continuous wave in all directions in the frequency band between 108 and 118 megahertz. The other's antenna pattern rotates at 30 times a second, sending out a narrow beam such that, when on north, the beam coincides with the steady wave, when on east it diverges by 90 degrees, when on south by 180 degrees, and so on. A receiver in the aircraft measures the divergence or phase difference as seen by it, and displays it on a radio magnetic indicator (RMI) as the bearing of the beacon.

NC—Abbreviation for no connection, or normally closed.

n-channel field-effect transistor—Field-effect transistor in which the source, drain, and the conductive channel between them are of n-type conductivity material. *Transistors.*

Nd—Symbol for neodymium (q.v.).

N-display—Radar display that is a combination of the K-display and the M-display (qq.v.). *Radar.*

Ne—Symbol for neon (q.v.).

near field—(See Fresnel region.)

near infrared—End of infrared band nearest to visible spectrum with wavelengths between 750 and 1500 nanometers. (See electromagnetic spectrum.)

NEC—Abbreviation for National Electric Code (q.v.).

NEMA—Abbreviation for National Electrical Manufacturers Association (q.v.).

nematic liquid crystal—Liquid crystal in which the molecules are oriented with their long axes parallel. Under the influence of an electric field they can be realigned, switching the material from clear to opaque, a property used in liquid crystal displays (q.v.).

neodymium—Silvery-white rare-earth metal used to color glass purple, for use instead of ruby in solid-state lasers (q.v.). Also used in an alloy called misch metal, from which cigarette-lighter flints are made. Symbol, Nd; atomic weight, 144.24; atomic number, 60.

neon—Noble (inert) gas used in electric signs and fluorescent lamps. Colorless, odorless, tasteless and lighter than air, it emits the familiar orange-red light when, under low pressure, an electric current is passed through it. Symbol, Ne; atomic weight, 20.183; atomic number, 10.

neon bulb—Small glass envelope containing neon gas at low pressure. When a voltage of 90 V d.c. or 65 V r.m.s. is applied between the two electrodes, the gas ionizes and conducts. A voltage of 65 V d.c. or 46 V r.m.s. is then required to maintain conduction. This is usually obtained by using a suitable resistance in series with the bulb, since it is generally used as an indicator for 115 V a.c. The bulb glows with a bright orange-red. (See also numerical-readout tube.)

neper—Alternative spelling of Napier. (See Napierian logarithms.)

network analysis—Evaluation of the overall performance of a circuit from the individual performances of its components.

neutralizing capacitor—Capacitor used to prevent oscillation by neutralizing the plate-grid capacitance of a triode used in a transmitter power amplifier.

neutrino—Subatomic particle with no charge and no mass, emitted accompanying the radioactive emission of beta decay. *Nuclear Physics.*

neutron—Constituent particle of every atomic nucleus, except ordinary hydrogen, that has no electric charge and a mass slightly less than that of a hydrogen atom. *Nuclear Physics.*

newton (N)—S.I. unit of force, being the force that, when applied to a body having a mass of one kilogram, causes an acceleration of one meter per second per second in the direction of application of the force. *S.I. Units.*

nF—Symbol for nanofarad (q.v.).

Ni—Symbol for nickel (q.v.).

nichrome—Nickel-chromium alloy used in making thin-film resistive components, wire-wound resistors, and heating elements. *Microelectronics, Resistors.*

nickel—Silvery-white tough metal, harder than iron, used for electrodes in electron tubes, alloyed with iron in magnetic cores, and in Edison storage cells (q.v.). Symbol, Ni; atomic weight, 58.71; atomic number, 28.

nickel-cadmium cell—Rechargeable alkaline cell (q.v.) in which the positive electrode is nickel and nickel oxide, the negative electrode is cadmium, and the electrolyte is potassium hydroxide. The voltage of a nickel-cadmium cell is 1.25 V and the recharging rate is 45 mA for 14 hours (slow), or 150 mA for 4 hours (fast).

nickel silver—(See German silver.)

nixie tube—(See numerical-readout tube.)

NMOS—MOS technology using electron majority carriers. *Microelectronics.*

noble gases—Family of six elements making up Group 0 in the periodic table of elements (q.v.): helium, neon, argon, krypton, xenon, and radon; called "noble" because they are extremely unreactive.

noble metals—Metals with high resistance to oxidation: rhenium, ruthenium, rhodium, palladium, silver, osmium, iridium, platinum, and gold.

node—1. Junction point of a network, such as the input.   2. Points of minimum voltage in standing waves. *Transmission Lines.*

noise—Random unpredictable and undesirable signals, or changes in signals, that mask desired information content. Noise in radio transmission is called static, or atmospherics, and in television it is called snow, or confetti, from its appearance on the picture screen. *Radio Noise and Interference.*

nondestructive readout—Memories that retain their stored data regardless of how many times it is read are said to have nondestructive readout (NDRO).

NOR gate—Logic circuit type that gives a true (1) output only when all inputs are false (0). *Computer Hardware, Microelectronics.*

normal distribution—Found wherever a large number of independent random causes produces additive effects, the graph of which is a typical bell-shaped curve. If the mean, or average, is designated $x_0 - \delta$ to $x_0 + \delta$, where δ is the standard deviation, x_0 is also the highest point of the curve. Also called Gaussian distribution. See also standard deviation.

north pole—In a magnet, that part from which magnetic lines of force emanate. In a compass needle, the north pole is the end which points toward the Earth's magnetic north pole, so is sometimes called the north-seeking pole. Another term used is positive (+) pole. Since unlike poles attract, the Earth's magnetic north pole is really a south pole.

notch filter—Band-reject filter. *Active Filters, Passive Filters.*

NOT gate—An inverter. Its output is always the opposite of its input. *Computer Hardware, Microelectronics.*

ns—Symbol for nanosecond (q.v.).

NTSC—Abbreviation for National Television System Committee (q.v.).

n-type—Semiconductor material in which electrons are the majority carriers.

nuclear physics—Scientific discipline concerned with structure of the atomic nucleus. *Nuclear Physics.*

NUCLEAR PHYSICS

According to present knowledge, an atom—defined as the smallest unit of a chemical element that retains its elemental identity—consists of a tiny but massive central core, called the *nucleus*, surrounded by a number of electrons. As far as mass is concerned, the atom is 99.95 percent nucleus.

The nucleus consists mainly of two kinds of *baryons: protons*, which bear a positive charge; and *neutrons*, which have no electrical charge.

Understanding what goes on in an atomic nucleus is far from complete, and research continues at a high level of intensity. At present *four classes* of subatomic particles are recognized, acted upon by *four forces*. In the following table the classes of particles are listed in the left-hand column. The forces shown heading the other five columns go from left to right in the order of their strength from weakest to strongest. The x's indicate the forces acting upon each class of particle. The terms in parentheses are the *field quanta* involved in the interaction.

Classes of Particles	Interacting Forces				
	Gravitational (graviton)	Weak (boson)	Electromagnetic (photon)	Strong (gluon)	(pion)
Hadrons	x				x
Quarks	x			x	
Electrons	x		x		
Leptons	x	x			

Hadrons are further divided into two classes: the heavy particles called baryons which, as mentioned above, are protons and neutrons (or other hadrons that can transform into protons or neutrons); and the not-so-heavy particles called *mesons*, which are unstable, decaying into electrons, neutrinos, and photons.

All hadrons are assumed to be composed of combinations of *quarks*. There are four kinds of quark, designated u, d, s, and c. (The c quark is also called a "charmed quark.") These basic types are further subdivided by a quantum property called "color." According to the current theory of *quantum chromodynamics* each baryon is composed of a unique combination of three quarks or antiquarks (quarks with opposite electrical charge), and each meson is made up of a quark-antiquark pair. These quarks must have color-quantum properties that cancel out. For instance, a proton's three quarks are u, u, and d. They must have the properties designated "red," "blue," and "green," which together make up "white" (no color), so it also follows that the two u quarks cannot have the same colors. Similarly, a neutron's quarks (udd) must have colors that together make up white, so the two d quarks cannot be identical also. Conversely, a meson's quark and antiquark must be of the same color, say red, so that the red-antired cancel out. A meson with a very high mass may consist of a charmed quark-charmed antiquark pair ($\overline{c}c$). This is a short-lived "atom" of an imaginary element called "charmonium", by analogy with "positronium," another "atom" composed of an electron-antielectron (positron) pair.

Leptons are the light particles. They include the ordinary negative electron and heavy negative electron, or muon, together with their antiparticles, the positron and antimuon. They also include the electrically neutral and massless neutrino and antineutrino. The ordinary negative electrons orbiting atomic nuclei are bound by the electromagnetic force existing between them and the protons in the nucleus, but are not affected by the weak nuclear force, so they are listed separately in the table.

The forces interacting with the particles are conceived as minute packages of energy called *quanta*. In some ways they behave as if they also were particles. These quanta are being exchanged continuously between the particles. The aggregate of all the exchanges constitutes a force field.

The length of time during which a quantum can exist separately depends upon its mass. A quantum with a large mass has taken a large amount of energy from the particle that emitted it, and must be reabsorbed either by that particle or another much sooner than if its mass and borrowed energy were smaller. For this reason a quantum with a larger mass has a lesser range than one with a smaller mass. The mass (or rest energy) of a boson is in excess of 50 gigaelectron-volts (GeV), while that of a pion is only 0.14 GeV, so the boson's range is much shorter (10^{-15} centimeter) than that of the pion (10^{-13} cm). This explains why the boson's force field is termed weak and the pion's field strong.

On the other hand, the graviton, photon, and gluon do not have any mass. Their range is therefore unlimited, and they travel at the speed of light. However, the influence of gravitons and photons declines as the square of the distance between the two interacting particles, whereas the gluon's influence remains constant regardless of distance. This accounts for the gluon being ranked with the pion in strength of interaction.

In its normal condition an atom is electrically neutral. The number of protons (+) is exactly equal to the number of electrons (−). The number of protons is called the *atomic number* (Z).

In light elements the number of neutrons is approximately equal to the number of protons, but for atomic numbers above 20 the number of neutrons averages 1½ times the number of protons.

The mass of a proton is almost equal to that of a neutron. These particles cling together in the nucleus because of the strong interaction. This force overcomes the electrical repulsion between individual protons due to their like charges. The balance between these forces keeps the particles evenly spaced in orbits about a common center in the nucleus. These orbits have various energies and are arranged in shells according to the rules of quantum mechanics. The radius of a nucleus is approximately one ten-thousandth of the radius of an atom, which is typically about 10^{-8} centimeter.

The masses of nuclei differ from one another very nearly by multiples of the *nuclear mass unit*, which is very slightly less than the mass of a proton. Each kind of nucleus is therefore assigned a *mass number* (A). For instance, oxygen with eight protons and eight neutrons, 16 baryons altogether, has a mass number of 16. A nucleus with charge Z (the atomic number) and mass A is made up of Z protons, and A–Z = N neutrons.

There are nuclei with all values of Z (except 43 and 61) from 1 to 105. Those from 1 to 92 occur naturally; the others have been created artificially. For each value of Z there may be several values of A. Nuclei with the same Z but different A are called *isotopes* of each other. Most naturally occurring chemical elements are a mixture of isotopes in which one is very much more abundant than the others. For example, there are about 90 atoms of carbon with A of 12 for every one with A of 13, so that the chemical atomic weight of the mixture is 12.011, instead of 12.000, the atomic weight of pure carbon-12.

The value of the mass unit has been arbitrarily adjusted to make it exactly 1/12 of the mass of the carbon-12 atom. A proton has a mass of 1.00728 mass units. A neutron therefore "weighs" 0.00139 mass unit more than a proton. Because of this extra mass the neutron is unstable. Alone in free space it has a maximum lifetime of approximately 17 minutes before it disintegrates into a proton, an electron, and a neutrino. This process is called *beta decay*, and when neutrons decay in large numbers the stream of electrons is termed *beta particles* or *beta rays*.

Whether or not this happens to a neutron depends upon its environment. If it is bound tightly in a stable nucleus it can last forever. Furthermore, it is far more likely to encounter another atom before it disintegrates and will react with it to increase its mass. If the atom it enters is a heavy atom with an unstable nucleus, the penetration of a neutron may be sufficient to disrupt the latter to produce nuclear fission.

Beta decay is one of the three kinds of radioactive decay of heavier nuclei, the others being termed *alpha* and *gamma*. These heavy nuclei are unstable and spontaneously dissipate excess energy by emitting a particle or a photon. In alpha decay a particle consisting of two protons and two neutrons is ejected at high speed (about 1/10 the speed of light) for a distance of from one to four inches. The actual energy varies according to the parent nucleus. The composition of the alpha particle is that of a helium-atom nucleus, so as soon as it has captured the two electrons required it becomes a helium atom.

Gamma decay takes place by the emission of penetrating electromagnetic radiation of the same nature as X-rays but of higher frequency. Photons, when emitted in radioactive decay, are gamma rays. X-rays are emitted by the acceleration or deceleration of charged particles. Gamma-ray photons eventually lose energy by being scattered by collisions with free electrons, being absorbed by other atoms or by the creation of an electron-positron pair in the field of the nucleus. (A positron is a positive electron. It ultimately reacts with a free electron to emit two more photons of lower energy.)

These particles, and also electrons, are used in the laboratory to study nuclear structure and reactions. Accelerators are used to give the particles energies from 100 kiloelectron-volts to over 100 gigaelectron-volts. With high enough energy, some of it can even be converted into matter in the form of other particles with extremely short lives.

The electron-volt is the unit for the kinetic energy acquired by a particle bearing one unit of electric charge (1.6×10^{-19} coulomb) when it is accelerated through a potential difference of one volt. The usual nuclear unit of energy is a million electron-volts, or one megaelectron-volt (MeV), which is the energy involved in transporting an electron through a potential difference of one million volts. The energy imparted by an accelerator is usually given in these units.

Constant-voltage accelerators. The simplest type of accelerator works on the same principle as the television picture tube. The potential difference between the ion source and the target accelerates the ions in a vacuum. Suitable electrodes may be used to form the ions into a beam. The energy gained by the particle is equal to eV, where e is the charge on the particle and V is the accelerating potential. Although such forward-action accelerators comprise more than half of all accelerators in use, they are limited by the high-voltage power supplies that can be provided. Although higher potentials can be obtained by voltage multiplication, corona discharge in the ambient air becomes a problem. By enclosing the generator in a high-pressure container, beams of ions with energies to 4 MeV and power to 20 kW have been built.

Another type of power generator is the *Van de Graaff* or electrostatic generator. As shown in Figure N-3, this generator has a traveling belt which conveys charged particles to the high-voltage electrode, where they accumulate. This electrode is in the form of a hollow sphere, so the charge builds up on the outside. The electrical field inside is negligible, so it does not exert any back pressure against the stream of charges arriving on the conveyor belt. Contemporary electrostatic generators produce voltages that can accelerate ions to energies of 20 MeV. In another version, where two stages of acceleration are used in tandem, heavy ions can reach energies greater than 100 MeV.

Betatrons are used for accelerating electrons ("beta particles"). An electromagnetic coil wound on a laminated steel core when excited by an alternating current produces an annular magnetic field in a circular cavity between the poles of the magnet. An electron beam injected into this doughnut-shaped vacuum chamber can be accelerated to energies as high as 300 MeV in a large betatron. Most, however, used in medical and radiographic applications, accelerate the electrons to 15 to 25 MeV, during which the particles make up to a quarter of a million revolutions, gaining about 100 eV per revolution. (See Figure N-4.)

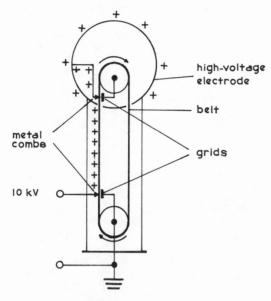

Figure N-3
Van de Graaff Accelerator

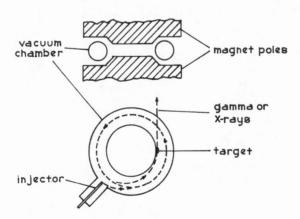

Figure N-4
Betatron: (top) cross-section; (bottom) plan view

Cyclotrons also accelerate particles in circular orbits. As shown in Figure N-5, a circular vaccum chamber is located between the poles of a magnet, so that it is permeated by a unipolar magnetic field. In this chamber there are two hollow metal electrodes called dees, which are connected to a radio-frequency oscillator. At the center is an ion source from which ions are accelerated in a spiral path within the dees by the combined effect of the electric and magnetic fields, until they acquire energies up to 25 MeV. This may be increased in two ways. Either the oscillator frequency may be modulated to track the circulation frequency of the accelerated particles (synchrocyclotron, or phasotron), or the magnetic field strength made to increase from the center to the outer perimeter of the vacuum chamber (sector-focussing, or isochronous cyclotron).

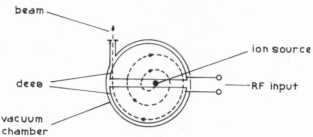

Figure N-5
Cyclotron (plan view). The magnetic field is vertical to the page.

The Berkeley synchrocyclotron attains acceleration energy of 730 MeV, while the Oak Ridge sector-focussing cyclotron (called the Mc² cyclotron) achieves 900 MeV. Others are planned that will exceed 1 GeV.

Linear resonance accelerators are long waveguides. High-power radio-frequency energy introduced into the waveguide produces a linear electical field with standing waves, which are able to alternately accelerate and decelerate the particles as they travel along the tube. Since only acceleration is wanted, metal drift tubes are located along the waveguide to shield the particles from deceleration at the points where it would occur. In a standing-wave accelerator, protons can be given energies up to 68 MeV. (See Figure N-6.)

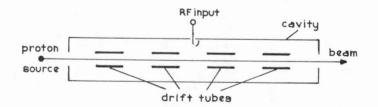

Figure N-6
Standing-wave linear proton accelerator, consisting of a long circular wave guide with drift tubes, frequently used for injection of protons into a synchroton. (May be as much as 1500 meters long for energy up to 1 gigaelectronvolt.)

Another type of linear accelerator is used to accelerate electrons. This is called a traveling-wave linear accelerator. A traveling electromagnetic wave is propagated through a waveguide. This waveguide has metallic diaphragms or irises mounted internally (see Figure N-7) at quarter-wave intervals, to reduce the phase velocity of the waves to equal the velocity of the electrons, which then "ride the waves" in bunches. The largest linear accelerator of this type was built at Stanford in 1966, with a length of 10,000 feet (3048 m) and a diameter of 4 inches (~ 10 cm). The radio frequency is 2.86 gigahertz, or a wavelength of 10.5 cm, in the form of microsecond pulses at a repetition rate of 136 Hz. The accelerator can accelerate electrons to 21 GeV.

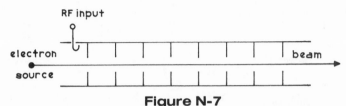

Figure N-7

Traveling-wave linear electron accelerator, consisting of a long circular waveguide with metal irises to slow down the RF waves to the proper velocity.

Synchrotrons are the ultimate in accelerators at present. There are two types: electron and proton. The electron type shown in Figure N-8 consists of a circular evacuated channel of oval cross-section. Typical dimensions range from 5.5 × 2.5 cm to 15 × 4 cm. The radius of the circular channel, or orbit, will be typically from 25 to 125 m. Electromagnets are placed around the orbit, which passes between their pole pieces, and resonators are placed in the gaps between the magnets. A linear accelerator forms an electron beam of about 10 MeV traveling at close to the speed of light, which is injected into the synchrotron orbital channel. The magnets center the electrons in the channel, and the resonators accelerate them further until they are as close to the speed of light as possible without a relativistic increase in mass. The maximum energy obtainable is limited by the generation of electromagnetic radiation by the circular orbit. The electron synchrotron at Ithaca, N.Y. develops 10 GeV, which is about the practical limit. At the end of the acceleration cycle (hundreds of thousands of revolutions in a microsecond or two), the beam of electrons is deflected out of the synchrotron, or directed against a target placed inside.

The proton synchrotron differs from the electron synchrotron in that the protons are injected at a lower velocity (0.3 to 0.6 times the velocity of light). The device is usually constructed like a racetrack rather than in a circle, and the accelerating, focussing, and beam-ejecting structures are mounted on the straight sections. Because of the initial lower velocity of the protons, the magnetic field is made to increase in step with their acceleration until they reach 0.98 times the velocity of light. The frequency of the accelerating voltage increases in a similar way to match the increase in velocity. The proton synchrotron at Serpukhov, built in 1967, with a radius of 236 m, develops 76 GeV. A proton accelerator at Batavia, Illinois, with an ultimate energy of 500 GeV, has a radius of one kilometer. Progress on further improvements in traditional methods, and discoveries of new principles, are expected to result in energies in the teraelectron-volt range.

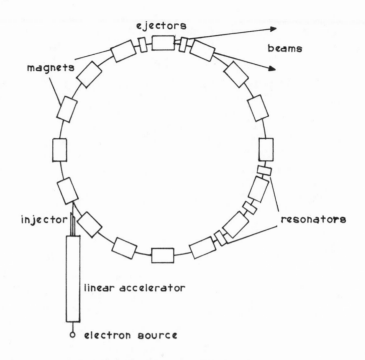

Figure N-8
Electron Synchrotron (plan view)

Radiation detection is performed by observing the interaction of accelerated particles with matter. Charged particles as they pass through a medium excite or ionize atoms in the medium. Uncharged radiation produces secondary charged particles, which make possible detection and analysis of the primaries.

Electrons not only lose energy by ionization but also by *bremsstrahlung,* which is the conversion of the particle's energy to radiation as it loses velocity in the vicinity of the strong electric fields of atomic nuclei. For heavier charged particles ionization is the most important mechanism.

Gamma rays, being photons, produce any of three effects. Electrons or ions may be liberated by the photoelectric effect. The photon may collide with a free electron and undergo elastic scattering, called Compton scattering. Or the photon may create an electron-positron pair in the field of the nucleus; the positron ultimately meets with a free electron, resulting in mutual annihilation and the emission of a pair of photons.

Energetic particles are detected by various devices. An *ion chamber* is a tube containing a cathode, grid, and collector (anode) in a gas atmosphere. These electrodes are charged and then isolated from the charging source. Exposure to radiation causes leakage of the charges and a consequent change in voltage. A *geiger counter* (Figure N-9) consists of a thin-walled metal cylinder containing argon and ethyl alcohol. Particles enter a window in the metal cylinder and temporarily ionize the gas, causing a brief pulse discharge. This is amplified and gives a visible and audible indication. A *scintillation counter* is a photomultiplier tube (q.v.) containing

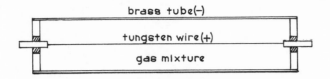

Figure N-9
Heart of a geiger counter is a thin-walled brass tube filled with a mixture of gases. The tungsten wire in the middle of the tube is positively charged; the tube is negatively charged. Particles passing through the gas ionize it, and this process continues until an avalanche of negative ions (electrons) falls to the positively charged wire and, similarly, an avalanche of positively charged ions falls to the negatively charged tube. This results in an electrical pulse, which is counted electronically by a counting circuit, and the result displayed. The amplified pulses may also be applied to a speaker for audible indication.

a scintillator. This is a chamber in which a substance such as sodium iodide, anthracene, and the like emits a small flash of light when a gamma ray passes through it and causes a photoelectric, Compton, or pair-production event. This is detected by the photocathode and amplified by the photomultiplier. A *Cerenkov counter* uses a transparent dielectric material in which a fast electron or proton emits visible light in a narrow cone. This is also detected and amplified by a photomultiplier. A *semiconductor detector*, consisting of an intrinsic region between p-n regions, works in the same way as an ion chamber, but is much more sensitive. A *bubble chamber* contains a superheated liquid which boils wherever an ionized particle passes, leaving an evanescent trail of bubbles. Stereo photographs are taken of the bubble trails and studied afterward to determine what caused the tracks. A *spark chamber* consists of two sets of interleaved, thin metal plates, parallel to each other and separated by small gaps, enclosed in a container filled with an inert gas such as neon, as shown in Figure N-10. A very high voltage pulse is applied between the two sets of plates whenever a charged particle enters the chamber, resulting in a trail of sparks along its path where gas molecules were ionized. This can be photographed and analyzed in the same way as in the bubble chamber.

nucleus—Tiny, positively charged, dense core at the center of an atom, comprising 99.95 percent of its mass. The nucleus consists of elementary particles (nucleons) called neutrons and protons bound together by nuclear force. *Nuclear Physics.*

null—Condition of balance in a bridge circuit in which no current flows through the indicating device connected between the two sides ("known" and "unknown").

null detector—Sensitive galvanometer (q.v.) used with a bridge to indicate when a condition of balance (null) has been obtained.

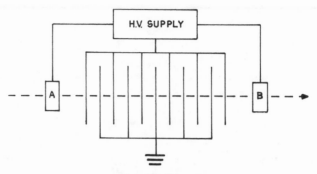

Figure N-10

In a spark chamber, when an ionizing particle travels along the path indicated by the dashed line, the radiation detectors at A and B trigger a very high voltage pulse from the power supply. The gas between the plates that was ionzed by the particle causes sparks to jump between the charged plates and the grounded plates. A photograph of the trail of sparks reveals the trajectory of the particle.

number—One of the positive integers. Rational numbers can be expressed as integers or the quotient of integers; irrational numbers are those that cannot be so expressed.

numeric—Consisting only of numerals.

numerical readout tube—Gas-discharge diode with a common anode and ten cathodes made of fine wire with the shapes of the numerals 0 through 9. Whichever is selected glows brightly, the others remaining invisible. (See Figure N-11.)

Figure N-11
Numerical Readout Tube

nV—Symbol for nanovolt (q.v.)

nW—Symbol for nanowatt (q.v.).

O—Symbol for oxygen (q.v.).

O-attenuator—Balanced π attenuator. Also called an O-pad. *Attenuators.*

octave—Interval between two frequencies, the higher of which is twice that of the lower, e.g., the band of frequencies from 1 kilohertz to 2 kilohertz is an octave. In Western music the octave encompasses eight notes of the scale, so that A (220 Hz) to A′ (440 Hz) is an octave.

oersted—Unit of magnetic field strength in the Centimeter-Gram-Second System. Defined as the intensity of a magnetic field in a vacuum in which a unit magnetic pole experiences a mechanical force of one dyne in the direction of the field. In S.I. units magnetic field strength is measured in amperes per meter (A/m). To convert from oersteds to amperes/meter, multiply by $7.957\ 747\ 2 \times 10^1$.

ohm (Ω)—Unit of electrical resistance, defined as the resistance between two points on a conductor which does not contain any source of electromotive force when a constant potential difference of one volt maintained between those points results in a current of one ampere in the conductor. *S.I. Units.*

ohmic contact—Junction between two materials in which resistance is proportional to the current, as in Ohm's law $R = \frac{E}{I}$, as opposed to semiconductor junctions where other properties are involved. *Semiconductors.*

ohmmeter—Electrical instrument for measuring resistance. Most ohmmeters consist of a permanent-magnet moving-coil movement (q.v.) and a current source. The movement responds to the current passing through it and the resistance in series, so the greater the resistance the lower the current. Consequently the values in ohms on the meter scale are highest at the left-hand end. In a shunt type ohmmeter the resistance to be measured is connected in parallel with the meter, so that more current flows through the meter for higher resistance values. Its scale, therefore, has the zero reading at the left-hand end.

Ohm's laws—For a d.c. circuit, E/I = R; for an a.c. circuit, E/I = Z. *Electrical Equations.*

ohms per volt—Rating of a voltmeter, the reciprocal of its full-scale current in amperes. The higher the ohms-per-volt rating, the lower the loading effect on any circuit to which the voltmeter is connected.

Omega—Hyperbolic navigation system using high-power transmitters with time-shared continuous-wave signals at around 10 kilohertz, and having baselines and service range of 5000 miles. *Navigation Aids.*

omega—Twenty-fourth and final letter of the Greek alphabet (Ω, ω), corresponding to the English long o. Ω is used as the symbol for ohm (q.v.), and ω is used as the symbol for angular velocity ($\omega = 2\pi f = 6.28 \times$ the frequency in hertz). See *Greek Alphabet.*

omnidirectional—All-directional, having no direction of emission or reception that is better or worse than another. Used of antennas, loudspeakers, microphones, etc.

omnirange—Omnidirectional range, a facility providing bearings equally well in all directions. VHF omnirange is called VOR; another popular term is "omni." *Navigation Aids.*

op amp—Abbrevation for operational amplifier (q.v.).

open—Said of a circuit, switch, etc., where there is an interruption in the current path.

open loop—Having no feedback path. *Operation Amplifiers.*

open-wire transmission line—Transmission line consisting of two parallel uninsulated wires supported a fixed distance apart by means of insulating rods called spacers. *Transmission Lines.*

operating angle—Electrical angle during which plate current flows in an amplifier, which varies for different classes. *Electron-Tube Circuits.*

operating code—First byte (8 bits) of an instruction, consisting of a pattern of 1's and 0's that serves as a code for the command to be performed. Also called op code. *Computer Hardware.*

operational amplifier—Dual-input differential amplifier, followed by one or more direct-coupled gain stages, characterized by very high gain and input impedance, very low output impedance, and very broad bandwidth. *Operational Amplifiers.*

OPERATIONAL AMPLIFIERS

An operational amplifier is a very high-gain, direct-coupled amplifier that uses feedback for control of its response characteristics. Originally used mainly in analog computers to perform various mathematical functions such as differentiation, integration, analog comparison and summation, it is now also used for numerous other applications, as diverse as wideband video amplifiers and active filters (q.v.). This diversity makes the op-amp the most versatile of linear integrated circuits available to modern designers.

Most op-amps consist of two direct-coupled cascaded differential amplifiers together with an appropriate output stage, as shown in Figure M-28. This is a stable configuration which lends itself well to the monolithic diffusion process used in microelectronics (q.v.) fabrication, and also provides high common-mode rejection at least 20 dB greater than the differential gain. There are two input terminals, the inverting input (II or $-$) and the noninverting input (NII or $+$). The input signal and feedback are usually (but not invariably) connected to the inverting input, so that the output signal is of opposite phase to the input signal and the feedback is negative.

This basic configuration is shown in Figure O-1. The load resistor R_L is of a high-enough value so that its effect on the transfer characteristic can be neglected (in other words, $I_o \simeq 0$). In this case, the gain is given by $\dfrac{R_f}{(R_1 + R_s)}$. To avoid any d.c. offset, d.c. paths for each input must be equal, so R2 is made equal to $(R_i + R_s)$.

While a simple resistive feedback circuit has negligible phase shift at low frequencies, this is not the case at higher frequencies, so manufacturers generally provide phase-compensation terminals to which may be connected external components that modify the performance of the amplifier. This is called internal phase compensation. In one method, called straight roll-off, a suitable RC network is connected across an internal resistor. In another, called Miller-effect roll-off, the phase-compensating network is connected between the input and output of an inverting-gain stage.

Several more op-amp circuits are given in Figure O-2. These are simplified by the omission of R_s, which is assumed to be included in R_i. (R_L is also not shown, but its value must always be such that I_o is virtually zero.) Amplifier performance may be calculated from the formulas accompanying each amplifier circuit.

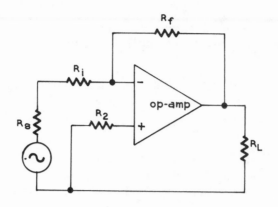

Figure O-1

Basic Op-Amp External Circuit

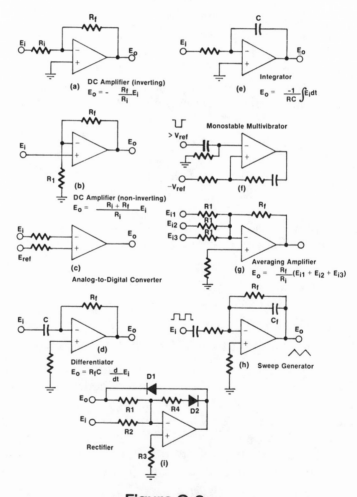

Figure O-2

Various Op-Amp circuits

optical communication—Communications system consisting of an optical source, a means of modulating the source, and an optical receiver. *Optical Communications.*

OPTICAL COMMUNICATIONS

Optical communication is made with modulated signals whose carrier frequencies are in the optical range instead of in the radio range. At present, wavelengths in the visible and near infrared portions of the spectrum are being used. Transmission may be through the atmosphere or via optical-fiber waveguides. The advantages and disadvantages of each system are shown in Table I.

TABLE I

Comparison of Atmospheric and Optical-fiber Waveguide Systems

System	Advantages	Disadvantages
Atmospheric	No license needed	Limited to line of sight
	No radio-frequency interference	Needs careful alignment
	Fairly private	Affected by atmospheric interference
	Fairly high data rate	Affected by external lighting, etc.
Waveguide	No interference of any kind	Special handling, splicing, and termination required
	Compact	
	Completely private	
	Very high data rate	
	Eventually will be cost effective	

An optical communications system requires a transmitter or optical source, a means of modulation, and a receiver or an optical detector, in addition to the transmission medium.

Optical sources comprise light-emitting diodes and lasers. There are two types of LED: the gallium arsenide type with emission in the near infrared (750-950 nm), and the super-luminescent diode (SLD). The structure of the LED (see *Semiconductors*) is modified so as to project the light in a parallel-ray beam, or into a glass fiber according to which type of transmission medium is being used.

The SLD is a modified stripe-geometry injection laser (see Figure O-3). It differs from a normal solid-state laser in that the top electrode is incomplete, so as to inhibit laser action, but it retains the laser characteristic of stimulated emission. It has a higher output with a narrower bandwidth than a conventional LED.

Lasers (see *Lasers*) are the most powerful optical sources. For communications use, the solid-state Nd-doped YAG can emit up to 10^3 watts CW, and is more efficient than its two nearest rivals, the HeNe and Ar gas lasers. For comparison, the wavelength and output power of certain lasers that may be used for communications are given in Table II.

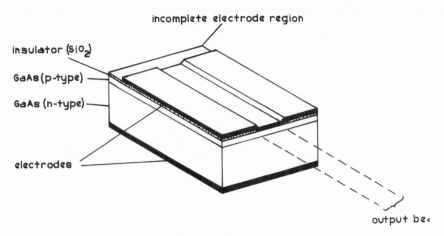

insulator (SiO₂)

GaAs (p-type)

GaAs (n-type)

electrodes

incomplete electrode region

output bee

Figure O-3
Superluminescent Diode

TABLE II
Laser Optical Sources

Laser	Wavelength	Output Power
Nd³: YAG (solid state)	1 065 nm	$\leqslant 10^3$ W
HeNe (gas)	633 nm	0.001 - 0.1 W
Ar (gas)	450 - 530 nm	0.02 - 100 W
Co₂ (gas)	10 600 nm *	1.0 - 100 W

*Only special optical materials and detectors can be used at this wavelength.

Modulation may be direct or external. LED's and SLD's are direct-modulated by varying the forward bias. These diodes have a very linear response, so they are suitable for analog signals. The modulation bandwidth may exceed 100 MHz. HeNe lasers are direct-modulated by varying the discharge current of the plasma tube.

External modulation can be done in three different ways. Each one consists of passing the laser beam through a transparent medium that is altered in accordance with the modulation signal applied to it. Electro-optic modulation uses lithium niobate ($LiNbO_3$) and lithium tantalate ($LiTaO_3$). When an electric field is applied to either of these, the index of refraction of the medium is altered, causing either phase, intensity or amplitude modulation, depending on the setup. Acousto-optic modulation is done with GaAs, $LiNbO_3$, or As_2Se_3, a type of glass. The modulating signal creates a phase grating by propagating an acoustic wave in these materials, which are called photoelastic. The phase grating diffracts the beam to a greater or lesser extent, resulting in intensity modulation. Magneto-optic modulation may be performed by passing the beam through certain magnetic iron garnets. Application of the modulating signal to the medium causes Faraday rotation, or rotation of the plane of polarization, so if the beam has been prepolarized, and afterwards is passed through a polarizing filter, more or less of it will be absorbed according to the degree of rotation.

Optical detectors can be photomultiplier tubes (q.v.) or solid-state devices, such as phototransistors, photodiodes, solar cells, and so on (qq.v.). The most important solid-state detectors are PIN photodiodes and avalanche photodiodes. Table III compares the characteristics of a photomultiplier tube and a PIN photodiode. The tube is much more sensitive, but requires a higher operating voltage.

TABLE III
**Comparison of Photomultiplier Tube
with Solid-State Photodiode**

Characteristic	Photomultiplier Tube	Photodiode
Bandwidth	200-1000 nm	200-1100 nm
Sensitivity	0.003-0.1 A/W	0.3-0.6 A/W
Minimum detectable light level	1 photon	5×10^{-15} W
Response time	5 ns	5 ns
Operating voltage	600-3000 V	3-90 V

Transmission media, as stated above, are the atmosphere and optical-fiber waveguides. The atmosphere can be expected to provide reasonable service up to two kilometers (1.25 miles), but cannot be relied upon beyond this, especially in areas with frequent precipitation

However, many of the limitations imposed by the atmosphere can be overcome by using optical-fiber waveguides. These are made of ultrapure fused silica, glass, or transparent plastic. The loss per kilometer of each type is given in Table IV. Contaminants are the cause of the high losses in the glass and plastic types, but they are much cheaper to manufacture.

TABLE IV
**Loss Per Kilometer of Three Types
of Optical Fiber**

Material	Loss in dB/km
Ultrapure fused silica	2 (1050 nm)
Multicomponent glasses	800
Plastic (e.g., Dupont PFX)	500 (580-680 nm)

An optical-fiber waveguide core may be as fine as 0.1 mm in diameter and has a cladding, or outer covering, as shown in Figure O-4. However, an optical ray will also propagate through an unclad fiber. The purpose of the cladding, which must have a lower refractive index than the core, is to protect the optical surface of the core to maintain total internal reflection. Optical-fiber waveguides should not be confused with the optical fibers used to transmit light to otherwise inaccessible areas, or used in medical procedures to observe the human body internally. These are usually 0.005-0.1 mm in diameter and packed in bundles of several thousand each.

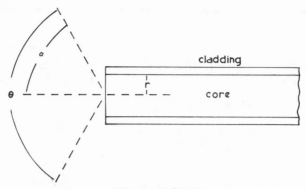

Figure O-4
Optical-Fiber Waveguide

The refractive index of the core is denoted n_1 and that of the cladding n_2. In the formula

$$V = \frac{2\pi r}{\lambda} \, (n_1^2 - n_2^2)^{1/2} \tag{1}$$

V — the normalized frequency,
r — the radius of the core,
λ — the wavelength of the optical rays.

The number of modes[*] that will be propagated is given by

$$N = \frac{V^2}{2} \tag{2}$$

and the maximum acceptance angle θ (see Figure O-4) is given by

$$\sin \quad \alpha = (n_1^2 - n_2^2)^{1/2} \tag{3}$$
$$\theta = 2\alpha$$

A problem that occurs in multimode transmission is called pulse broadening, a form of distortion that limits the maximum data rate. However, pulse broadening can also occur in single-mode operation if the frequency varies, since this changes the waveguide properties. Very accurate control of all parameters is necessary for high rates.

(See also articles on *Waveguides and Resonators, Lasers, Semiconductors, Optoelectronics.*)

optical-fiber waveguide—Fiber of glass or plastic used as a transmission line for signals with wavelengths between 950 and 550 nanometers (visible and near infrared radiation). *Optical Communications.*

---

[*]If V does not exceed 2.405, only a single mode will be propagated.

optoelectronics—That branch of electronics which covers optical radiation, the interaction of light with matter, radiometry, photometry, and the characteristics of various sources and sensors. *Optoelectronics.*

OPTOELECTRONICS

Optoelectronics is that portion of the science of electronics that deals with the interaction of optical waves with matter and includes such topics as optical radiation, radiometry, photometry, and devices that sense optical radiation or emit it.

Optical radiation, by convention, consists of ultraviolet, visible, and infrared radiation. Ultraviolet (UV) is subdivided into extreme (10-200 nm), far (200-300 nm), and near (300-370 nm). Visible light is that radiation which is perceived by the human eye (370-750 nm). Infrared (IR) is also subdivided into near (750-1500 nm), middle (1600-6000 nm), far (6100-40 000 nm), and far-far ($4.1 \times 10^4 - 10^6$ nm). It is clear that while the UV and visible light spectra are almost equal, the IR spectrum is some 1350 times as great as the other two put together.

Radiometry is the science of measuring radiant energy throughout the spectrum using the unit of power, the watt. *Radiant flux* is the rate of flow of radiant energy per second. *Radiant incidence* is density of radiant flux falling on a surface, in watts per square meter. Radiant power reflected from a surface is called *radiant exitance*. If a point source of radiation is located at the center of a hollow sphere, all parts of the inner surface of the sphere will be irradiated equally. An area of the surface equal to that of a square with sides the same length as the radius of the sphere, but actually circular, can be visualized as the base of a cone whose apex is at the point source. This cone is a solid angle called a steradian. The amount of radiation per steradian is called *radiant intensity*. If a surface on which radiation is falling is "viewed" from the side instead of straight down on it, it will appear narrower because of the angle of view. This apparently reduced area is called the projected area. Dividing the radiant intensity from this surface by the apparent or projected area gives the *radiant sterance*, expressed in watts per steradian per square meter.

Photometry is the science of measuring visible light in terms of the response of the human eye. Since the eye is not uniformly sensitive to all colors and is itself an extension of the forebrain, photometric measurements are affected by the suggestibility of the mind instead of being based on reality alone, as is the case with radiometric measurements. Photometric measurements are almost always based on the photopic luminosity curve, which is a curve showing the relative spectral luminous efficiency as a function of wavelength of a standard observer under normal conditions of illumination. Another curve shows the response at very low levels of illumination, when night vision takes over. This curve is called the scotopic curve.

The photopic curve is shown in Figure O-5. It represents the response of the photoreceptors in the eye, called cones, that are sensitive to color. Each cone is characterized by the presence of one of three different pigments, so that it absorbs light preferentially in the red, green, or blue part of the spectrum as the case may be, a

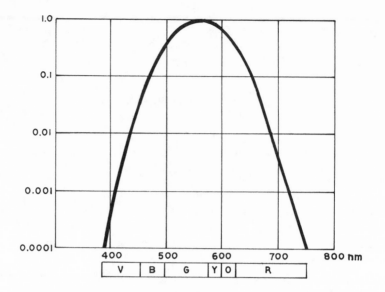

Figure O-5

Relative Sensitivity of Cones to Color (maximum is 1.0)

principle very like color television. The photopic response is greatest at 555 nm (green-yellow). These cones are concentrated in the center of the field of vision, a tiny area only 0.3 mm in diameter called the fovea (Latin for a small pit). This minute depression contains about 10,000 cones. Surrounding the fovea, the rest of the retina contains a high density $(1.5 \times 10^5/\text{mm}^2)$ of rods, which are photoreceptors that are not able to differentiate colors but are more sensitive to weak light than cones. They are most sensitive to light of 510 nm (green-blue).

Instead of the watt, the photometric unit is the lumen (673 lm = 1 watt). For comparison, radiometric and photometric units are compared in Table I.

TABLE I

Radiometric Term	Expressed in watts	Photometric Term	Expressed in lumens
Radiant flux	W	Luminous Flux	lm
Radiant incidence	W/m²	Luminous incidence	lm/m²
Radiant intensity	W/sr	Luminous intensity	cd*(lm/sr)
Radiant exitance	W/m²	Luminous exitance	lm/m²
Radiant sterance	W/sr/m²	Luminous sterance	lm/sr/m²

*candela (see *S.I. Units*)

Optical sensors fall into two categories: thermal detectors and quantum detectors. Thermal detectors respond to heat generated by radiation and include the bolometer, thermocouple, thermopile, thermopneumatic cell, and pyroelectric detector (qq.v.). Quantum detectors react to incident photons and include the photovoltaic cell, photoconductive cell, photoelectromagnetic cell, and photomultiplier tube (qq.v.).

Optical emitters, or sources of optical radiation, include tungsten filament lamps, fluorescent lamps, glow discharge lamps, light-emitting diodes, and lasers (qq.v.).

(See also articles on *Semiconductors, Transistors, Optical Communications, Electron Tubes, Lasers, S.I. Units.*)

ordinate—One of two dimensions used in fixing a point on a geometric graph.

OR gate—Logic gate in which a true (1) signal at any input gives a true (1) output. *Computer Hardware, Microelectronics.*

orthicon—Type of camera tube most widely used for live commercial television. Also called an image orthicon. *Electron Tubes.*

oscillation—Periodic to-and-fro motions, such as a swinging pendulum, a vibrating guitar string, sound waves in air, or alternating electric current.

oscillator—Device for producing alternating electric current, commonly employing tuned circuits and amplifier stages. Oscillators are of two main types: L-C oscillators, in which the frequency of oscillation is determined by an inductance-capacitance network, and R-C oscillators, in which the frequency is determined by a resistance-capacitance network. The best-known L-C oscillators employ the Hartley and Colpitts circuits; R-C oscillators are exemplified by various types of multivibrators (qq.v.).

oscillograph—Instrument for indicating and recording time-varying electrical quantities. A galvanometer causes a pen to trace an ink line on a moving paper chart, or reflects a beam of line from a small mirror on to a photographic film. Also called a recorder. (See Figure O-6.)

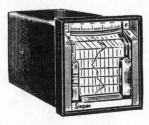

Figure O-6
Oscillograph

oscilloscope—A modern oscilloscope consists basically of the circuits shown in the block diagram of Figure O-7. In the usual mode of operation the signal to be observed is applied to the Y input and processed in the vertical amplifier. An attenuator in this section allows the operator to adjust the degee of amplification to give a display of convenient size on the screen of the cathode-ray tube (q.v.). The push-pull output of the vertical amplifier is applied to the upper and lower deflection plates in the CRT. In oscilloscopes used for higher frequencies, the signal also passes through a delay line which retards the signal slightly so that the trigger and time-base generators and horizontal amplifier can get their signal to the horizontal deflection plates at the same time.

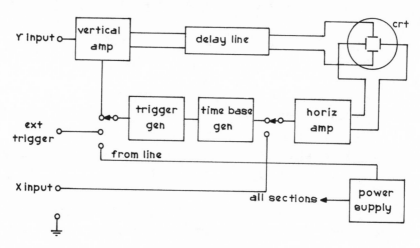

Figure O-7

Principle Circuits of an Oscilloscope

A small sample of the input signal is taken internally from the vertical amplifier, or externally directly from the circuit being examined, and applied to the trigger generator, where a trigger pulse is produced. This pulse is of the proper shape and amplitude for application to the time-base generator, regardless of the shape or amplitude of the sample. It causes the time-base generator to turn on so that a sawtooth-voltage ramp is started in synchronism with the signal. The slope and duration of this ramp are according to the selected RC combination, which is called the time base. After amplification in the horizontal amplifier, the push-pull output signal is applied to the left- and right-hand deflection plates in the CRT.

In the CRT an electron gun directs a beam of electrons toward the phosphor screen, which glows where the beam strikes it. On its way to the screen, the beam passes between the two pairs of deflection plates and is deflected according to the potentials on them. The signal on the horizontal plates causes the beam to sweep in an arc from left to right (as viewed by the observer), so that it would trace a straight line across the screen if there were no variation of potential on the vertical plates. The rate at which the beam travels is in accordance with the time base selected. At the end of the sweep the electron beam is blanked so no trace appears on the screen until the next trace is triggered.

Meanwhile, during each sweep any fluctuating signal applied to the vertical plates causes the beam to be deflected up and down according to its amplitude at each instant, so that a wavy line is traced across the screen. Since this line is the result of signal amplitude and time, the waveform of the signal is displayed.

The X input allows a second signal to be applied to the horizontal deflection plates instead of a sweep. This signal, taken from a different part of the circuit being examined, combines with that on the vertical plates to produce a Lissajous figure (q.v.), useful for determining phase shift or distortion.

In addition to the study of electrical signals, oscilloscopes are used for studying a great many non-electrical physical phenomena by converting them to electrical signals by means of suitable transducers.

output—End result or product of a piece of electronic equipment. In a television receiver, the output is both picture and sound; in a radio, or hi-fi system, sound only; in an amplifier, the amplified signal; and so on.

overcoupled—In inductive coupling, the condition when the coupling is tightened beyond the critical point, resulting in a flat-topped or double-humped resonance curve. (See Figure O-8.)

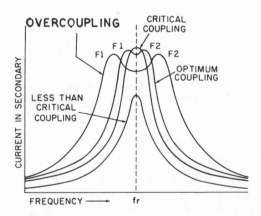

Figure O-8

Overcoupling (frequency response of interstage transformer)

overdamping—(See aperiodic damping.)

overload—When a circuit is subjected to a higher current than it is designed for it is overloaded. This can happen because of an internal defect, the wrong output load, or an excessive input. Circuits should be provided with overload protection, such as fuses, circuit breakers, and the like.

overmodulation—Modulation exceeding 100 percent. *Modulation.*

overshoot—1. Pulse top distortion which may occur when a pulse with a fast rise time causes ringing in an amplifier.   2. The amount by which the pointer overshoots when the energy applied to a meter is changed.

Owen bridge—Bridge used for measuring parallel inductance. (See also Maxwell bridge.) (See Figure O-9.)

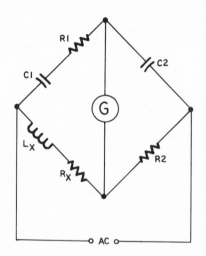

Figure O-9
Owen Bridge

From *Complete Guide to Reading Schematic Diagrams*, First Edition, page
211. ©1972. Reprinted with permission of Parker Publishing Company, Inc.

oxidation—Class of chemical reaction in which the number of electrons associated with an atom or group of atoms is decreased. The electrons lost by the substance oxidized are taken up by some other substance which is then said to be reduced. The reaction does not necessarily involve chemical combination with oxygen.

oxygen—Colorless, odorless, tasteless gas, the most plentiful element in the Earth's crust; its most important compound is water. Oxygen in the lower atmosphere consists almost entirely of molecules of two atoms (O_2), but in the upper atmosphere ozone (O_3) and monatomic oxygen (O) are more predominant. Ozone shields the Earth from most of the Sun's ultraviolet radiation. Symbol, O; atomic weight, 15.9994; atomic number, 8.

P—1. Symbol for permeance (q.v.).   2. Symbol for chemical element phosphorus (q.v.).   3. Symbol for power.   4. In a schematic diagram, class letter for plug.

p—1. Abbreviation for per, as in rpm.   2. Symbol for S.I. unit prefix pico- (q.v.).

Pa—1. Symbol for S.I. unit pascal (q.v.).   2. Symbol for chemical element protactinium (q.v.).

pA—Symbol for picoampere (q.v.).

packaging—Art and technology of physically locating, connecting and protecting elements, circuits, components, modules, subsystems, etc.

pad—Resistive network that attenuates a signal without introducing appreciable distortion or impedance mismatch. (See attenuator.)

padder—Small trimming capacitor (15-30 pF) used in the tuning circuit of a receiver to make the sections of a ganged capacitor track (tune to the same frequency at each setting).

PAM—Abbreviation for pulse-amplitude modulation. *Modulation.*

panel—In a rack, console or cabinet, a steel or aluminum plate 1/8 or 3/16-inch thick, with a standard width of 19 inches and height varying from 1-3/4 inches to 21 inches, used to cover the front of the enclosure. Panels may be blank or drilled to provide for control knobs, connectors, meters, etc.

paper capacitor—Foil-type capacitor with paper dielectric. *Capacitors.*

parabola—Open curve, a conic section produced by the intersection of a right circular cone and a plane parallel to an element of the cone. (See Figure P-1.)

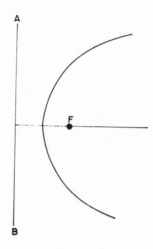

Figure P-1
*Parabola curve in which each point is the same distance
from the focus F and the directrix AB.*

parabolic antenna—Antenna with paraboloid reflector. *Antennas.*

paraboloid—Open surface generated by revolving a parabola (q.v.) about its axis.

parallax—Apparent displacement of an object with respect to its background when viewed from different points. Some meters have mirror scales to avoid parallax errors: the user keeps the pointer and its reflected image in line when taking a reading.

parallel—1. In geometry, lines are parallel if they lie in the same plane but do not intersect. 2. In a circuit, parallel paths that are connected to the same pair of terminals, so that the current between the terminals divides between the paths. Also called shunt.

parallelogram—Four-sided plane figure with the property that the opposite sides are parallel.

parallelogram law—Commonly used to determine the result of two vectors by constructing a parallelogram. The diagonal then gives the vector sum of the two adjacent sides.

parallel-resonant circuit—Circuit consisting of an inductance and a capacitance in parallel. At the resonant frequency, given by $f_r = \frac{1}{2\pi\sqrt{LC}}$, the parallel-resonant circuit offers a high impedance to signals at that frequency.

paramagnetism—Kind of magnetism found in materials weakly attracted by a strong magnet, not to be confused with ferromagnetism (iron, cobalt, nickel, etc.). Strong paramagnetism is exhibited by compounds containing iron, palladium, platinum, and rare-earth elements. Weak paramagnetism is found in sodium and the other alkali metals.

parameter—A variable with a range of possible values. For instance, the z, y, and h parameters of transistor circuit configurations, which characterize their behavior. When values are assigned to the parameters, they are no longer parameters because they are specific values.

parametric amplifier—Low-noise microwave electron tube, generally a linear beam (O type) device, using a fast space-charge wave or fast cyclotron wave. *Electron Tubes.*

paraphase amplifier—Phase splitter (q.v.).

parasitic element—Element of a directional antenna that receives power by induction or radiation from the drive element (which receives power directly from the transmitter or is directly connected to the receiver). *Antennas.*

parity—Method of detecting errors in computer encoding. An extra bit is added to each word. In even parity the total of all the parity bits should be even (in odd parity the total should be odd), and the parity status flag is set accordingly. *Computer Hardware.*

parsec—Astronomical unit of distance equal to 3.262 light-years. Defined as the distance at which the radius of the Earth's orbit subtends an angle of one second of arc.

particle—Body of negligible dimensions but definite mass. *Nuclear Physics.*

particle accelerator—Device for accelerating particles to high kinetic energy. *Nuclear Physics.*

passband—Frequency band passed by a filter. *Active Filters, Passive Filters.*

passive element—Circuit element that does not require an external source of power in order to function (e.g., resistors, capacitors, inductors, etc.).

passive filter—Filter network constructed only of passive elements. *Passive Filters.*

PASSIVE FILTERS

The four basic filters are the low-pass, high-pass, band-pass, and band-stop filters. Although many refinements are possible, the four constant-K circuits given below are practical for most purposes and have the advantage of not requiring advanced mathematics to design them.

Low-Pass and High-Pass Filters

The low-pass filter is shown in Figure P-2 and its response curve in Figure P-3. The high-pass filter and its response curve are shown in Figures P-4 and P-5. Values of C and L are calculated from the following formulas:

$$C = 1/\omega_c R \qquad\qquad L = R/\omega_c$$

where:

C = capacitance in farads
L = inductance in henries
R = nominal terminating resistance [$=\sqrt{L/C}$]
ω_c = cutoff frequency $\times$ 6.28

Band-Pass Filter

The band-pass filter is shown in Figure P-6 and its response curve in Figure P-7. Values of C_1, C_2, L_1, and L_2 are calculated from the following formulas:

$$C_1 = (\omega_2 - \omega_1)/R\omega_o^2 \qquad\qquad L_1 = R/(\omega_2 - \omega_1)$$
$$C_2 = 1/R(\omega_2 - \omega_1) \qquad\qquad L_2 = R(\omega_2 - \omega_1)\omega_o^2$$

where:

C_1 = series capacitance in farads,
C_2 = shunt capacitance in farads,
L_1 = series inductance in henries,
L_2 = shunt inductance in henries.
R = nominal terminating resistance [$= \sqrt{L_1/C_2} = \sqrt{L_2/C_1}$],
ω_o = midband frequency $\times$ 6.28 [$= \sqrt{\omega_1\omega_2}$],
ω_1 = lower cutoff frequency $\times$ 6.28,
ω_2 = upper cutoff frequency $\times$ 6.28.

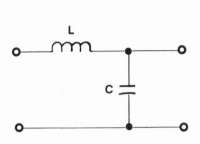

Figure P-2

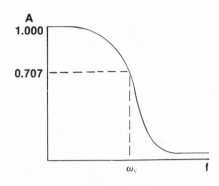

Figure P-3

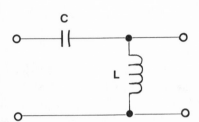

Figure P-4

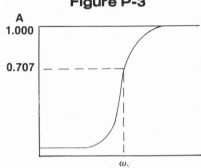

Figure P-5

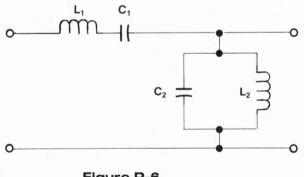

Figure P-6

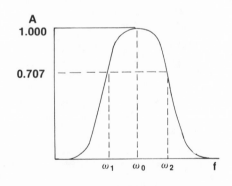

Figure P-7

Band-Stop Filter

The band-stop filter is shown in Figure P-8 and its response curve in Figure P-9. Values of C_1, C_2, L_1, and L_2 are calculated from the following formulas:

$$C_1 = 1/R\,(\omega_2 - \omega_1) \qquad\qquad L_1 = R(\omega_2 - \omega_1)/\omega_1\omega_2$$
$$C_2 = (\omega_2 - \omega_1)\omega_1\omega_2\,R \qquad\quad L_2 = R/(\omega_2 - \omega_1)$$

where:

ω_0 = midband frequency $\times\ 6.28\ [=\sqrt{\omega_1\omega_2} = 1/\sqrt{L_1C_1} = 1/\sqrt{L_2C_2}]$;
all other expressions are the same as for high-pass filters.

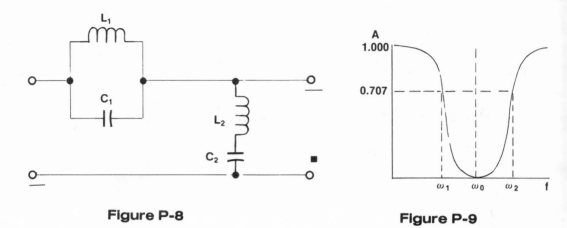

Figure P-8   **Figure P-9**

More Advanced Filters

The foregoing examples belong to a class of filters called "image parameter filters" which, as mentioned above, require only simple arithmetic for their design. More advanced filters use "modern network theory" which, while giving much sharper filtering, unfortunately require the solution of complex network equations by computer programs or precomputed tables. Such filters include the Chebishev, Butterworth and others, which are not covered here (for a lucid introduction to this subject see Walter H. Buchsbaum, "Buchsbaum's Complete Handbook of Practical Electronic Reference Data," Second Edition, © 1978, Prentice-Hall). However, the introduction of integrated-circuit operational amplifiers has made it possible to design some modern network theory filters by simpler calculations, as explained under *Active Filters.*

PA system—Public-address system (q.v.).

patch—Temporary connection, often using a patch board or panel with jacks, and patch cords (short connecting cables with plugs).

P-band—Microwave frequency band between 0.225 and 0.390 gigahertz. *Frequency Data.*

pC—Symbol for picocoulomb (q.v.).

p-channel field-effect transistor—Field-effect transistor in which the source, drain, and conducting channel between them are of p-type conductivity material. *Semiconductors, Transistors.*

PCM—Abbreviation for pulse-code modulation (q.v.).

P-display—Radar indicator in which range is measured radially from the center and bearing as the angle around the center. Also called plan position indicator (PPI). *Radar*.

PDM—Abbreviation for pulse-duration modulation (q.v.).

peak detector—Detector used for detection of transients.

peaking coil—Coil used to extend the high-frequency response of an amplifier by forming a resonant circuit with the distributed capacitance of the circuit to compensate for the leakage to ground of the higher frequencies. Most commonly used in video amplifiers. (See Figure P-10.)

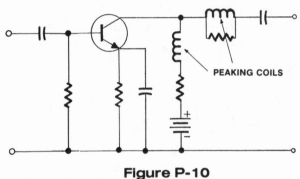

Figure P-10

Peaking Coils (in video amplifier)

From *Complete Guide to Reading Schematic Diagrams,* First Edition, page 38. ©1972. Reprinted with permission of Parker Publishing Company, Inc.

peak inverse voltage—In a rectifier, the peak voltage which appears across the rectifier during the time it is not conducting.

peak-to-peak value—The overall voltage or current difference between the maximum positive and maximum negative excursion of a sine-wave voltage or current from its mean value, equal to 2 × the peak value, or 2.828 × the r.m.s. value.

peak value—The maximum excursion of a sine-wave current or voltage from its mean value, equal to 1.414 × the r.m.s. value, or 1.57 × the average value (for sinusoidal waves). Also called crest value.

pentagrid converter—Superheterodyne converter stage, using an electron tube with five grids, which serves as both local oscillator and mixer.

pentode—Electron tube with five electrodes; cathode, control grid, screen grid, suppresser grid, and anode. *Electron Tubes.*

periodic damping—In a meter, where the pointer oscillates about the reading before coming to rest.

periodic table of the elements—Table that shows a recurring pattern in the properties of the elements when arranged in order of increasing atomic number. (See element.)

peripheral device—Input-output device used with a computer. *Computer Hardware.*

permalloy—Metallic core materials composed mainly of iron and nickel. Permalloy 45 is 45 percent nickel, permalloy 65 is 65 percent nickel, permalloy 78 is 78 percent nickel and 0.6 manganese. Other permalloys are Hipernik (50% nickel with small amounts of silicon and manganese) and Supermalloy (79% nickel, 5% molybdenum). All have much higher permeability (q.v.) than iron alone and are widely used for transformer laminations. *Transformers.*

permanent magnet—Piece of hardened steel or steel alloy that has been magnetized and retains its magnetism indefinitely. Naturally occurring lodestone (Fe_3O_4) is also a permanent magnet.

permanent-magnet moving-coil (PMMC) meter movement—The most commonly used direct-current operated meter movement, in which a small coil is suspended between the poles of a permanent magnet. When a current flows through the coil its magnetic field interacts with that of the permanent magnet, causing the coil to rotate until the opposition of the hairsprings just balances the magnetic torque. The attached pointer then indicates a value on the dial scale. Also called d'Arsonval meter movement. (See Figure P-11.)

permanent-magnet (PM) speaker—(See loudspeaker.)

permeability—The ratio between the magnetizing field (H) produced by electric-current flow in an air-core coil and the magnetic flux density (B) inside a material when it is substituted for air as the core of the coil, given by $\mu = B/H$ (assuming the permeability of air to be the same as a vacuum, equal to 1).

permeability tuning—(See slug tuning.)

permeance—Ratio of the flux in maxwells through any cross-section of a magnetic

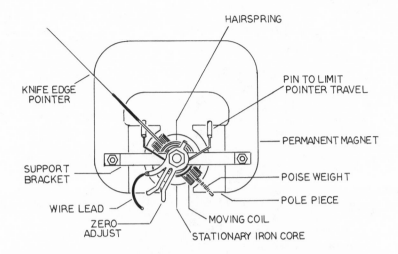

Figure P-11
Permanent-Magnet Moving-Coil Meter Movement

From *Complete Guide to Electronic Test Equipment and Troubleshooting Techniques*, page 33.

circuit to the magnetomotive force in gilberts impressed on the same cross-section. Permeance is the reciprocal of reluctance.

persistence—Duration of the phosphorescence of a phosphor. *Electron Tubes.*

pF—Abbreviation for picofarad (q.v.).

pH—Measure of the acidity or alkalinity of a solution. A solution with a pH of less than 7 is acidic; one with a higher number is alkaline.

phantastron—Type of Miller rundown sweep generator that also generates its own gating signal, used in oscilloscopes, etc. A negative trigger at IN1 drives V1's control grid momentarily negative, and screen grid current is cut off, resulting in a positive-going pulse being coupled via C2 to V1's suppresser grid. This allows plate current to flow, lowering the potential on C1's left-hand plate and causing the right-hand plate to discharge through R1, thus developing a negative-going sawtooth, until the plate voltage is so low that the screen grid takes over, and the cycle is ready to be repeated when the next trigger is received. Triggering with positive triggers can also be applied at IN2. (See Figure P-12.)

phase—In wave motion, the fraction of the time required to complete a full cycle that a point on the wave completes after last passing through the reference, or zero, position. This fraction is generally given in degrees of arc (phase angle), a full cycle being 360 degrees. (2π radians). Symbol, ϕ (Greek letter phi).

phased array—Array of dipoles in which the signal feeding each has a small phase shift with respect to the others, in such a way that an antenna beam is formed and scanned without mechanical movement.

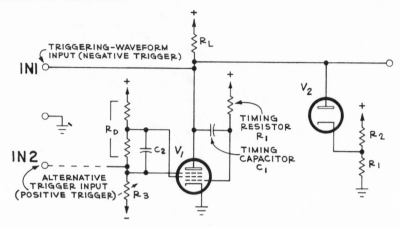

Figure P-12
Phantastron Sweep Generator

From *Practical Oscilloscope Handbook,* page 96.

phase detector—Circuit which compares the phase of an oscillator signal with a reference signal and generates an error voltage if any difference exists. Equal amplitudes of the reference signal, but of opposite phase, are applied to the anode of D1 and the cathode of D2. Equal amplitudes of the oscillator signal, but of the same phase, are applied to the cathode of D1 and the anode of D2. If the oscillator frequency varies, the relative conductions of the two diodes alter. The difference between the d.-c. voltages caused by the different conduction currents in R1 and R2 is applied to the oscillator, raising or lowering its frequency until the phase difference disappears. (See Figure P-13.)

phase distortion—Unequal reproduction of the various phase relationships in a signal. *Distortion.*

phase inversion—In an amplifier or other circuit, where the output waveform is inverted with respect to the input waveform.

phase inverter—Phase splitter (q.v.).

phase-locked loop—Type of phase detector (q.v.).

phase modulation (PM)—Modulation in which the carrier phase angle is made to vary by the message function. *Modulation.*

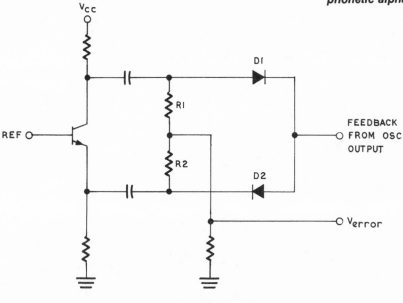

Figure P-13
Phase Detector

phase shift—Difference in phase angle between input and output signals of an amplifier or other circuit, expressed in degrees of lag or lead (qq.v.). (See also phase.)

phase-shift keying (PSK)—Form of modulation in which one phase of the carrier is used to represent one binary state, and a second phase (usually 180 degrees different) is used for the other binary state.

phase splitter—Circuit which produces two output signals of opposite phase from a single-phase input signal. Used to drive push-pull stages, etc. (See Figure P-14.)

phenol—Any of a family of organic compounds in which at least one hydroxyl (OH) group is attached to a carbon atom forming part of a benzene ring. Phenolic materials are usually thermosetting and have high mechanical strength and electrical resistance. One of the most well-known is Bakelite (phenol-formaldehyde resin).

Phillips screw—Type of crossed-slot screw used in many appliances. It requires a Phillips screwdriver.

phone jack, plug—Connector, female and male, used mainly with audio equipment (headphones, microphones, etc.). (See Figure P-15.)

phonetic alphabet—Standard word list used by radio-telephone operators to spell out doubtful words or expressions to avoid confusion between similar sounding letters of the alphabet. *Phonetic Alphabet.*

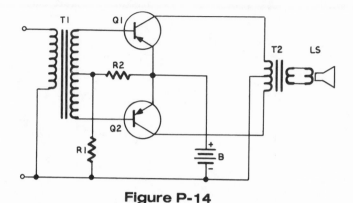

Figure P-14

Using a Transformer as a Phase Splitter

From *Complete Guide to Reading Schematic Diagrams*, First Edition, page 66. ©1972. Reprinted with permission of Parker Publishing Company, Inc.

Figure P-15

Phone Jack and Plug

PHONETIC ALPHABET

Alfa	Juliet	Sierra
Bravo	Kilo	Tango
Charlie	Lima	Uniform
Delta	Mike	Victor
Echo	November	Whiskey
Foxtrot	Oscar	X-ray
Golf	Papa	Yankee
Hotel	Quebec	Zulu
India	Romeo	

phonograph—Instrument for reproducing sounds by means of the vibration of a stylus, or needle, following a spiral groove on a revolving disk or cylinder. The modern instrument is generally called a record player (q.v.) and forms part of a high-fidelity system. *Recording.*

phonograph cartridge—(See pickup.)

phono jack, plug—Coaxial female and male connectors for use with shielded cables. (See Figure P-16.)

Figure P-16
Phono Jacks and Plugs

phosphor—Solid material that emits light (luminesces) when exposed to radiation such as an electron beam. *Electron Tubes.*

phosphorescence—Second stage of luminescence, an afterglow that remains after initial fluorescence. Electrons excited by radiation almost instantly fall back to a lower energy level. One such level is the metastable level. Transition from this level to others is forbidden, so electrons can be trapped in it. Their eventual escape, if back to the ground state (some are returned to the excited state), is accompanied by further emission of energy.

phosphorous—Nonmetallic chemical element of the nitrogen family, a colorless waxy solid that glows in the dark and ignites spontaneously when exposed to air. Symbol, P; atomic number, 15; atomic weight, 30.9738.

photocathode—Element of a phototube that emits electrons when struck by light. *Electron Tubes.*

photocell—(See photoelectric effect.)

photoconductive cell—(See photoconductivity.)

photoconductivity—Decrease in the resistance of a material when struck by light. The energy absorbed from the light increases the number of electrons available for conduction. This property is found in semiconductors (q.v.) and is used in photoconductive transducers or cells, such as light meters in cameras, in which the flow of current from a small battery varies with the light intensity, with corresponding variation in the indication given by a light meter, etc.

photodiode—Light-sensitive diode. (See photoelectric effect.)

photoelectric effect—Interaction of radiation with matter causing dissociation of matter into electrically charged particles. A current will flow across the junction of two dissimilar materials when light falls on it, producing equal numbers of positive and negative charges, which migrate to a place where charge separation can occur. This is normally at a potential barrier between two semiconductor regions. This is the principle of solar cells, photocells, photodiodes, phototransistors, and similar photovoltaic devices.

photoelectromagnetic cell—Detector that produces an output voltage which varies in response to incident optical radiation. The voltage is obtained by separating charge carriers with a magnetic field.

photoflash—(See flash bulb.)

photomask—Photographic mask, similar to a negative, placed over a silicon wafer coated with photoresist in microelectronic fabrication. *Microelectronics.*

photomultiplier tube—Electron tube with photocathode and series of dynodes (electrodes), each at a successively higher positive potential so that it will attract electrons given off by the previous dynode. Light striking the photocathode causes emission of primary electrons, which strike the first dynode, causing the emission of a larger number of secondary electrons. These strike the second dynode with a similar result, and so on until the final diode is reached. Usually there are nine diodes, and total multiplication may reach a factor of 10^6. *Electron Tubes, Nuclear Physics.*

photon—Subatomic particles, photons are bosons, having no electric charge, or rest mass, and one unit of spin. They are field particles, or carriers of the electromagnetic field, traveling at the speed of light, with energy given by $h\mu$, where h is Planck's constant and μ is the frequency. *Nuclear Physics.*

photoresist—Photosensitive organic polymeric material used to selectively protect substrate surfaces against subsequent plating or etching. *Microelectronics.*

phototransistor—(See photoelectric effect.)

phototube—Electron tube with photocathode that emits electrons when illuminated and an anode for collecting the electrons. The electron current is proportional to the intensity of the illumination. *Electron Tubes.*

photovoltaic cell—(See photoelectric effect.) *Semiconductors.*

pickup—Device that converts some form of intelligence, visual or aural, into corresponding electrical signals. In the broadest sense, television cameras and microphones are pickups, but the term is most commonly used for the phonograph pickup that converts vibrations caused by the sound track in a record groove into an electrical signal. *Recording.*

pico- —S.I. prefix meaning one millionth millionth (10^{-12}) of the unit to which it is prefixed. Symbol, p. *S.I. Units.*

picoampere—One millionth millionth (10^{-12}) of an ampere. Symbol, pA. *S.I. Units.*

picocoulomb—One millionth millionth (10^{-12}) of a coulomb. Symbol, pC. *S.I. Units.*

picofarad—One millionth millionth (10^{-12}) of a farad. Symbol, pF. *S.I. Units.*

picosecond—One millionth millionth (10^{-12}) of a second. Symbol, ps. *S.I. Units.*

picowatt—One millionth millionth (10^{-12}) of a watt. Symbol, pW. *S.I. Units.*

picture carrier—Main carrier of a television signal. *Modulation, Broadcasting.*

picture tube—Cathode-ray tube containing a screen of luminescent materials on which are produced visible images, used mainly in television receivers. *Electron Tubes.*

picture tube brightener—Adapter connected between the base of a picture tube and its socket. It contains a small transformer which increases the filament voltage from 6.3 to 7.7, thereby raising the temperature of the cathode and stimulating emission of electrons. Used to restore brightness to a picture tube and extend its life, instead of replacing it immediately.

Pierce oscillator—Crystal oscillator in which the crystal is connected between plate and grid, collector and base, or drain and gate, as the case may be. (See Figure P-17.)

piezoelectric effect—When certain crystals are subjected to mechanical pressure, positive and negative charges appear on opposite sides, because the physical distortion due to the pressure slightly separates the center of positive charge from that of negative charge. The converse effect occurs when a potential difference is applied between opposite faces of the crystal, the crystal being mechanically deformed by the electric field. This property is made use of in various ways. Crystals can be "cut" to oscillate at a certain frequency, within very close limits. Other crystals are used in pickups (q.v.) to convert mechanical vibrations to electrical signals. High-frequency signals can also be converted to ultrasonic waves of the same frequency, but with much slower velocity, and then back again to electro-magnetic waves.

pi filter—Type of passive filter in which the series and shunt elements are usually diagrammed in the form of the Greek letter π (pi). *Rectifiers and Filters.*

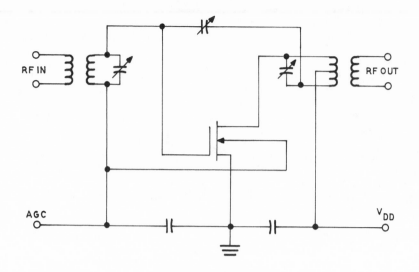

Figure P-17
Pierce Crystal Oscillator Using a Field-Effect Transistor

pigtail—Short length of wire, as for instance, one of the leads of a small component.

pile—1. Old name for a battery; e.g., Voltaic pile. 2. Nuclear reactor.

pillbox antenna—Antenna with reflector shaped like a semi-circular section of a pillbox, used to produce a fan beam. Also called a cheese antenna. *Antennas.*

pilot light—Indicator lamp of various colors to show the condition of a circuit, light dials, etc.

pilot subcarrier—Pilot signal of 19 kilohertz which modulates the main carrier of an FM broadcast station transmitting a stereophonic program. Since the stereophonic subcarrier of 38 kilohertz is suppressed, the pilot subcarrier is doubled and reinserted in the receiver. *Broadcasting.*

pinch effect—Self-constriction of gaseous plasma when a current passes through it, due to the current's magnetic field, which forces the current-carrying particles together. Self-constriction is generally unstable and must be augmented by an external magnetic field to produce a stable magnetic bottle (magnetic field that confines a plasma as if it were in a bottle).

pinch-off voltage—Gate potential required to cut-off conduction in the channel of a depletion-mode field-effect transistor. *Transistors.*

pincushion distortion—Geometric distortion that may occur in a cathode-ray or picture tube where vertical and horizontal lines near the borders of the display are bowed inward.

pin diode—Diode consisting of an intrinsic or lightly doped region between two heavily doped p^+ and n^+ end regions. *Semiconductors.*

pin jack—Small jack (receptable), panel-mounted, that accepts a tip plug, meter plug, probe tip, etc. Also called a tip jack (q.v.).

pink noise—White noise (q.v.), after passing through a "pink" filter that progressively attenuates the higher frequencies, becomes pink noise, which is noise whose energy level halves each time the frequency doubles.

planar device—Device formed in one plane and essentially two-dimensional; the term planar is applicable to most microelectronic devices that are fabricated in the surface of a substrate and do not penetrate it by more than a few micrometers. *Microelectronics.*

Planck's constant—The energy of a quantum or photon is given by $E = h\nu$, where $E =$ the energy in joules: $h =$ Planck's constant, 6.626196×10^{-34} joule-second; $\nu =$ the radiation frequency in hertz. *Constants, Nuclear Physics.*

plan position indicator (PPI)—(See P-display.)

plasma—Collection of positive and negative charges, about equal in number, usually in the form of an ionized gas. This condition is called the fourth state of matter, because its properties are not those of a gas, liquid, or solid. Nearly all the matter in the universe exists in the plasma state, in bright stars, including the Sun. The gas in a gaseous discharge tube, such as a neon light, is also in the plasma state when glowing.

plate—Common name for the anode in an electron tube. *Electron Tubes.*

plate current—In an electron tube, the electron current flowing to the plate. *Electron Tubes.*

plate modulation—Amplitude modulation by modulating the plate supply voltage to the final stage of a transmitter. *Modulation.*

platinum—Very heavy, precious, silver-white metal, soft and ductile, with high melting point and good resistance to corrosion and chemical attack. Commonly alloyed with small amounts of iridium to improve its hardness and strength, platinum is used for electrodes, crucibles and dishes, electrical contacts, etc. Symbol, Pt; atomic number, 78; atomic weight, 195.09.

playback—Reproduction of a recording by a tape player, phonograph, etc. *Recording.*

PLL—Abbreviation for phase-locked loop (q.v.).

plotter—Device for converting an electric function, such as a waveform, to a drawing. Plotters include graphic plotters, X-Y recorders, strip-chart recorders and oscillographs, and use pen-and-ink or thermal writing methods. Because of the mechanical nature of the latter, the use of plotters/recorders is restricted to very low frequencies. Many are designed for computer control.

plug—Connector, generally male, usually at the end of a cable, that mates with a corresponding, generally female, connector, which may be wall or panel mounted, as well as at the end of another cable.

plug-in—Subassembly designed to be inserted into a cell in the main instrument. By having a variety of plug-ins, the versatility of the main instrument is greatly enhanced.

plutonium—Silvery metal that takes on a yellow tarnish when exposed to air, warm because of energy released in alpha decay. All isotopes are radioactive. The most important is plutonium-239 because it is fissionable and readily produced from uranium-238, which is plentiful. It has a half-life of 24,360 years. It was used in the original atom bomb. Symbol, Pu; atomic number, 94.

PM—Abbreviation for permanent magnet and phase modulation (qq.v.).

PMOS—Abbreviation for p-channel enhancement MOS (metal-oxide-semiconductor). *Transistors.*

pn junction—Boundary between p and n regions. *Semiconductors.*

pnp transistor—Transistor with an n-type base between p-type collector and emitter. *Transistors.*

point-contact diode—Diode in which a metal-semiconductor junction is formed by means of a metal point that contacts the surface of the semiconductor. Because of its low capacitance it is used for video detectors and microwave mixers. *Semiconductors.*

point-to-point communication—Two-way communication between fixed stations.

point-to-point wiring—"Hand wiring" as opposed to printed circuits boards.

polar coordinates—System of locating points in a plane with reference to a fixed point by giving their distance (r), and angle (θ) between the vector r and the axis drawn through the reference point.

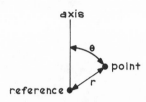

Figure P-18
Polar Coordinates

Polaris—The Pole Star, or North Star, also known as Alpha Ursae Minoris, is a triple star with combined magnitude of 2.04. It will actually come nearest to the north celestial pole around AD 2100, and in about 12,000 years from now will have been replaced by Vega, assuming the present precession of the equinoxes continues undisturbed.

polarity—Property of matter or force analogous to that of a magnet (of possessing magnetic poles), or of being electrically positive or negative.

polarization—1. Electrical polarization is a slight relative shift of positive and negative electric charge in opposite directions within an insulator or dielectric, induced by an external electric field. *Capacitors.* 2. Optical polarization is where the direction of vibration of waves of light is in one plane only. 3. In electromagnetic waves, the electric field is oriented with respect to the ground in accordance with the orientation of the transmitting antenna. Consequently, reception is best when the receiving antenna is oriented similarly. *Antennas.*

polarized capacitor—Electrolytic capacitor. *Capacitors.*

pole—Either end of any axis, such as that of the Earth; or either of two opposed or differentiated forces, such as the ends of a magnet, the terminals of a battery, etc.

polycrystalline material—Any solid object composed of randomly oriented crystalline regions. *Semiconductors.*

polyester—Organic substance with molecules linked in chains or forming a network. The chain type are represented by Dacron, Fortrel, Terylene, and Mylar. The net type are the alkyd resins used in paints, varnishes, etc.

polyethylene—Synthetic fiber, the basis of Herculon, Polycrest, Vectra, etc., often used as dielectrics and insulators.

polygraph—Instrument that simultaneously records blood pressure, pulse rate, respiration, and psychogalvanic skin reflex. Used in police interrogation to test the veracity of a suspect's answers to questions, the polygraph is popularly called a lie detector.

polymer—Substance composed of molecules joined in chains or networks. Such substances range from natural proteins to synthetic man-made materials.

polyphase—More than one phase. Said of electric systems in which voltages and currents are out of step with each other by equal phase angles. The most common is the three-phase system, in which alternating currents separated by phase angles of 120 degrees are transmitted over the same three wires instead of three different pairs, making for greater economy. Polyphase generators and motors used with such a system operate more smoothly because of the constant flow of power. In the three-phase system, circuits are Y, or star-connected, or delta (Δ)-connected. (See three-phase system.)

polystyrene—Synthetic organic polymer composed of styrene, often with the addition of rubber latex to make it impact resistant, used for making housings for appliances, etc.

polyvinyl chloride (PVC)—Synthetic resin made by polymerizing vinyl chloride and used for insulation, phonograph records, and plumbing fittings.

positive feedback—Returning a portion of the output power of an amplifier to the input, where it is in phase with the input signal. *Feedback.*

positive ion—Anion, an atom that has lost one or more electrons so that it has an overall positive charge.

positive logic—Where the more positive voltage level represents the 1 state and the less positive level represents the 0 state. Used by the majority of computers.

positron—Subatomic particle with the same mass as an electron, but positively charged; in other words, the antiparticle of a negative electron. *Nuclear Physics.*

post-deflection acceleration (PDA)—In a cathode-ray tube, a means of increasing display brightness without increasing deflection potentials, in which the electron beam is accelerated only moderately before passing between the deflection plates, and then progressively accelerated by the rising strength of the electric field produced by the high voltage applied across a narrow ribbon of resistive material deposited in a close spiral on the inside of the tube between the deflection plates and the screen.

pot—Short for potentiometer (q.v.).

potassium—Soft white metal with a silvery luster and a low melting point, it is a good conductor but has very little commercial application. Symbol, K; atomic number, 19; atomic weight, 39.102.

potential—Property of each point in an electric field that is equal to the amount of work or energy per unit charge required to transport a small positive test charge to a specific point in the field from an arbitrary reference point beyond the influence of the charge causing the electric field. The potential difference between two such points is the difference resulting from subtracting the lesser from the greater. The potential drop between two points in a circuit is equal to the current between the two points multiplied by the resistance between the two points.

potentiometer—1. Precision measuring instrument used to compare directly and accurately through a nulling technique unknown d.c. voltages with reference-derived voltages.   2. Type of rheostat (q.v.) consisting of a resistor between two fixed terminals and a third terminal connected to a variable contact arm, used as a volume control, etc. *Resistors.*

potting—Process in which the space between a component (e.g., a transformer) and its case is filled with a potting compound which hardens to provide an airtight, moisture-proof, insulating seal. Also called embedment.

powdered-iron core—Core used in coils in high-frequency circuits, made by pulverizing iron filings and mixing them with a binder to hold them together and to insulate the particles from each other. The "slug" is inserted by means of a screw thread so that it can be adjusted in or out; therefore this type of coil is called a slug-tuned coil.

power—Rate at which electric energy is delivered to or absorbed by a circuit, expressed in watts (symbol, W). In a d.c. circuit, power (P) is given by $E \times I$, where E is the voltage and I the current (in volts and amperes). In an a.c. circuit, $P = EI \cos \theta$, where $\cos \theta$ is the power factor (q.v.). *Electrical Equations.*

power amplifier—Current amplifier, generally used in an output stage to drive a

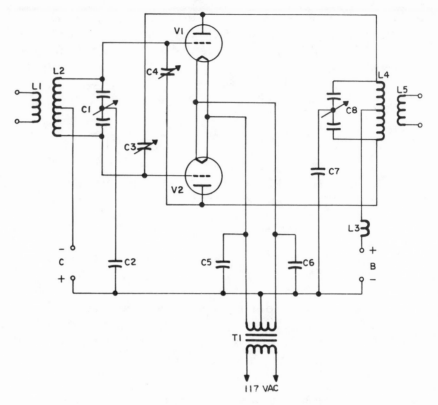

Figure P-19
RF Power Amplifier

From *Complete Guide to Reading Schematic Diagrams,* First Edition, page
70. ©1972. Reprinted with permission of Parker Publishing Company, Inc.

transducer, such as a loudspeaker, antenna, cathode-ray tube, etc., which is the load.
(See Figure P-19.)

power dissipation—Some of the power fed to an active device such as an electron tube
or transistor is wasted as heat within the device. To prevent damage to the device the
heat must not be allowed to build up. It is dissipated by convection, radiation, or
conduction. The maximum power that can be safely dissipated is the power rating of
the device.

power factor—Ratio of power consumed to apparent power, usually expressed as a
percentage. It is given by:

$$PF \quad = \frac{EI \cos \theta}{EI} = \cos \theta$$

where PF = circuit load power factor,
EI cos θ = true power in watts,
EI = apparent power in volt-amperes,

E = applied potential in volts,
I = load current in amperes,
θ = phase angle.

power gain—1. Of an antenna, ratio of power required to produce a given field strength with an isotropic or half-wave dipole to the power required to produce the same field strength with a specified type in its most favorable direction, expressed in decibels. *Antennas.* 2. In an amplifier, ratio of output signal power to input signal power.

power output rating—Maximum continuous r.m.s. power that an amplifier can deliver to a load (e.g., a loudspeaker), expressed in watts.

power supply—Source of power for an electric circuit, either a battery or an electronic power supply. The latter is a rectifier-filter circuit for converting a.-c. 60-Hz line voltage to d.c. *Rectifiers and Filters.*

power transformer—Magnetic-core transformer for operation at 60 hertz, with nearly zero source impedance, to transfer power from line voltage to some required voltage. *Transformers.*

power transistor—Transistor with dissipation rating higher than about one watt. It usually requires a heat sink. *Transistors.*

PPI—Abbreviation for plan-position indicator (q.v.).

PPM—Abbreviation for pulse-position modulation (q.v.).

ppm—Abbreviation for parts per million.

preamplifier—Separate amplifier used "ahead" of the main amplifier. Frequently has special circuits for processing various kinds of input signals (e.g., equalizing, mixing, differential, dual-trace, etc.).

precession—Comparatively slow rotation of the axis of rotation of a spinning body about a line intersecting the spin axis, caused by a force applied to the latter. In the case of the Earth, precession (of the equinoxes) is caused by the gravitational attraction of the Sun and Moon. In a gyroscope, precession is the tendency of the rotor's axis to move at right angles to any perpendicular force applied to it.

preemphasis—In FM broadcasting, emphasis of higher frequencies to improve signal-to-noise ratio in transmission. (See also deemphasis.)

preferred values—Orderly progression of parts values, each value differing from its predecessor by a constant multiplier based on the tolerance. *Preferred Values.*

PREFERRED VALUES

In order to limit the quantities of parts that must be stocked and to standardize their values, preferred numbers are used. These are calculated according to their tolerances, each nominal value being separated from the next by a constant multiplier. For small electronic components, such as fixed composition resistors and fixed ceramic, mica and molded capacitors, the following values are used:

TABLE I
ANSI Standard C83.2-1971

± 20%	± 10%	± 5%
10	10	10
		11
	12	12
		13
15	15	15
		16
	18	18
		20
22	22	22
		24
	27	27
		30
33	33	33
		36
	39	39
		43
47	47	47
		51
	56	56
		62
68	68	68
		75
	82	82
		91
100	100	100

A slightly different set of preferred values is used for fixed wire-wound power-type resistors and for time-delay fuses:

TABLE II
ANSI STANDARD Z17.1-1973

Series "5" (± 24%)	Series "10" (± 12%)
10	10
	12
16	16
	20
25	25
	32
40	40
	50
63	63
	80
100	100

These tables give the two significant figures of each value, which could therefore be, for example, 33, 330, 3300, 33,000 330,000, or 3,300,000, and so on. Those numbers that appear in more than one column are values available in more than one tolerance, the tolerance of each column in which they appear.

Not all manufacturers adhere to the preferred values, especially in the area of wire-wound resistors, so catalogs should be consulted. "Mil-spec" resistors, which are often available from surplus outlets, have the following values if their tolerance is ±1 percent (multiply by multiples of 10 for higher values):

Ohms	Ohms	Ohms	Ohms	Ohms	Ohms	Ohms	Ohms
1.00	1.33	1.78	2.37	3.16	4.22	5.62	7.50
1.02	1.37	1.82	2.43	3.24	4.32	5.76	7.68
1.05	1.40	1.87	2.49	3.32	4.42	5.90	7.87
1.07	1.43	1.91	2.55	3.40	4.53	6.04	8.06
1.10	1.47	1.96	2.61	3.48	4.64	6.19	8.25
1.13	1.50	2.00	2.67	3.57	4.75	6.34	8.45
1.15	1.54	2.05	2.74	3.65	4.87	6.49	8.66
1.18	1.58	2.10	2.80	3.74	4.99	6.65	8.87
1.21	1.62	2.15	2.87	3.83	5.11	6.81	9.09
1.24	1.65	2.21	2.94	3.92	5.23	6.98	9.31
1.27	1.69	2.26	3.01	4.02	5.36	7.15	9.53
1.30	1.74	2.32	3.09	4.12	5.49	7.32	9.76

presence—Term used in high-fidelity sound reproduction to describe the naturalness of the sound, "as if the instrument, etc., were actually present."

pressure—Perpendicular force per unit area, variously expressed in pounds per square inch (psi), newtons per square meter or pascals, dynes per square centimeter, atmospheres, and bars. Atmospheric pressure is measured in millibars, millimeters of mercury, and inches of mercury. The standard, or nominal pressure of the atmosphere at sea level is 1013.25 millibars, 760 millimeters of mercury, 29.92 inches of mercury, one bar, one atmosphere, 14.7 pounds per square inch, or 101.325 kilopascals. All except the last named are being gradually phased out.

pressure pad—Felt pad that presses the tape in a tape recorder against the head gap. *Recording.*

pressure roller—Rubber roller that presses the tape in a tape recorder against the capstan. Also called pinch roller. *Recording.*

primary battery—Battery consisting of primary (not rechargeable) cells.

primary colors—Those colors which given the widest range of other colors when mixed in various proportions. In additive mixing (as used in television) the primary colors are red, green, and blue. In subtractive mixing (as in color printing) the primary colors are magenta, yellow, and cyan.

primary winding—In a transformer, the winding connected to the source of energy. *Transformers.*

printed circuit board—Insulating substrate on which a thin pattern of copper conductors has been formed by any of several graphic arts procedures (such as photo-etching, silk-screening, etc.) to connect together discrete components also mounted on the board. *Printed Circuits.*

PRINTED CIRCUITS

Printed circuits originally were introduced to reduce costs of manufacturing electronic assemblies by eliminating hand wiring. It was soon realized that there were other advantages, such as greater uniformity, use of printed components instead of discrete ones in many cases, savings in weight and bulk, and new methods of fabrication that had not been possible before. Today, practically all radio and television sets use printed circuits for most of their wiring, and microelectronic devices are usually mounted on printed circuit boards.

A printed circuit board starts out as a laminate with a thickness varying from 1/64 to 1/2 inch. The usual materials are paper-base phenolic or epoxy, and glass-fabric-base epoxy, polytetrafluoroethylene°, or fluorinated ethylene propylene. The glass-fabric-

°PTFE, or Teflon.

base boards are superior to the paper-base boards in all respects except their "punchability," which is a serious disadvantage, since drilling holes is more expensive. There are some other materials, including even rubber and ceramic, but the most widely used board material is paper-base phenolic (NEMA type XXXP).
type XXXP).

Most boards are copperclad on one side, but boards with copper on both sides are available for applications where there will be a high densityof components with many crossovers. The copper foil is commonly supplied in two thicknesses, called "1 oz." (0.0014 inch), and "2 oz." (0.0028 inch). When there is copper on one side only, that side is called the conductor side, and the other is called the component side. The copper may be "sensitized" or "nonsensitized," according to whether it has been coated with photo resist. Sensitized boards are used for photoetching, unsensitized for silk screening.

However, before fabrication can begin, the master artwork must be prepared. This has to be done with care, because components for use on printed circuit boards have leads spaced to match a standard grid. There are three grids with spacings of 0.100, 0.050 and 0.025 inch, the first being the most widely used. The material used for the master is generally Mylar, and the scale is two to five times oversize. The draftsman uses black pressure-sensitive tape for the conductor pattern, with adhesive black disks for the terminals, or draws the design with opaque, permanent black ink. The master is then photographed to obtain a negative the same size as the board.

This negative can be used in various ways. Best results are obtained by exposing a sensitized board to light (often ultraviolet) through the negative. The conductor pattern on the negative is white, of course, so the photo resist is exposed only where illuminated by the pattern. This portion now becomes insoluble, so that when the board is treated with a solvent such as trichloroethylene, the pattern remains in place while the rest of the photoresist is washed away. The board is then placed in an etching bath containing a mordant, such as ferric chloride, which eats away all of the unprotected metal. After the etching has been completed the photo-resist pattern is removed with a stripping agent.

Because photoetching is rather slow and requires a darkroom (or special lighting), the silk-screen process is favored for work not requiring high resolution. The silk screen is made in a similar manner to the negative described in the previous paragraph. But once made, it can be used to fabricate a large number of printed circuit boards with greater simplicity. The screen is mounted in a frame, and as each unsensitized board is placed under it an acid-resist ink or lacquer is squeegeed through it on to the copper. After this dries, the board is placed in the etching bath as before.

Etching away the unwanted copper is called a subtractive process. An opposite process, called additive, is a plating process in which the circuit pattern is printed on the unclad board with conductive ink. The conductor pattern is then deposited in a manner similar to electroplating. This method has the advantage that if the holes are punched or drilled first they can be "plated through" for making connections to the other side of the board. Other additive processes include metal spraying and die stamping.

As the boards may have to be stored for a while before they are used, some method of finishing is necessary to protect the conductor pattern and to make soldering easier. The most widely used method is to dip or wave-solder the boards so that the copper is coated with a layer of solder. In some special applications silver plating is used, or solder may be applied by hot rolling or plating. Plating with gold, or rhodium and nickel, is generally only for high-reliability circuit boards used in military or space equipment. These boards are also treated later with acrylic or other insulating coatings to resist humidity and increase the firmness of attachment of parts.

Holes for component leads are punched or drilled. As already mentioned, some boards must be drilled because they cannot be punched. Special drill bits that leave a smooth hole must be used if holes are to be plated through. Sometimes eyelets or tubelets are inserted in the holes. The diameters of punched holes should be not less than 2/3 the thickness of the board and should not exceed by more than 0.020 inch the diameter of the lead to be inserted.

All components are mounted on the side opposite the conductor pattern, since the latter is to be dip or wave-soldered. Any component which must be mounted on the conductor side has to be hand-mounted afterward, so this is avoided if at all possible. In mass production, components are mounted by machines that bend the leads, cut them to the proper length, and insert the components in the board. They then crimp the lead wires on the bottom of the board to ensure that the components do not fall out before soldering.

Dip soldering consists of applying flux to the conductor pattern and placing the board in contact with 60/40 tin-lead solder at a temperature of 230°C (450°F) for five seconds, while slightly agitating it. Wave soldering is a refinement where, instead of placing the board in contact with the solder, it is passed horizontally over the surface of the bath. A pump forces the solder upward so that a wave or ridge is formed across the bath. This wave is about one inch high and extends the full width of the solder *pot*. As the board moves along, it is in contact with this wave for several seconds (eight to ten for a four-inch board). The advantage of this method is that the entire board is not exposed to the maximum heat all the time, the solder temperature is not lowered where it is in contact with the board, because the solder is moving rapidly, and for the same reason oxidation is eliminated.

Solder masks are frequently used for these types of soldering to restrict the areas of the board the solder will wet. This prevents bridging between closely spaced conductors and concentrates the heat of the solder around the joints to be soldered. The mask material is applied with a silk screen in the same way as acid resist.

printer—Output device that prints in alphanumeric characters on accordion-folded paper the data output of a computer. The simplest kind operates like a typewriter; faster versions, called line printers, print a line at a time (each line can have 132 characters, at speeds of up to 1285 lines per minute); wire matrix printers form each character from a pattern of dots, the ends of small wires arranged in a five-by-seven rectangle. Selected wires are pressed against an inked fabric ribbon to print the characters on paper. The output of any type of printer is called a print-out. *Computer Hardware.*

probe—Assembly consisting of test lead and convenient manual connecting device for touching or temporarily latching to a measuring point, so that the meter, oscilloscope, or other instrument to which the test lead is connected may make a reading. A probe should: (1) not load the circuit; (2) not distort a.c. signals; (3) not pick up unwanted signals. Some probes contain circuitry for demodulating r.f. signals, for attenuation of high voltages, for measuring a.c. current without breaking the circuit (clamp-on or clip-on probe), and so on. Where the probe's internal circuitry requires a power supply it is called an active probe; if no external power is needed it is called a passive probe. (See Figure P-20.)

program—An unambiguous, ordered sequence of instructions to a computer to enable it to perform a computational task. *Computer Software.*

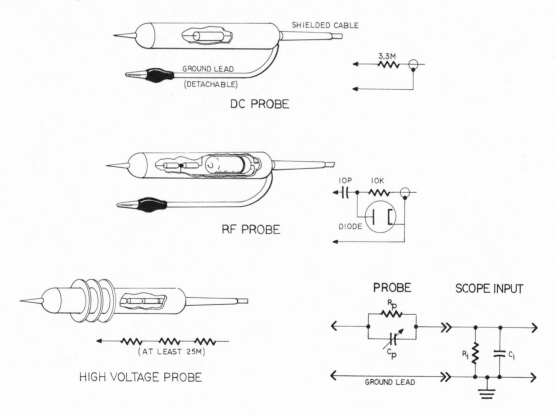

Figure P-20

Probes: (a), (b), and (c) would be used with voltmeters; (d) is a compensated probe for use with an oscilloscope.

programmable calculator—Electronic calculator with a programming facility. A long series of program steps can be entered, and then the function to which they relate can be performed repeatedly without having to enter the operational instructions again.

projection television—Television reception system in which the picture is projected on to a large screen.

PROM—Abbreviation for programmable read-only memory. ROM with program specified by customer; ROM programmed by customer. Usually can be done only once, but there are PROM programmer circuits which allow instructions to be stored and restored in some PROM's. However, this is far slower than normal storage in a RAM. *Computer Hardware.*

propagation—Transmission of radio waves through the atmosphere, along the ground, or out into space. *Electromagnetic-Wave Propagation.*

properties of materials—Data relating to metals, semiconductors, insulators, and ferromagnetic materials used in electronics. *Properties of Materials.*

PROPERTIES OF MATERIALS
Metals Commonly Used in Electronics

Metal	Density at 20° C g/cm^3	Melting Point ° C	Coefficient of Linear Expansion at 20° C $\times 10^{-6}$/° C	Resist-ivity $\mu\Omega$/cm	Modulus of Elasticity kg/mm^3	Thermal Conduct-ivity at 20° C W/cm/° C/s
Aluminum	2.70	660	22.90	2.62	7 250	2.18
Beryllium	1.82	1 278	12.00	10.00	30 000	1.64
Brass	8.55	900	18.77	3.90	13 200	*
Bronze	8.15	1 040	18.45	6.50	16 500	*
Copper	8.96	1 083	16.50	1.67	11 000	3.94
Gold	19.30	1 063	14.20	2.19	7 300	2.96
Iridium	22.40	2 410	6.50	5.30	52 500	1.40
Iron (wrought)	7.87	1 535	11.70	9.71	20 000	0.79
Lead	11.34	327	28.70	21.90	1 800	0.35
Magnesium	1.74	651	25.20	4.46	4 600	1.55
Manganese	7.44	1 244	23.00	5.00	16 000	
Mercury	13.55	−39		95.80		0.08
Molybdenum	10.20	2 610	4.90	4.90	35 000	1.46
Monel	8.90	1 400		42.00		
Nickel	8.90	1 453	13.30	6.84	21 000	0.90
Osmium	22.48	3 000	5.00	9.50	57 000	0.61
Palladium	12.00	1 552	11.80	10.80	12 000	0.70
Platinum	21.45	1 769	8.90	9.83	15 000	0.69
Rhodium	12.00	1 966	8.10	4.51	30 000	1.50
Ruthenium	12.20	2 250	9.10	7.60	42 000	

Metal	Density at 20° C g/cm³	Melting Point ° C	Coefficient of Linear Expansion at 20° C × 10⁻⁶/° C	Resist- ivity μΩ/cm	Modulus of Elasticity kg/mm³	Thermal Conduct- ivity at 20° C W/cm/° C/s
Silver	10.49	961	18.90	1.59	7 200	4.08
Tantalum	16.60	2 996	6.60	12.40	19 000	0.54
Tin	7.30	232	23.00	11.40	41 100	0.64
Titanium	4.54	1 675	8.50	80.00	8 500	0.20
Tungsten	19.30	3 410	4.30	5.50	35 000	1.99
Zinc	7.14	419	29.39	6.00	8 400	1.10
Zirconium	6.40	1 852	5.60	41.00	7 500	1.40

*Varies with composition, somewhat lower than copper

Principal Semiconductors

Semiconductor	Density (g/cm³)	Melting Point (° C)	Coefficient of Linear Expansion (× 10⁻⁶/° C)	Energy Band Gap at 300K (eV)	Electron Mobility Light Mass (cm³/V.s)	Electron Mobility Heavy Mass (cm³/V.s)	Hole Mobility Light Mass (cm³/V.s)	Hole Mobility Heavy Mass (cm³/V.s)
AlSb	4.28	1 065		1.60	180-230			420-500
B	2.34	2 075		1.40	1	1		2
C (diamond)	3.51	3 800	1.18	5.30	1 800	1 800		1 600
GaAs	5.32	1 238	5.70	1.43	8 600-11 000	1 000	3 000	426-500
GaSb	5.62	706	6.90	0.70	5 000-40 000	1 000	7 000	700-1 200
GaP	4.13	1 450	5.30	2.25	120-300			420-500
Ge	5.32	937	6.10	0.66	3 900	3 900	14 000	1 860
InAs	5.67	942	5.30	0.33	33 000-40 000		8 000	450-500
InP	4.79	1 062	4.50	1.27	4 800-6 800			150-200
InSb	5.78	530	5.50	0.17	78 000		12 000	750
Se (amorphous)	4.82			2.30	0.005			0.15
Se (hexagonal)	4.79	217	36.9	1.80				1
Si	2.33	1 417	4.20	1.09	1 500	1 500	1 500	480
Te	6.25	432	16.80	0 38	1 100	1 100	10 000	700

Carrier mobilities are at 300 K

Electrical Properties of Commonly Used Insulators

Material	Resistivity (Ω/cm at 25° C)	Dielectric Constant (at 1MHz at 25° C)
Air	—	1.0
Asbestos	—	3.1
Bakelite	10¹¹	4.4
Beeswax	—	2.5
Glass	—	8.3
Gutta-percha	10¹⁵	2.5
Mahogany	—	2.3
Nylon	8 × 10¹⁴	3.1
Paper	—	3.0
Phenol (formaldehyde, 50% paper laminate)	—	4.6
Plywood (Douglas fir)	—	1.9
Porcelain	—	5.1
Polyvinylchloride (PVC)	10¹⁴	2.9

Table of Standard Annealed Bare Copper Wire
Using American Wire Gauge (B&S)

Gauge (AWG) or (B & S)	DIAMETER INCHES			AREA	WEIGHT	LENGTH	RESISTANCE AT 68° F			Current* Capacity (Amps)— Rubber Insulated
	Min.	Nom.	Max.	Circular Mils	Pounds per M'	Feet per Lb.	Ohms per M'	Feet per Ohm	Ohms per Lb.	
0000	.4554	.4600	.4646	211600.	640.5	1.561	.04901	20400.	.00007652	225
000	.4055	.4096	.413/	167800.	507.9	1.968	.06180	16180.	.0001217	175
00	.3612	.3648	.3684	133100.	402.8	2.482	.07793	12830.	.0001935	150
0	.3217	.3249	.3281	105500.	319.5	3.130	.09827	10180.	.0003076	125
1	.2864	.2893	.2922	83690.	253.3	3.947	.1239	8070.	.0004891	100
2	.2550	.2576	.2602	66370.	200.9	4.977	.1563	6400.	.0007778	90
3	.2271	.2294	.2317	52640.	159.3	6.276	.1970	5075.	.001237	80
4	.2023	.2043	.2063	41740.	126.4	7.914	.2485	4025.	.001966	70
5	.1801	.1819	.1837	33100.	100.2	9.980	.3133	3192.	.003127	55
6	.1604	.1620	.1636	26250.	79.46	12.58	.3951	2531.	.004972	50
7	.1429	.1443	.1457	20820.	63.02	15.87	.4982	2007.	.007905	
8	.1272	.1285	.1298	16510.	49.98	20.01	.6282	1592.	.01257	35
9	.1133	.1144	.1155	13090.	39.63	25.23	.7921	1262.	.01999	
10	.1009	.1019	.1029	10380.	31.43	31.82	.9989	1001.	.03178	25
11	.08983	.09074	.09165	8234.	24.92	40.12	1.260	794.	.05053	
12	.08000	.08081	.08162	6530.	19.77	50.59	1.588	629.6	.08035	20
13	.07124	.07196	.07268	5178.	15.68	63.80	2.003	499.3	.1278	
14	.06344	.06408	.06472	4107.	12.43	80.44	2.525	396.0	.2032	15
15	.05650	.05707	.05764	3257.	9.858	101.4	3.184	314.0	.3230	
16	.05031	.05082	.05133	2583.	7.818	127.9	4.016	249.0	.5136	6
17	.04481	.04526	.04571	2048.	6.200	161.3	5.064	197.5	.8167	
18	.03990	.04030	.04070	1624.	4.917	203.4	6.385	156.5	1.299	3
19	.03553	.03589	.03625	1288.	3.899	256.5	8.051	124.2	2.065	
20	.03164	.03196	.03228	1022.	3.092	323.4	10.15	98.5	3.283	
21	.02818	.02846	.02874	810.1	2.452	407.8	12.80	78.11	5.221	
22	.02510	.02535	.02560	642.4	1.945	514.2	16.14	61.95	8.301	
23	.02234	.02257	.02280	509.5	1.542	648.4	20.36	49.13	13.20	
24	.01990	.02010	.02030	404.0	1.223	817.7	25.67	38.96	20.99	
25	.01770	.01790	.01810	320.4	.9699	1031.	32.37	30.90	33.37	
26	.01578	.01594	.01610	254.1	.7692	1300.	40.81	24.50	53.06	
27	.01406	.01420	.01434	201.5	.6100	1639.	51.47	19.43	84.37	
28	.01251	.01264	.01277	159.8	.4837	2067.	64.90	15.41	134.2	
29	.01115	.01126	.01137	126.7	.3836	2607.	81.83	12.22	213.3	
30	.00993	.01003	.01013	100.5	.3042	3287.	103.2	9.691	339.2	
31	.008828	.008928	.009028	79.7	.2413	4145.	130.1	7.685	539.3	
32	.007850	.007950	.008050	63.21	.1913	5227.	164.1	6.095	857.6	
33	.006980	.007080	.007180	50.13	.1517	6591.	206.9	4.833	1364.	
34	.006205	.006305	.006405	39.75	.1203	8310.	260.9	3.833	2168.	
35	.005515	.005615	.005715	31.52	.09542	10480.	329.0	3.040	3448.	
36	.004900	.005000	.005100	25.00	.07568	13210.	414.8	2.411	5482.	
37	.004353	.004453	.004553	19.83	.06001	16660.	523.1	1.912	8717.	
38	.003865	.003965	.004065	15.72	.04759	21010.	659.6	1.516	13860.	
39	.003431	.003531	.003631	12.47	.03774	26500.	831.8	1.202	22040.	
40	.003045	.003145	.003245	9.888	.02993	33410.	1049.	0.9534	35040.	
41	.00270	.00280	.00290	7.8400	.02373	42140.	1323.	.7559	55750.	
42	.00239	.00249	.00259	6.2001	.01877	53270.	1673.	.5977	89120.	
43	.00212	.00222	.00232	4.9284	.01492	67020.	2104.	.4753	141000.	
44	.00187	.00197	.00207	3.8809	.01175	85100.	2672.	.3743	227380.	
45	.00166	.00176	.00186	3.0976	.00938	106600.	3348.	.2987	356890.	
46	.00147	.00157	.00167	2.4649	.00746	134040.	4207.	.2377	563900.	

*Note: Values from National Electrical Code.

Mica, ruby	5×10^{13}	5.4
Shellac	—	3.5
Silicon dioxide	$>10^{19}$	3.8
Silicone-rubber	—	3.2
Teflon	10^{17}	2.1
Vaseline	—	2.2
Water (distilled)	10^{6}	78.2

Properties of Ferromagnetic Materials
Representative Core Materials

Material	Permeability		Coercivity (A/m)	Retentivity (T)
	Initial	Maximum		
Ferroxcube 3 (Mn-Zn-Ferrite)	1.26×10^{-3}	1.88×10^{-3}	7.96×10^{-2}	0.10
Ferroxcube 101 (Ni-Zn-Ferrite)	1.38×10^{-3}		1.43×10^{-1}	0.11
HyMu 80 (Ni 80%, Fe 20%)	2.51×10^{-2}	1.26×10^{-1}	3.98×10^{-2}	
Iron, silicon (transformer) (Fe 96%, Si 4%)	6.28×10^{-4}	8.80×10^{-3}	2.39×10^{-1}	0.70
Mumetal (Ni 77%, Fe 16%, Cu 5%, Cr 2%)	2.51×10^{-2}	1.26×10^{-1}	3.98×10^{-2}	0.60
Permalloy 45 (Fe 55%, Ni 45%)	3.14×10^{-3}	3.14×10^{-2}	2.39×10^{-1}	
Permendur 2V (Fe 49%, Co 49%, V 2%)	1.01×10^{-3}	5.65×10^{-3}	1.59×10^{0}	1.40
Rhometal (Fe 64%, Ni 36%)	1.26×10^{-3}	6.28×10^{-3}	3.98×10^{-1}	0.36
Sendust (high-frequency powder) (Fe 85%, Si 10%, Al 5%)	3.77×10^{-2}	1.51×10^{-1}	3.98×10^{-2}	0.50
Supermalloy (Ni 79%, Fe 16%, Mo 5%)	1.26×10^{-1}	1.26×10^{0}	1.59×10^{-3}	

Properties of Ferromagnetic Materials
Representative Permanent Magnetic Materials

Material	External Energy (B_dH_d)	Coercivity (A/m)	Retentivity (T)
Alnico V (Fe 51%, Co 24%, Ni 14%, Al 8%, Cu 3%)	35 810	45 757	1.20

Material	External Energy (B_dH_d)	Coercivity (A/m)	Retentivity (T)
Alnico VI (Fe 48.75%, Co 24%, Ni 15%, Al 8%, Cu 3%, Ti 1.25%)	27 852	59 683	1.00
Carbon steel (Fe 98.5%, C 1%, Mn 0.5%)	1 432	3 820	0.86
Chromium steel (Fe 95.5%, Cr 3.5%, C 1%)	2 308	5 013	0.90
Cobalt steel (Co 36%, Cr 35%, Fe 25.15%, W 3%, C 0.85%)	7 448	16 711	0.90
Cunife I (Cu 60%, Ni 20%, Fe 20%)	15 597	4 775	0.58
Iron oxide powder (4.96 g/cm^3) (Fe_3O_4 92%, Fe_2O_3 8%)	—	31 433	0.75
Platinum alloy (Pt 77%, Co 23%)	30 239	159 155	0.45
Tungsten steel (Fe 94%, W 5%, C 1%)	2 546	5 570	1.03
Vectolite (sintered) (Fe_3O_4 44%, Fe_2O_3 30%, Co_2O_3 26%)	4 775	79 577	0.16

GALVANIC SERIES IN SEAWATER

If any two metals from the following list are connected by a conductor and immersed in sea water they form a galvanic cell. If they are in different groups the metal coming first in the list will be corroded by the other metal. It is therefore described as anodic. The other metal is therefore cathodic, or most noble. If the two metals are in the same group corrosion will be negligible.

> Magnesium
> Magnesium alloys
>
> Zinc
> Galvanized steel
> Galvanized wrought iron
>
> Aluminum:
> 52SH, 4S, 3S, 2S, 53ST
> Aluminum clad
>
> Cadmium
>
> Aluminum:
> A17ST, 17ST, 24ST

Mild steel
Wrought iron
Cast iron

Ni-resist

13% chromium stainless steel

50-50 lead-tin solder

18-8 stainless steel (active)
18-8-3 stainless steel (active)

Lead
Tin

Muntz metal
Manganese bronze
Naval brass

Nickel (active)
Inconel (active)

Yellow brass
Admiralty brass
Aluminum bronze
Red brass
Copper
Silicon bronze
Ambrac
70-30 copper nickel
Comp. G bronze
Comp. M bronze

Nickel (passive)
Inconel (passive)

Monel

18-8 stainless steel (passive)
18-8-3 stainless steel (passive)

proton—Stable subatomic particle of unit positive charge that is a constituent of every atomic nucleus. *Nuclear Physcis.*

PRR—Abbreviation for pulse repetition rate (q.v.).

ps—Symbol for picosecond (q.v.).

p-type material—Semiconductor material in which a small quantity of impurity atoms provide freely moving positive charges (holes). *Semiconductors.*

public-address system—System consisting basically of a microphone, amplifier, and at least one loudspeaker, to increase loudness of speech or music. *Electroacoustics.*

ulling—In an oscillator, a slight deviation from the proper frequency caused by a strong signal coupled in to the circuit. If this signal is very close in frequency it takes control, making the oscillator change to its frequency.

pulse—A wave that rises from zero or some other reference line to a peak amplitude and then returns to zero, always with the same polarity. (See Figure P-21.)

pulse-amplitude modulation (PAM)—Series of pulses modulated in amplitude. *Modulation.*

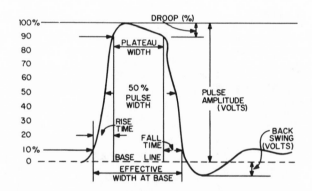

Figure P-21

Pulse ("Pulse amplitude" shown is also called "peak amplitude." In this case pulse amplitude is the amplitude of the midpoint of the pulse.)

pulse-code modulation (PCM)—Amplitude of modulating signal is represented by code groups of pulses. *Modulation.*

pulse generator—Generator for producing pulses, singly or in trains, at selected repetition rates, intervals, amplitudes, etc.

pulse modulation—Modulation of some parameter (amplitude, duration, time of occurrence, or shape) of a series of regularly recurrent pulses (the carrier). *Modulation.*

pulse-position modulation (PPM)—Modulation by varying the position in time of a pulse relative to its unmodulated time of occurrence. Also called pulse-phase modulation. *Modulation.*

pulse-repetition rate (PRR)—The number of pulses per second, expressed in hertz.

pulse-time modulation (PTM)—Variation of the time of occurrence of some parameter of the carrier pulses. *Modulation.*

pulse-width modulation (PWM)—Modulation in which the time of occurrence of either the leading or trailing edge (or both) of each pulse is varied from its unmodulated position. Also called pulse-duration modulation (PDM), and pulse-length modulation (PLM). *Modulation.*

punched card—Card that stores data to be used with a computer system by the presence of small rectangular holes in specific locations. See IBM card. *Computer Hardware.*

punched tape—Paper or other tape that stores data to be used with a computer system by the presence of small round holes along the tape. *Computer Hardware.*

purple plague—Purple compound that may appear where gold and aluminum are bonded, causing a mechanically weak joint which is unreliable.

push-button switch—Switch operated by depressing a button instead of moving a lever.

push-pull amplifier—Two transistors or electron tubes connected oppositely in a balanced circuit, so one is amplifying the positive excursion of the input signal at the same time that the other is amplifying the negative. The positive and negative outputs are then coupled together. (See Figure P-21.)

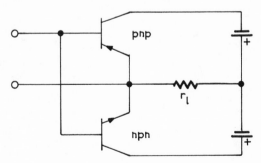

Figure P-22
Push-pull Amplifier Using Complementary Symmetry (Class B large-signal amplifier)

PVC—Abbreviation for polyvinyl chloride (q.v.).

pW—Symbol for picowatt (q.v.).

PWM—Abbreviation for pulse-width modulation (q.v.).

pyroelectric detector—Temperature sensing device consisting of a thin wafer of ferroelectric crystal (e.g., lithium tantalate or triglycine sulfate) acting as the dielectric of a capacitor whose capacitance varies with temperature variations induced by radiation.

pyrometer—High-temperature thermometer in which radiation from the hot object generates a voltage in a thermopile connected to a recording instrument.

Q—1. Quantity of electric charge, expressed in coulombs.   2. Quality of a reactive circuit, given by the ratio of the reactance of the circuit to the total resistance of the circuit, or $Q = X/R$. 3. In a schematic diagram, the class letter for a transistor.

Q Band—Microwave band containing frequencies from 36 through 46 gigahertz. *Frequency Data.*

Q meter—Instrument for measuring Q.

Q signal—In color television, the modulating signal carrying the green and magenta information. *Broadcasting.*

Q-switch—Technique for obtaining brief, intense pulses of power from a laser. *Lasers.*

quadrant—Quarter circle, one of four segments into which the Cartesian plane is divided. (See Figure Q-1.)

quadraphonic sound system—Extension of stereophonic sound system using four channels and speakers. Also called quadriphonic, quadrisonic, quad.

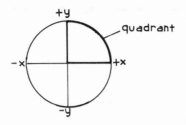

Figure Q-1
Quadrant

quantum—Discrete natural unit, or packet of energy, charge, angular momentum, or other physical property. Quanta of electromagnetic radiation are called photons. *Nuclear Physics.*

quartz—Widely distributed mineral consisting of silica or silicon dioxide (SiO_2). Transparent to ultraviolet light and piezoelectric (q.v.), quartz has numerous applications in electronics. Quartz crystals are used for frequency control in oscillators. (See Pierce oscillator.)

q.v.—Abbreviation for quod vide, Latin for "which see," for which there is no convenient English equivalent.

R—1. Symbol for Réaumur (q.v.). 2. Symbol for resistance (q.v.). 3. In a schematic diagram, the class letter for a resistor.

r—1. Symbol for radius. 2. Symbol for roentgen (q.v.).

Ra—Symbol for radium (q.v.).

rad—Unit of absorbed dose of ionizing radiation (e.g., X-rays) equal to the amount of radiation that releases an energy of 100 ergs per gram of matter. In S.I. units a rad $= 1 \times 10^{-2}$ J/kg. (See also rem, roentgen.)

radar—Acronym derived from "radio detecting and ranging," a system that determines the direction and distance of an object by radio echoes, using a highly directional radio beam and measuring the time taken by a transmitted pulse to go to the object and return to the receiver. Nearly all radar operation is at microwave frequencies. *Radar.*

RADAR

The word radar is an acronym of "radio detecting and ranging." A basic radar system is shown in block form in Figure R-1. A short pulse of microwave energy is fed from

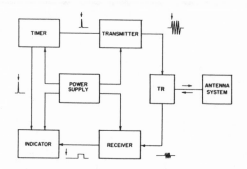

Figure R-1

*Functional Block Diagram of a Fundamental Pulse-
Modulated Radar System*

the transmitter to a directional antenna. The antenna is then switched to the receiver. If an echo is received from a target it is detected, amplified, and displayed on a cathode-ray tube indicator. Then another pulse is fed by the transmitter to the antenna, and so on.

Any radio wave striking an object will have a certain amount of its energy reflected back toward the transmitter. Echoes from desired targets are target signals; echoes from nontargets, such as the ground, the ocean, precipitation, hills and the like, are clutter. The range of the target is determined by the time interval betwen the transmitted pulse and the received target signal. Since radio waves travel at 2.99776×10^8 meters per second, the range of the target will be equal to half the time interval in seconds multiplied by this velocity. Its direction is given by the direction to which the antenna is pointing.

If the transmitter did not send pulses, with intervals between them for reception of target signals, it would not be possible to associate a target signal with the part of the transmitted signal of which it is an echo, and therefore its range could not be determined. (However, continuous-wave radar, which gets round this difficulty by changing frequency continuously, is used in one type of Doppler radar, described below.) Figure R-2 illustrates two successive pulses. The pulse width is designated by the Greek letter τ, and the pulse power is designated by its height P_p. The pulse repetition rate f_r is the reciprocal of the pulse interval $1/f_r$. The duty cycle, which is the ratio of the pulse width to the pulse interval, is therefore given by f_r, and the average output power $P_A = P_p \tau f_r$. This means among other things that the transmitter can operate at much higher peak power than would be possible if it were operating continuously.

The pulse repetition rate (PRR) has to be such that echoes from all detectable targets for one pulse will appear before the next pulse is transmitted; otherwise a target signal may be received from a distant target after the next pulse, and be incorrectly interpreted as a close-range target: if no targets will be detected beyond 25 miles the pulse interval can be 300 microseconds; but if targets may be detected up to 50 miles the pulse interval must be at least 600 microseconds.

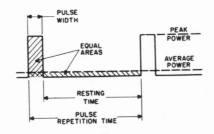

Figure R-2

Transmitter Pulses Showing Peak and Average Power

The frequency of the pulse-modulated carrier wave is in the microwave region of the electromagnetic spectrum, between 1 and 35 gigahertz. The higher the frequency the narrower the beam that can be transmitted, thus permitting more accurate direction determination. This is the case especially with laser radars, the beam width of these being less than a milliradian (0.057 degree).

When the target is moving toward the transmitter, the frequency of the target signal is increased, and when it is moving away it is decreased, relative to the frequency of the transmitted signal. The difference between the transmitted frequency and that of the target signal is called the Doppler frequency. The velocity of the target is given by:

$$v = \frac{f^d \lambda}{89.4} \text{ miles per hour,}$$

where f_d = the Doppler frequency in hertz, and
λ = the transmitted wavelength in centimeters.

The range of a radar depends on four things: the peak power of the transmitter, the size of the antenna, the sensitivity of the receiver, and the reflective characteristics of the target.

There are two basic types of transmitter. One is a self-excited oscillator that converts d.c. directly to microwaves, using a magnetron. The other has a low-power oscillator followed by one or more amplifiers, the final being a high-power klystron. Until recently the magnetron was the more powerful, but modern klystrons produce power in the megawatt range, and although they are not as efficient as magnetrons, the klystron transmitter is also more stable, which makes it more suitable for Doppler radar. Traveling-wave tubes are also used as microwave amplifiers. (These tubes are all described in the article *Electron Tubes.*)

Very small solid-state radar systems with integrated circuitry are presently being developed. Singly, they could be used as collision-avoidance devices for automobiles, intrusion alarms, traffic monitors, and so on. Stacked in phased arrays, they provide light reliable systems of considerable combined power that can be installed in airplane wings, fuselages, and other locations where they take up little room and do not interfere with the aerodynamic efficiency of the structure.

Standard radar antennas consist of microwave horns with parabolic reflectors which focus the beam in one direction. (These are described in the article *Antennas.*) The antenna is moved so as to scan the area being searched in a methodical manner. The operator may or may not know if there is a target in this space. Since there is a large volume of space, a search radar frequently uses two fan-shaped beams, one sweeping horizontally and one vertically. A tracking radar uses a pencil-shaped beam that is pointed at the position of the target revealed by the search radar, or otherwise known in advance (an orbiting satellite, for instance). Having been acquired, if the target moves off the center of the beam an error signal is generated that adjusts the antenna to keep the beam on the target.

Two types of receiver are used in radar systems: the superheterodyne and the video. The first-named is much more sensitive, but the video is simpler and is suitable for use with short ranges and strong target signals. The limiting factor in a radar receiver is its signal-to-noise ratio. Noise is always generated at the input of the first stage, and any incoming signal must be strong enough to override it. The noise in a receiver fluctuates continuously, with some of the noise peaks exceeding the signal levels of some target signals. To reduce this undesirable condition as much as possible, a range gate is used. This is a circuit that turns the receiver on just before the target signal is due to arrive and turns it off just afterwards. This reduces the number of noise peaks appearing on the display as signals. (See also article on *Radio Noise and Interference.*)

The radar cross-section (σ) is a figure of merit of the microwave reflection efficiency of a target. It varies with frequency. If the target is small compared to the radar wavelength, it is said to be in the Rayleigh region, where the radar beam is subject to backscattering. This occurs in rain and other types of precipitation. When the target dimensions are approximately equal to the wavelength, it is said to be in the resonance region. This gives strong reflections. (Weather radars operate on wavelengths that get strong reflections from rain.) The optical region is where the target dimensions are much greater than the radar wavelength. In this region the size, shape, and aspect ratio of the target are important. A ship broadside on gives a much stronger reflection than one end on, and a large airplane reflects better than a small one.

Discrimination between fixed and moving targets is often necessary, especially when the echoes from low-flying aircraft can be masked by stationary objects, as in ground-controlled approach. This is overcome by use of the Doppler principle mentioned above. However, simple continuous-wave Doppler radar cannot measure range, so this type of radar is suitable only for detection of the presence of a moving target. Transmitting pulses gives the range information, and a moving target produces pulses that vary in amplitude periodically, but this still does not give the discrimination required. However, the moving-target-indication (MTI) radar exploits the phase difference between a reference signal and the target. This remains constant for fixed targets but is continuously changing for a radially moving target.

There are many variations of the ways in which radar information can be displayed to the operator. All use a cathode-ray tube and make use of one or other of two characteristic methods of using the CRT: deflecting the electron beam or modulating it. (For more information on CRT's see article on *Electron Tubes.*)

In the first method, when a pulse is transmitted it triggers the beginning of a sweep of the electron beam horizontally across the CRT screen, as shown in Figure R-3. The pulse itself appears as a large "blip" at the beginning of the trace. The electron beam sweeps across the screen at a rate that will bring it to the right-hand side in the time interval between pulses. It is then blanked to prevent it making a return trace as it returns to the starting point. A new sweep begins with the next pulse, and so on. A target signal appears as a second, smaller blip at a distance along the trace corresponding to the time interval between its receipt and the transmission of the pulse. The range can be read from a horizontal scale marked in miles, meters, etc. This type of display is called an A-scope presentation. It shows the range only. The bearing is given by the direction in which the antenna is pointing, as indicated by a synchro dial, or similar device. Variations of the A-scope presentation using parallel sweeps as on a TV screen can also show azimuth or elevation. A more accurate reading of range can be made by using moveable markers controlled by a turns-counting dial. The marker blip is moved to the exact position of the target blip, and the range is read from the dial.

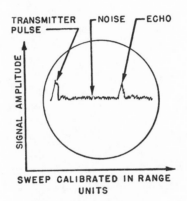

Figure R-3
A-Scope Presentation

The other type of display (Figure R-4) is obtained by having the horizontal sweep of the A-scope rotated radially about a point in the center of the screen, using deflection coils revolving around the neck of the CRT in synchronism with the antenna. Each trace starts at the center of the screen and sweeps out to the edge at an angle from the zero reference at the top that corresponds to the direction the antenna is facing when the pulse is transmitted. Since the movement of the antenna is infinitesimal between the pulse and its echo, an accurate bearing of the target is obtained. The target signal intensifies the trace at a distance from the center corresponding to its range, so that a bright spot appears. By using a screen phosphor with suitable persistence, this spot remains visible until the antenna completes its rotation and comes back to the same bearing, when a new target signal is received. By adjusting the brightness of the display the trace lines can be dimmed so that they do not interfere with the presentation. Since all objects in the circular area around the antenna give echoes that appear on the screen, the effect resembles a map. Moving targets are identified by the

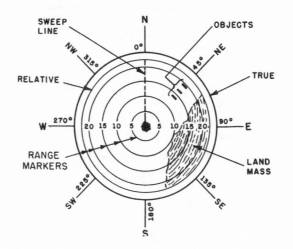

Figure R-4
PPI Presentation

change in position of their blips on successive sweeps. This type of display is called a plan position indicator (PPI), and is unsurpassed in its ability to keep track of all the aircraft flying in the vicinity of an airport. However, it is too slow for military use unless used in conjunction with a computer, because of the high speeds of missiles and enemy aircraft.

Laser radars are limited by optical visibility conditions, but are extremely accurate on account of their short wavelength. Most at present are used as rangefinders. The best-developed types use the helium-neon laser (see *Lasers*) and a photomultiplier receiver. The laser beam is intensity-modulated, and the range is determined by comparing the phase of the transmitted signal with that of the target signal. This is then displayed numerically by a suitable readout device.

radian measure—Alternative way of measuring angles, preferred in mathematics and physical science. An angle of one radian is defined as the angle between two radii of a circle when the arc between them is equal to the radius r. Since the circumference of a circle is given by $2\pi r$ it follows that an angle of 360 degrees is equal to 2π radians. One radian, therefore, equals $360/2\pi$, or 57.3 degrees approximately.

radiant flux—Rate of flow of radiant energy per unit time, expressed in watts. *Optoelectronics.*

radiation—1. Process of emitting electromagnetic energy or subatomic particles. 2. Energy or particles emitted. *Electromagnetic-Wave Propagation, Nuclear Physics.*

radiation resistance—Equivalent resistance that would dissipate the power radiated by *Physics.*

radiation hardening—Techniques use to minimize damage to electronic circuits by radiation.

radiation pattern—Diagram showing the relative field strength of an antenna at different angles but the same distance. The solid line shows the pattern of a horizontal half-wave antenna at a vertical angle of 15 degrees. Dashed lines are patterns for the angles shown. The double-pointed arrow indicates the orientation of the antenna. *Antennas.* (See Figure R-5.)

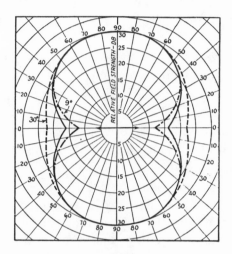

Figure R-5
Radiation Pattern

radiation pyrometer—(See pyrometer.)

radiation resistance—Equivalent resistance that would dissipate the power radiated by an antenna when the current flowing in it is the same as in the antenna. At the center of a half-wave dipole the radiation resistance is approximately 70 ohms, the same at the impedance at that point. It is not the same as the ohmic resistance, which is ordinarily small enough to be neglected. *Antennas.*

radio—Transmission and reception of communication signals carried by electromagnetic waves. *Broadcasting, Electromagnetic-Wave Propagation.*

radioactivity—Spontaneous emission of energy and subatomic particles by certain elements. *Nuclear Physics.*

radio astronomy—Study of electromagnetic emissions in the radio spectrum from celestial bodies, which has led to many surprises: nonthermal emission, solar bursts, quasars, pulsars, complex interstellar molecules, etc.

radio beacon—Fixed transmitter from which aircraft and ships can obtain navigational bearings by means of their own directional receiver systems. *Navigation Aids.*

radio beam—(See radio range.)

radio bearing—Angle in the horizontal plane with respect to a reference, usually North, measured clockwise in degrees from the reference, obtained by any of several radio methods. *Navigation Aids.*

radio direction finding (RDF)—Technique in which a direction-sensing receiver determines the direction in which a transmitter lies with respect to the receiver. *Navigation Aids.*

radio fix—Position of aircraft or ship determined by the intersection of two or more position lines (radio bearings) obtained by radio direction finding. *Navigation Aids.*

radio frequency (r.f.)—1. Electromagnetic waves used for radio communication in the frequency spectrum from 30 hertz to 3 terahertz. 2. Commonly used to designate signals in the input section of a receiver or the output section of a transmitter, as distinct from intermediate, audio, or modulation frequencies.

radio frequency amplification—1. In a receiver, selectivity and sensitivity are improved by use of a stage or stages, before the converter stage, to amplify r.f. signals to which the receiver is tuned, and to discriminate against r.-f. images, which become a greater problem at higher frequencies. 2. In a transmitter, an r.-f. power amplifier stage or stages are required to raise the power level of the oscillator output signal before feeding it to the antenna. *Electron Tube Circuits, Transistor Circuits.*

radio-frequency heating—Method of raising the temperature of materials by the application of electric currents of frequency above 70 kilohertz. In induction heating, used for conductive materials, eddy currents are induced by an inductor, within which the material is placed. In dielectric heating, used for poor conductors, the material is placed between two metal plates so that it forms the dielectric of a "capacitor," or placed in an electromagnetic field. Magnetrons operating at frequencies in the gigahertz range with continuous output power up to a kilowatt are used for dielectric heating. (A microwave oven is a low-power example of dielectric heating.)

radio-frequency interference—Noise and interference picked up by a receiver that limit its useful operating range, originating in the atmosphere, the galaxy, other equipment, or the receiver itself. *Radio Noise and Interference.*

radio-frequency signal generator—Oscillator which produces signals in the range from 10 kilohertz through 40 gigahertz that can be amplitude, frequency, or pulse-modulated. Popular service instruments are usually built to operate between 100 kilohertz and 50 to 60 megahertz, and provide an internal modulation frequency of 400 or 1000 hertz.

RADIO NOISE AND INTERFERENCE

Man-made noise is the strongest form of interference experienced in radio reception and is due chiefly to electric motors, neon signs, power lines, and ignition systems that are not more than a few hundred yards from the receiving antenna. High-voltage transmission lines and radio-therapy devices may affect reception over greater distances. Obviously these sources are at their worst in urban areas, somewhat less in suburban areas, and very much less in rural areas.

Generally, man-made noise decreases as the frequency of reception is increased. The principal exception is ignition noise from automobiles, which causes most interference in the VHF area (e.g., television). Most countries have legislation requiring the use of some means of preventing the radiation of this type of noise. A combination of radio-resistance type spark-plug wires and a capacitor in parallel with the ignition-coil primary is standard in U.S. automobiles as an ignition-noise suppression filter.

Propagation of man-made noise is mostly by transmission over power lines and by ground wave. Radio power supplies should therefore be provided with a line filter if powerline conducted noise is to be suppressed. A typical filter is shown in Figure R-6, where C1 and C2 are each 0.01 microfarad and R1 is 100 kilohms.

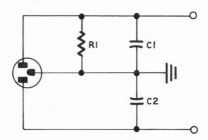

Figure R-6
Typical Line Filter at Input to Power Supply

Another source of interference is from undesired signals that overlap the channel reserved for the desired channel. This problem is met with in communications receivers mostly, and the better types are provided with selectable crystal filters to narrow their bandwidth so as to exclude the interfering signal. The use of a directional antenna also helps to discriminate against the undesired signal.

Galactic Noise

Galactic noise originates outside the Earth and comes from the Sun, various individual cosmic sources, and a general background radiation concentrated along the galactic plane. It has a frequency spectrum from 15 to 100,000 megahertz, although it is strong only in the region between 18 and 1000 megahertz. In this area it is slightly higher than rural man-made noise.

The dominant extraterrestrial radio source is the Sun. Lying above its surface are regions of tenuous gas and magnetic fields in which intense radio signals are generated. At times when sunspots are at a minimum ("quiet Sun"), the steady emission of radio radiation at about ten centimeters wavelength causes little trouble on Earth, but when sunspots are many ("disturbed Sun"), more violent radio events take place. Intense energy in the form of light, x-rays, and protons is radiated from the vicinity of sunspots, and the resulting noise storms may last from hours to days. Their effect is to disturb the ionosphere and interfere with radio communication (see *Electromagnetic-Wave Propagation*).

Individual cosmic sources lie mainly within ten degrees of the galactic plane. The most intense of these is called Sagittarius A and is believed to be the nucleus of the Galaxy. Others are supernova remnants, of which the strongest is Cassiopeia A, which exploded in 1702 and is expanding at 4000 miles per second.

Atmospheric Noise

The atmosphere, which excludes galactic noise below 15 megahertz, generates its own interference at frequencies below 20 megahertz. This noise is produced mostly by lightning. It therefore most affects regions subject to thunderstorm activity. Another type of atmospheric noise is called precipitation static and is produced by rain, hail, snow, or dust storms in the vicinity of the receiving antenna.

Receiver Noise

The ultimate sensitivity of a receiver is set by the noise inherent to its input stage. This is caused by random movements of electrons in the resistors and other components. When there is no other signal present it is heard as a hiss in the loudspeaker or seen as snow on the television screen.

The signal-to-noise ratio of a receiver is the difference, in decibels, between a specified signal reference level and the level of noise measured in the absence of a signal. The higher the ratio the better the receiver.

radiography—Process of obtaining x-ray photographs.

radiology—Branch of medicine that deals with the use of x-rays, radioactive isotopes, and nonionizing radiation in the diagnosis and treatment of disease.

radiometry—Science of measuring optical radiation of any wavelength. *Optoelectronics.*

radio range—Radio navigational aid such as "omnirange," or A-N range, in which a ground transmitter broadcasts a signal that an aircraft can use for guidance or navigational information. *Navigation Aids.*

radio receiver—Equipment for reception of radio signals and recovery of the information encoded in them. Practically all modern receivers are superheterodyne receivers (q.v.).

Radio Shack—Originally designated that part of a ship, usually topside, where the radio transmitter and receiving apparatus were located. (More recently associated with the Tandy Corporation and its network of retail outlets for electronic products.)

radiosonde—Balloon-borne package of meteorological instruments for measuring the temperature, pressure, and humidity of the atmosphere at altitudes up to 150,000 ft., and radioing the data to a ground station. By tracking the horizontal vector of the sonde's movement a determination of wind velocities may also be made.

radio spectrum—Range of frequencies from 30 hertz to 3 terahertz. *Frequency Data.*

radio telephone—Telephone set using a radio link instead of a cable. Term includes transmitter, receiver, and other necessary equipment.

radio telescope—Usually a steerable paraboloid antenna that picks up radio signals from extraterrestrial sources and feeds them to a radio receiver. The amplified signals are recorded for subsequent analysis. The radio sources include the Sun, planets, radio stars, interstellar gas, the galactic center, other galaxies, pulsars, and quasars.

radio transmitter—Equipment consisting of an oscillator, power amplifier, and power supply that generates a radio-frequency signal and feeds it to an antenna for propagation through space. Provision is also made for attaching information to this signal by code (wireless telegraphy) or modulation (radio telephony). Additional stages found in many radio transmitters include buffers, doublers, and modulators. (See Figure R-7.)

radium—Radioactive, alkaline-earth metal, silvery white, highly reactive; its principal uses have been in medicine (radiology) and luminous paint (now superseded by cobalt-60 and cesium-137 for cancer therapy, and phosphor activated by tritium for dial markings). Symbol, Ra; atomic number, 88; atomic weight, 226.05.

radix—Base number; e.g., 10 in the decimal system, 2 in the binary system.

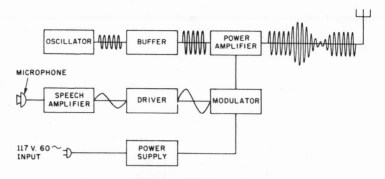

Figure R-7

Block Diagram of an A-M Radiotelephone Transmitter

radome—Dome to give environmental protection to a radar antenna (radar + dome = radome).

radon—Radioactive gas generated by the radioactive decay of radium; colorless, odorless, tasteless, it is 7½ times heavier than air. Symbol, Rn; atomic number, 86; atomic weight, 222.

rainbow generator—Type of generator that provides color-difference bars (keyed rainbow generator), or a display in which the colors merge as in a rainbow, on the television screen. Used for adjustment of TV color circuits. (See also color-bar generator.)

random-access memory (RAM)—Read/write memory, either volatile or nonvolatile, whose contents may be altered at will or read out without alteration, and which may be randomly addressed. *Computer Hardware.*

Rankine temperature scale—Absolute temperature scale in Fahrenheit degrees. Degrees Rankine = °F − 459.72.

rated—Designed. Said of a parameter (voltage, power, etc.). The rate value is usually the maximum value at which a device can be operated while still maintaining its specified performance.

ratio detector—FM detector and limiter. Frequency variations appear across the transformer output winding as voltages that charge C3 and C4 so that the ratio of the voltages across these capacitors changes with respect to each other. The voltage at the point between them therefore varies at the audio rate. (See Figure R-8.)

ray—1. Stream of particles. 2. Straight line along which any part of an electromagnetic wave travels from its source to a given point.

Rayleigh wave—(See seismic wave.)

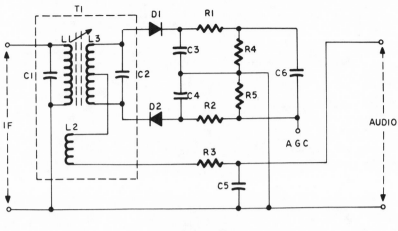

Figure R-8
Ratio Detector

From *Complete Guide to Reading Schematic Diagrams*, First Edition, page 141. ©1972. Reprinted with permission of Parker Publishing Company, Inc.

RC—Abbreviation for resistance-capacitance (q.v.).

RDF—Abbreviation for radio direction finding (q.v.).

R-display—Radar display consisting of an A-display (q.v.) with the added capability of magnification of the blip. *Radar*.

reactance—Measure of the opposition of a circuit or component to an alternating current, expressed in ohms. Inductive reactance (X_L) is a characteristic of coils and is given by $X_L{}' = 2\pi fL$, where f is the frequency and L the coil's inductance. Capacitive reactance (X_c) is a characteristic of capacitors and is given by $X_c = \frac{1}{2\pi fC}$, where f is the frequency and L the coil's inductance. Capacitive reactance (X_c) is a characteristic of capacitors and is given by $X_c = 2\pi fC$, where f is the frequency and C the capacitor's capacitance. Inductive reactance causes the voltage to lead the current, while capacitance reactance has the opposite effect. (See also resonance.)

reactance circuit—Circuit with an electron tube or transistor that "looks like" an inductance or capacitance to a capacitor or inductor connected in parallel with it. Varying the bias on the tube or transistor varies the apparent reactance, and so the resonant frequency. This type of circuit can be used for frequency modulation when the modulating signal is applied to the reactance circuit as shown. Capacitive reactance is provided by this circuit in which C2 at the resonant frequency has a reactance that is ten times the resistance of R2. This reactance is in parallel with L3 in a Hartley oscillator circuit. (See Figure R-9.)

read-only memory (ROM)—Nonvolatile memory whose contents are mask-programmed during microelectronic fabrication; consequently they cannot be

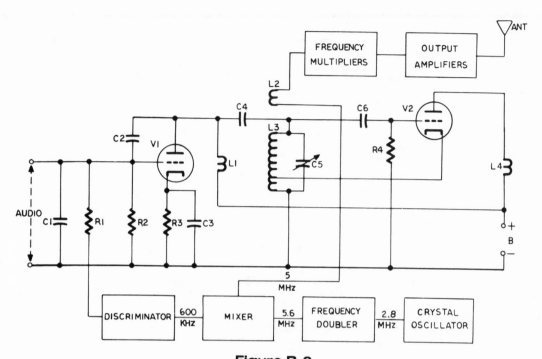

Figure R-9
Reactance Circuit and Oscillator for FM Transmitter

altered. This type of memory is used to store microinstructions, reference data, and various codes. *Computer Hardware, Microelectronics.*

readout—Output data of a computer. *Computer Hardware.*

real time—Actual time, or clock time, as opposed to fictitious time. A computer controlling some process must operate in real time, whereas many computer programs that are not dependent on or responsible for real-time events use an artificial time of their own.

Réaumur temperature scale—Obsolescent temperature scale in which 0° was the freezing point of water and 80° its boiling point.

receiver—(See radio receiver.)

receptacle—Any female type of connector, especially a wall socket or panel-mounted multicontact connector.

rechargeable—Capable of being recharged, as a secondary cell or battery.

recombination—1. The filling of a hole by an electron. *Semiconductors.*   2. In the ionosphere, the process by which free electrons recombine with charged atoms (ions) at night, neutralizing, or partially neutralizing, the ionized layers. *Electromagnetic-Wave Propagation.*

recording—Methods of storing for later reproduction any type of information that can be converted into electrical signals. *Recording.*

RECORDING

Modern recording is mainly divided into mechanical and magnetic. Optical recording is also used in the production of motion pictures, but will not be covered here (however, sound tracks on many motion-picture films today are magnetic, and the process is the same as described below).

Mechanical Recording

Phonograph records are made by cutting a master, usually a lacquer-coated aluminum disk, using a sharp, accurately-ground, heated stylus. This disk must rotate at either 33-1/3 or 45 revolutions per minute ± 0.1 percent, in a clockwise direction. The outer diameters of finished disks are nominally 12, 10 or 7 inches, no other sizes being made today.

The stylus is moved sideways at a steady rate from the outer margin of the revolving disk to where the diameter of the innermost recorded groove is approximately one half the diameter of the outermost groove. To go any further would result in an unacceptable level of distortion. (Distortion increases from the outside margin toward the center, as the actual speed of the groove past the stylus decreases.) There are between 200 and 300 grooves per radial inch.

The requirements in modern stereo recording are that the groove width shall be not less than 0.001 inch, with sides sloping inward at 45 degrees, so that they make a 90-degree angle with each other. (Actually the bottom of the groove is rounded to have a radius of not more than 0.0002 inch.) In recording two sound channels with one stylus, the right-hand channel is engraved on the outer wall of the groove by stylus vibrations perpendicular to that wall only. The same vibrations are parallel to the other wall, so do not engrave it. The left-hand sound channel, on the other hand, is engraved on the inner wall of the groove by vibrations perpendicular to it, which are therefore parallel to the other wall. The two motions, at right angles to each other, are imparted to the stylus by magnetic fields generated by the stereo audio signals in two electromagnets. To be compatible with either stereo or monophonic equipment, monophonic records are made in the same way, although both channels are identical.

To make copies of the master, dies are made by electro-forming, a process in which a layer of silver is deposited by a chemical process on the master to serve as an electrode in an electrolytic solution. When a suitable current is passed through the solution, a metal plate is formed, the surface of which is an exact negative replica of the original.

From this negative master a positive master, called a "mother," is then made by the same process. It is then used as a mold to produce a number of stampers, from which the records are made.

These are molded from vinyl plastic, which becomes soft and flows under pressure and heat, but the finished records are remarkably durable when handled and played with proper care.

The source of the audio signal that drives the record cutter is usually a microphone. Other sources include magnetic recordings (see below), but these have first to be recorded, also using microphones.

A microphone is a device for converting the energy of sound waves in the air into an essentially equivalent electrical signal. The largest class of microphones are those used in telephones (q.v.). This class—carbon microphones—is not suitable for recording, however, because the frequency response is not broad enough. The same objection applies to piezoelectric crystal microphones.

In the microphone, the sound waves impinge on a slightly flexible surface (diaphragm), causing it to vibrate in a manner corresponding to the movement of the air particles. This motion of the diaphragm causes a similar variation in some property of an electrical circuit. In a dynamic microphone, the diaphragm moves a coil to-and-fro in a magnetic field, inducing a current in the coil that is proportional to the movement. An electrostatic microphone employs the field variations produced by the movements of the diaphragm in relation to an adjacent fixed plate (condenser microphone) or electret (q.v.) to generate an electrical signal. These microphones have frequency-response ranges from 20 to 20,000 hertz, generally with an impedance of 600 ohms.

The ribbon, or velocity microphone does not have a diaphragm. Instead, a very thin and light aluminum-alloy ribbon is suspended between the poles of a powerful magnet. This ribbon is accordion pleated so that it has practically no resistance to movement. As sound waves move the air particles around it, it moves with them. As it moves in the magnetic field a voltage is induced in the ribbon proportional to the sound. This microphone is very sensitive and must be treated with great care to avoid damage, but is excellent for use in a recording studio.

Microphones are designed to pick up sound from all around (omnidirectional), or mostly from in front and from the sides (cardioid). The latter has very little pickup from the rear, so it is the preferred type for recording purposes. However, a velocity microphone has equal pickup from front and rear, but practically none from the sides. This can be modified by special acoustic chambers in the microphone housing.

The microphone's output signal must be amplified to enable it to drive the cutter, and the amplifier is also responsible for ensuring that the sound track complies with the required characteristics of the recording.

To convert the undulations of the record groove back to sound, a pickup, turntable, amplifier, and speaker or speakers, are required. These may all be parts of one unit such as a "stereo compact" (except that the speakers are separate), or they may be individual components in a high-fidelity system.

The pickup is a device actuated by the record groove. To minimize wear on the record and the stylus, the better pickup is designed to exert a tracking force of 3/4 to 1½ grams (0.026 to 0.053 ounce). Diamond styli with tip radii of between 0.0125 and 0.0175 millimeters (0.0005 and 0.0007 inch) are standard for modern equipment. These should perform satisfactorily for 100-500 hours of playing time (a saphire stylus for 20-100 hours). The tip is in the form of a cone often elliptical with a rounded apex. The angle of slope of the cone is between 40 and 55 degrees to fit the record groove. As the stylus wears, its sides become flat and it runs gradually deeper in the groove until it rides on the bottom, with noisy and distorted reproduction. As explained above, the groove has two tracks. The undulations of one track move the stylus at right-angles to the direction it is moved by the undulations of the other. These movements are converted by two electromagnetic coils into the left- and right-channel audio signals.

The pickup cartrige is mounted on the end of the tone arm, which is balanced so that only minimum tracking force is applied. The tone arm is part of the record player, which consists of a turntable driven by an electric motor. Turntables for the ultimate in high fidelity do not have record changers, so that they can use tracking force of under one gram, and the best use a 12-pole brushless d.c. servomotor, of which the motor shaft itself is the spindle, thus elminating all speed reduction gear. By using a feedback system, the speed of the motor is kept accurate to within 0.0005 percent of the mean speed. (The standard average deviation may not exceed 0.1 percent.)

However, most people prefer record changers which allow a stack of selected records to be played in sequence automatically. The modern mechanism, once the motor has been turned on, lifts the tone arm from its rest, drops the bottom record from the stack by retracting the lower retainer into the locking "umbrella" spindle (the rest of the stack is held by the upper retainer), moves the tone arm over and lowers it on to the edge of the record, where the stylus indexes on the lead-in groove, follows the grooves all the way to the "trip," or reject grooves, lifts the tone arm and sends it back to a position above its rest, and drops the next record. It then repeats the above procedure. When the last record has been played, the tone arm is replaced on its rest and the motor, and in some cases the amplifier also, is shut off. When changing a record there are more moving parts in action than are required to fire a machine gun. Since all these operations are controlled by the stylus tracking in the trip groove—moving the tone arm so as to actuate the mechanism—the stylus must track with somewhat more force if it is not to jump out of the groove and "skate." This shortens the life of both stylus and record to some extent.

Another feature that is not popular with audiophiles is the use of pulleys and belts, or rubber-tired wheels, for speed reduction. These are adjustable for 45, 33-1/3 and, but not always, the obsolete 78-rpm speed (since many of these older records are still in existence). The shaded-pole motor seldom causes problems, but pressure grooves on rubber wheels caused by disuse can give rise to a noticeable thumping. The predominant cause of trouble, however, is transfer of drive to the turntable rim where insufficient, or excessive, friction will result in there not being enough power to operate the change cycle, uneven turntable speed, or low motor speed. Uneven turntable speed produces "flutter" or "wow" (higher or lower cyclic deviations), while vibrations termed "rumble" may be generated by the turntable, pickup pivot, speed-reduction gear, or any combination of them.

The amplifier used to build up the minute signals from the pickup to the power necessary to actuate modern loudspeakers (as much as 120 watts for wall-shaking volume) is a remarkable piece of equipment. It has at least two, sometimes three, tone controls (bass, midrange, and treble), a balance control for adjusting the volume levels of the two stereo channels relative to each other, and many others, the total depending on whether it is a plain amplifier or contains also a preamplifier and FM- and AM-radio tuners. Since the same amplifier is often designed to record on tape, there are microphone inputs and outputs to the tape deck (see below). The frequency response of a good amplifier may span from 15 to 25,000 hertz, with total harmonic distortion of 0.1 percent and a signal-to-noise ratio of 70 decibels.

Loudspeakers are classified in much the same way as microphones, and the dynamic, or moving-coil type is the standard model. This has a coil mounted centrally on a conical diaphragm so that it is positioned in the powerful field of a permanent magnet. The audio output signal of the amplifier flows in this coil, and the current generates a magnetic field that interacts with the field of the permanent magnet. Since the diaphragm is flexible, the coil can move, though the permanent magnet cannot. Variations in the coil current cause it to move backward and forward in conformity with the resultant fluctuations in its magnetic field, and the diaphragm is accordingly made to move in and out. This movement creates low and high-pressure zones in the air in contact with the coil, which radiate outward as sound waves. Most loudspeakers used in high-fidelity systems have at least two such units mounted on a baffle board, which forms the front of a closed cabinet. This closed cabinet is sound insulated so as to prevent sound waves from the rear of the speakers getting around to interfere with sound waves from the front. Generally, there is a large heavy-duty speaker with a diameter from 8 to 15 inches, called a "woofer," to handle the bass, a medium-size midrange speaker with a smaller diameter (4 to 8 inches), and a very small speaker (2-4 inches) called a "tweeter." Each speaker handles a different frequency band:

Woofer	20-800 Hz
Midrange	800-8,000 Hz
Tweeter	8,000-25,000 Hz

These figures are for a good quality speaker. The average range overall is generally 50-18,000 Hz. The boundaries between the ranges of each speaker are called crossovers, and provision is made in better-class speakers for adjustment of crossover frequencies to suit room acoustics. (Loudspeakers from an acoustic point of view are dealt with more fully under *Electroacoustics*.)

A recent development has been the application of the phonograph recording principle to the recording of television programs. These records are called videodisks. There are two types, optical and capacitive, which are not compatible with each other.

The optical videodisk rotates at 1800 rpm. It has thousands of circular tracks forming a continuous spiral from the inside of the videodisk to the outside (opposite to a phonograph). However, these tracks are not in grooves. They are "dotted lines" of minute indentations in the videodisk material. Each track is 0.4 micrometer (15.7×10^{-6} inch) wide and separated by 1.6 micrometers (63.0×10^{-6} inch) from the adjacent tracks, but the length and width of the indentations vary with the information recorded on them.

The recorded signal is a composite signal consisting of an 8.10-megahertz carrier, frequency-modulated with the television video signal and two sound channels, so the audio can be played back on a stereo system, if desired. A built-in laser focuses a minute beam of light on the video track, and this beam is intensity-modulated by it and reflected back to a photodiode, from which suitable circuitry reconverts the recorded signal to the standard television format, so it can be applied to an ordinary television receiver. The laser beam is kept on the track by two secondary beams that fall on each side of it and are reflected back to two other photodiodes. These generate focus-error and radial tracking-error voltages.

Optical videodisks have two playing modes. The standard-play mode allows up to 30 minutes per side; the extended play mode, 60 minutes.

The capacitive videodisk has grooves like a phonograph record, and employs a pickup with a diamond stylus. The stylus is wedge shaped, with the sharp end leading. The flat rear face of the wedge has a thin metallic coating. The record is made of conductive plastic, and the recorded signal consists of a frequency-modulated carrier that produces a sequence of transverse furrows of varying width and spacing in the groove. As the stylus passes over these furrows the capacitance between the metallic coating on the stylus and the videodisk varies in accordance with the recorded signal, resulting in an electrical signal which is then reconverted to the standard television format and applied to a television receiver. This videodisk rotates at 450 rpm, allowing 60 minutes playing time on each side of a 12-inch record.

Magnetic Recording

Modern recording tape consists of a plastic-film base one-quarter inch wide (except video tape) coated with magnetic powder held in place by a binder. The base is commonly Mylar or polyester. The most widely used magnetic material is red iron oxide (Fe_2O_3), but chromium dioxide CrO_2), which has a 50-percent greater output and wider frequency range, is expected eventually to replace it for high-fidelity use (at present it tends to make the head dirty, requiring more cleaning after use).

The electrical signal to be recorded (from a microphone, or other source, as in mechanical recording) is applied to the tape by a head. This is a magnet consisting of soft iron pole pieces wound with a wire coil. The pole pieces are separated by a small gap of some 0.005 millimeter (0.0002 inch). This gap allows the flux induced in the core to spread out so that it penetrates the tape, which is in contact with it. As the tape travels past the gap the variations in the magnetic flux magnetize the particles in the tape so that it receives a magnetic record containing direction (north or south polarity), amplitude, and linear dimension (along the tape), corresponding to the polarity (positive or negative), amplitude, and time of the electrical signal. During recording the head is also biased with an alternating current of 60 to 100 kilohertz to improve the frequency response, distortion, and signal-to-noise ratio. This bias is much higher in amplitude than the signal current and several times the highest frequency to be recorded, and assists the magnetic particles to reorganize themselves in a truer reproduction of the electrical signal. It is, however, essential that the tape is first completely demagnetized, and comes to the recording head in a neutral condition. Accordingly, it is, in many tape recorders, made to travel first past an erase head, which subjects it to a powerful, high-frequency alternating magnetic field that diminishes as the tape passes away from the gap, leaving it demagnetized. This may also be accomplished by placing the entire reel or cassette in a bulk eraser, in which the tape is subjected to a strong alternating magnetic field, which is slowly reduced to zero. Either method allows tapes to be used over and over again.

To play the recording, the tape is passed over a head similar to that used to record it (the same one is used in many cases). The tape speed (usually 7½ inches per second, but other speeds that may be used are 15, 3-3/4, and 1-7/8 inches per second) must, of course, be exactly the same as that at which it was recorded. The magnetized portions of the tape passing over the gap in the reproducing head cause the magnetic flux in the core to change, generating a signal voltage in the coil, which is a close facsimile of the original signal. This is applied to the input of the amplifier as in the case of a phonograph signal.

High quality magnetic-tape recorders are used professionally for recording the master programs for phonograph records, tape cassettes, and motion-picture sound. Magnetic tape is also used to store television programs (see below). Tape cartridges marketed for tape players are generally eight-track stereo tapes; as stereo requires two tracks, these cassettes are played through four times, which gives a program duration of about an hour.

The most common method for driving a tape recorder or tape player is the capstan drive in which the tape is pulled forward by the friction between a rubber pinch-wheel and a metal spindle, the capstan driven by the motor. The feed reel and the take-up reel are usually also driven by the motor to maintain proper tension on the tape.

Cassettes and cartridges are designed to save the user the trouble of threading the tape. Cassettes contain a feed reel and take-up reel, with tape, enclosed in a flat rectangular case. The tape is narrower than the standard quarter-inch width, usually 0.15 inch (3.8mm), and travels from reel to reel via guides, being pulled along by a pinch-wheel and a capstan that are part of the tape player. The feed and take-up reels are also driven by the player. Cartridges, on the other hand, use a quarter-inch endless loop of tape, fed from the inside of a single reel. The tape is driven by friction between the tape player's capstan and a pinch-wheel built in the inside of the cartridge, around a guide, past the head, around a second guide, and back to the outside of the reel. In the most widely used cartridge size (4″ × 5½″), there are over 550 feet of tape. Two other sizes that are used in professional equipment are 6″ × 7″ and 7-5/8″ × 8½″.

Professional video tape recorders use a tape that is two inches wide, but otherwise no different from the standard quarter-inch tape. The reason for the greater width is the high-frequency components of the video signal. These require a far higher tape speed than do the audio signal components, but it is impractical to run the tape at some 1500 inches per second, because of tape handling problems and the size of the reels required to hold so much tape (close to half a million feet for an hour's program time). So the video portion is recorded in transverse tracks across the tape while it moves forward at 15 or 7½ inches per second. This is done with a headwheel, which has four recording heads mounted at equal distances around its rim. The headwheel has a diameter of two inches and is mounted in a plane perpendicular to the tape. As it spins at 240 revolutions per second, and as the tape advances at 15 inches per second, the video track consists of diagonal sections 10 mils wide and separated by guard bands of 5 mils. The actual distance traveled by each head in a second is 1562 inches, which gives the required speed for the video signal, while 15 inches per second is adequate for the audio signal, which is recorded along one edge of the tape, while control signals are recorded on the other edge. The speed of 7½ inches per second mentioned above is a secondary speed used with a headwheel that makes a 5-mil track.

A simpler version of video tape recorder, which cannot be used in standard broadcasting because of the lack of standardization among the different models, is the helical-scan type. This uses 1.0 or 0.5-inch tape that passes in a slanted path around a drum. A headwheel with one or two heads rotates inside the drum and traces slanted tracks across the tape in a similar manner to that described above.

Reels for holding quarter-inch tape come in various sizes, and hold lengths of tape as follows:

Reel Diameter	Hub	1.5 mil	Max. capacity 1.0 mil	0.5 mil	Playing time (mins.) 7.5 ips		
3"	1.75"	125'	200'	300'	3½	5⅓	8
5"	1.75"	600'	900'	*1200'	16	24	32
7"	2.25"	1200'	1800'	*2400'	32	48	64
10.5"	4.5"	2500'	3600'	—	66⅔	96	—
14"	4.5"	5000'	7230'	—	133⅓	192	—

*tensilized

Specifications for standard video-tape reels are:

Reel Diameter	Hub	Max. Capacity (1.0 mil)	Playing Time (mins.) 15 ips
6.5"	4.5"	750'	10
8"	4.5"	1650'	22
10.5"	4.5"	3600'	48
12.5"	4.5"	5540'	74
14"	4.5"	7230'	96

record player—In modern high-fidelity equipment, the unit that rotates a phonograph record and converts the mechanical impressions in the groove to electrical signals for amplification in the amplifier and conversion to sound in the loudspeaker. *Recording, Electroacoustics.*

rectangular waveguide—Rectangular metal tube with one side twice as wide as the other, used to channel electromagnetic waves of microwave frequency. *Waveguides and Resonators.*

rectifier—Device that converts alternating current to direct current. It may be a solid-state device, electron tube, vibrator, or mechanical device. *Rectifiers and Filters.*

RECTIFIERS AND FILTERS

The most convenient source of electrical power is the utility company's 117-volt, 60-hertz line. However, this must be changed to d.c. at other voltages for most purposes in electronic equipment. Rectifiers are used to turn the a.c. to pulsating d.c., and filters then smooth the latter until it is as close to pure d.c. as necessary for the equipment concerned.

Rectifiers today are mostly solid-state devices using the rectifying properties of semiconductors. Silicon, germanium, selenium, and copper oxide are those most often used. However, electron-tube rectifiers are also used for some purposes. For the majority of uses the silicon rectifier is preferred.

Figure R-10 shows a single-phase half-wave rectifier circuit. The diode conducts only on a positive-going excursion of the input a.c.; therefore the output of the circuit consists of only half of the input, resulting in low efficiency.

Figure R-11 shows a single-phase full-wave center-tap rectifier circuit. The efficiency is good since both excursions of the input a.c. are used. However, it requires a larger transformer with the added complication of a center tap.

Figure R-12 shows a single-phase full-wave bridge rectifier circuit. This is equally efficient, requires a simpler transformer, and with the low cost of solid-state rectifier diodes, is cheaper to make than the center-tap rectifier circuit.

Figure R-13 shows a conventional voltage doubler circuit. There is no transformer. Capacitors C1 and C2 are each charged, during alternate half-cycles, to the peak value of the alternating input voltage. The capacitors discharge in series into the load R_L, so their individual voltages add together.

Figure R-14 shows a cascade voltage doubler circuit. In this circuit the capacitor C1 is charged to the peak value of the a.-c. input via D2 during one half-cycle. During the other half-cycle it discharges in series with the a.-c. input through D1 to charge C2 to twice the a.-c. peak voltage.

Figure R-15 shows a bridge voltage doubler, which is converted from the circuit in Figure R-13 by adding two more diodes.

Figure R-16 shows how cascade voltage doublers (Figure R-14) can be added to obtain further voltage multiplication. The number of stages is limited, however, by the load current, which must be small, and deterioration in regulation as the number of stages increases.

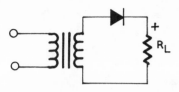

Figure R-10
Single-Phase Half-Wave Rectifier

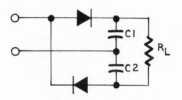

Figure R-11
Single-Phase Full-Wave Center-Top Rectifier

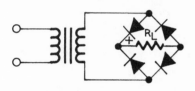

Figure R-12
Single-Phase Full-Wave Bridge Rectifier

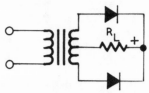

Figure R-13
Voltage Doubler Rectifier

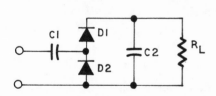

Figure R-14
Cascade Voltage Doubler Rectifier

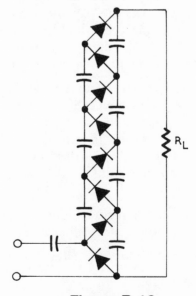

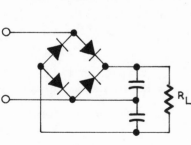

Figure R-15
Bridge Voltage Doubler Rectifier

Figure R-16
*Cascade Voltage Doublers Combined
to Make a Voltage Quadrupler*

The three types of rectifier filter are shown in Figure R-17. (A) is called an inductor-input filter; (B) is called a capacitor-input filter; and (C) is a resistor-input filter.

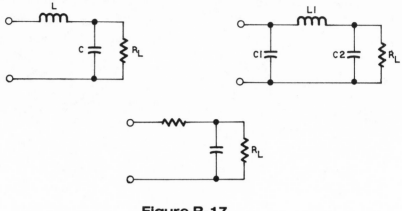

Figure R-17
Rectifier Filters

(a) inductor-input filter (one section shown)

(b) capacitor input filter (π filter)

(c) resistor-input filter

The percentage ripple from a single-section filter as in Figure R-17a, made up of any values of inductance and capacitance, may be determined closely enough for practical purposes for a full-wave rectifier from 100/LC, where L is in henries and C in microfarads. (In the case of a half-wave rectifier the values of inductance and capacitance must be doubled.) The minimum value of the choke in henries is determined by dividing the maximum load resistance in ohms by 1000, so once this is known the value of C can easily be calculated.

It is obvious from the above that the value of the choke is mathematically related to the total load resistance. Since this may vary considerably it is wiser to use chokes with twice the calculated value. Alternatively, a swinging choke may be used. This will have the lower value of inductance ($R_L/1000$) when there is no load (or only a bleeder resistor) and about twice this value at full load. (This also indicates that the value of the bleeder resistor should be 1000 times the maximum inductance of the swinging choke, in henries.)

Additional filter sections identical to Figure R-17a may be cascaded and each will reduce the ripple at its input in accordance with the factor 100/LC given above.

The value of C1 in a capacitor-input filter for a full-wave rectifier is determined from

$$r = 0.00188/C_1 \, R_L$$

where: r is the degree of filtering required, and is equal to the maximum r.m.s. ripple voltage appearing across C1 (E_r) divided by the d.-c. voltage across C1 (E_{dc}); C_1 is the capacitance of C1 in microfarads; R_L is the maximum value of total load resistance in megohms.

The maximum value of C1 is limited, however, by the maximum allowable peak-current rating of the rectifier. A minimum value of 4 microfarads is assumed in this calculation.

Greater filtering is obtained by adding the inductor L1 and capacitor C2 shown. Their values are calculated in the same way as for the inductor-input filter section, and the reduction of ripple is given by the same factor, 100/LC.

Since a resistor presents the same opposition to d.c. that it does to a.c., resistor-input filters are suitable only for low-current applications. The value of the resistor ideally should be about the same as the reactance of a choke in the same position in the circuit, but a lower-value compromise is often made so as not to reduce the output voltage unduly. This usually means that the resistor will be about one-tenth of the output load.

redundancy—Provision of duplicate circuits so that if one fails another will carry on. Redundancy is particularly important in unmanned space probes and the like, where there is no way of repairing the equipment, but the mission's reliability is extremely important.

reed relay—Type of relay in which two strips of magnetic material sealed inside a glass tube are caused to come into contact by the magnetic field of a surrounding coil whenever it is energized.

reference—Any standard, such as zero, 100 percent, one milliwatt, the threshold of hearing, etc., with which a quantity is compared to give its value above or below the reference.

reflected wave—In addition to the direct wave arriving at a receiver, waves may arrive which have been reflected by mountains, buildings, aircraft, the ionosphere, etc. Because of its longer path, the reflected wave may or may not be in phase with the direct wave, adding to it or subtracting from it as the case may be. Propagation beyond the radio horizon is made possible for high frequencies by waves reflected from the ionosphere. *Electromagnetic-Wave Propagation.*

reflector—1. Antenna parasitic element that reflects that part of the radiated field traveling in the opposite direction to the desired direction, reinforcing the field going in the desired direction.   2. A paraboloidal reflector that concentrates and directs the field emitted by an electromagnetic horn. *Antennas.*

refractive index—Measure of the bending of a ray of light when passing from one medium to another, given by $n = c/v$, where $n =$ the refractive index, $c =$ the velocity of light in empty space, and $v =$ the velocity of light in the medium. Also, $n = \sin i / \sin r$, where $i =$ the angle of incidence and $r =$ the angle of refraction. The refractive index varies with the wavelength and the medium. (See Figure R-18.)

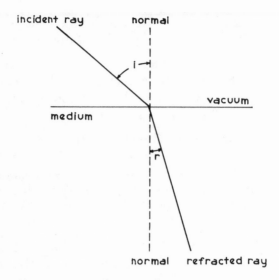

Figure R-18
Refractive Index

refreshing—In a dynamic RAM the data must read automatically at regular intervals of a few hundred nanoseconds, and new data bits written in before the previously stored bits fade away. *Computer Hardware.*

regeneration—Positive feedback. *Feedback.*

regenerative detector—Detector circuit with positive feedback that is very sensitive just below the point where oscillation begins. (See Figure R-19.)

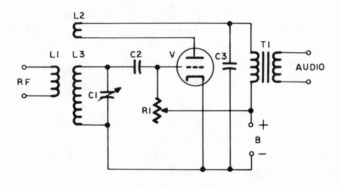

Figure R-19
Regenerative Detector

From *Complete Guide to Reading Schematic Diagrams*, First Edition, page 137. ©1972. Reprinted with permission of Parker Publishing Company, Inc.

register—Group of flip-flops used for temporary storage of, or operation on, a binary bit pattern (word or byte). *Computer Hardware.*

regulation—In a power supply, circuits that ensure a constant output regardless of variations in load or line voltage. If the output voltage should begin to rise or fall from the value set by the adjustment of the voltage-adjust potentiometer, the change is sensed by the comparator, which monitors it constantly against the voltage reference (frequently a zener diode). An error voltage is generated, amplified, and used to adjust the bias on the series regulator to restore the original level. (See Figure R-20.)

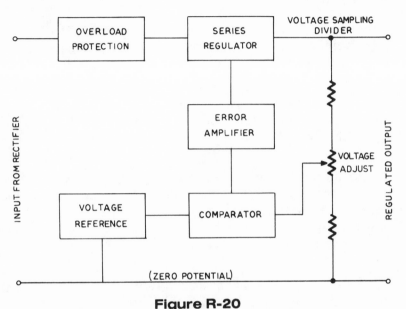

Figure R-20
Regulation (Series Regulator)

From *Complete Guide to Reading Schematic Diagrams,* First Edition, page 165. ©1972. Reprinted with permission of Parker Publishing Company, Inc.

relative humidity—Ratio of actual vapor pressure of water in the air at any given time to what it would be if the air were saturated, expressed as a percentage.

relaxation oscillator—Class of oscillator characterized by a large excess of positive feedback so that it operates in abrupt transitions between two end states. Examples are blocking oscillators and multivibrators (qq.v.). (Also called R-C or resistance-capacitance oscillators.)

relay—Type of switch actuated by current flowing in a coil wound around an iron core, which is thereby magnetized and attracts an armature, so opening or closing the switch contacts. Removal of the current returns the switch to its resting condition (normally open, or normally closed), except in the case of latching relays, which must be reset. (See also reed relay.)

reluctance—The geometrical and material properties of a magnetic circuit that offer opposition to the presence of magnetic flux, analogous to the resistance of an electric circuit, and given by $\mathcal{R} = F/\phi$, where $\mathcal{R}$ = the reluctance in ampere turns per weber, F = magnetomotive force (m.m.f.) in ampere turns, and ϕ = flux in webers.

rem—Unit of dosage of ionizing radiation that will cause the same amount of biological injury in a human being as one roentgen of x-rays or gamma rays. (See also rad, roentgen.) *Nuclear Physics*.

remanence—Flux density remaining in a magnetic material after the removal of a magnetic field which caused it to be saturated. Permanent magnets have very high remanence, core materials comparatively low. *Properties of Materials*.

remote sensing—In a regulated power supply, external voltage leads which are connected between the comparator circuit and the load so as to sample the voltage across the load, excluding voltage dropped in the power leads. (See also regulation.)

repeatability—Measure of the ability of a meter or other instrument to provide the same reading every time for the same input, given as a percentage of the full-scale value. For example, if the full-scale value is 100 volts and the repeatability is 0.5 percent, it means the reading will be within half a volt of the same scale reading each time for the same input.

repeater—1. Amplifier used in telephone lines or submarine cables to compensate for loss of signal strength over long distances. 2. Satellite repeaters for long-range communication. 3. Microwave relay stations which automatically receive, amplify, and transmit signals from one repeater to another to extend the range for as far as required.

repeller—In an electron tube, such as a reflex klystron, an electrode whose primary function is to reverse the direction of an electron stream. (Also called a reflector.) *Electron Tubes*.

repetition rate—(See pulse repetition rate.)

repulsion—Mutual action by which like charges or magnetic poles repel each other.

reset—To return a device to its original state, as in a latching relay, or to clear data from a register or similar storage device.

resist—(See photoresist.)

resistance—Property of a circuit or component that transforms electrical energy into heat energy (resulting from collisions between the charge carriers and fixed particles in the conductor) expressed in ohms (Ω), and given by R = E/I, where R is the resistance, E the voltage, and I the current (in amperes). *Resistors.*

resistance-capacitance (RC)—Combination of resistive and capacitive elements used for coupling between stages, filtering, or timing. The delay in charging a capacitor through a resistor connected in series with it is called the RC constant and is given by R × C, expressed in seconds, where R is the resistance in ohms and C is the capacitance in farads.

resistance-capacitance oscillator—(See relaxation oscillator.)

resistivity—Resistance of a conductor of unit cross-sectional area and unit length, symbolized by the Greek letter ρ (rho), and given by ρ = RA/1, where R is the resistance of, for example, a length of wire in ohms, A is its cross-sectional area and l its length, expressed as circular mils/foot, etc.

resistor—Component used to introduce resistance into a circuit. *Resistors.*

RESISTORS

Fixed Resistors

Resistors are the most ubiquitous components in electronics. They are used to protect, operate, or control circuits, to divide voltages and, in combination with other components, to shape waveforms, broaden the response of tuned circuits, and isolate parts from each other, so that one will not interfere with the action or operation of the other. All the various uses of resistors are based upon their opposition to a flow of current.

The opposition to a flow of current given by a resistor is its resistance, expressed in ohms (Ω). The ohm is the electrical resistance between two points on a conductor, which does not itself contain any source of electromotive force, when a constant potential difference of one volt maintained between those points results in a current of one ampere in the conductor.

The choice of the material used to make a resistor depends upon its resistivity, symbolized by the Greek letter ρ (rho), which is equal to the resistance of a specimen, such as a wire, divided by its length and multiplied by its cross-sectional area: ρ = RA/l. If R is expressed in ohms, A in square meters and l in meters, ρ is expressed in ohm meters. The resistivity of a good conductor such as copper is 1.77×10^{-8} ohm meter, carbon (graphite) 1400×10^{-8} ohm meter, and fused quartz $> 10^{21}$ ohm meters.

These resistivity values are all for a temperature of 20 degrees Celsius (68 degrees Fahrenheit). Resistivity of metallic conductors generally increases with a rise in temperature, but resistivity of semiconductors such as carbon and silicon decreases.

Materials used for resistors are composition, wire, and film. Composition resistors (see Figure R-21) are made with carbon particles mixed with a binder. This mixture is molded into a cylindrical shape and hardened by baking. Leads are attached axially to each end, and the assembly is encapsulated in protective insulation. Colored bands on the tan outer covering indicate value and tolerance (see *Color Coding*). Composition resistors have the advantage of cheapness, but have a high noise level, especially in values above one megohm. They are rated for maximum wattage at an ambient temperature of 70°C. This decreases with higher temperature. The maximum wattage ranges from 1/8 watt to 2 watts and is the maximum power that the resistor can dissipate. It is usual to select resistors with twice the wattage rating that will be required in the circuit. Composition resistors have an end-to-end shunted capacitance that may be noticeable at high frequencies, particularly in resistors with values over 0.3 megohm, which begin to be affected at about 100 kilohertz. Lower values are affected above one megahertz.

Figure R-21
Composition Resistor

Wire-wound resistors are made by winding wire of a nickel-chromium alloy on a ceramic tube and covering it with a vitreous coating. Power wire-wound resistors (see Figure R-22) are fairly large and are furnished in power ratings of from one to 210 watts. The connections for these may be axial or radial leads, or lugs. Brackets for mounting the larger sizes must be used since they are fairly heavy. As might be expected, the spiral winding has the inductive and capacitive characteristics of a coil of similar construction, making such resistors unsuitable for use above 50 kilohertz. The frequency limit can be raised by a noninductive winding, in which the magnetic fields produced by its two parts cancel each other, but these resistors should not be used in VHF circuits since the cancellation is not perfect and residual reactance becomes excessive. (These resistors are manufactured to have a maximum inductance of 0.5 microhenry at a test frequency of one megahertz.)

Figure R-22
Power Wire-Wound Resistor

Wire-wound resistors are also furnished as low-power insulated and precision types. The former type has values from 0.1 to 30,000 ohms. The individual resistors may be color coded, but if so the first band is twice the width of the other bands. Precision resistors have tolerances within 1, 0.5, 0.25, 0.1, and even 0.05 percent, as opposed to the tolerances of power resistors (within 1 and 5 percent) and composition resistors (within 5, 10, and 20 percent). Power ratings and noise levels are low, and temperature

coefficients (change of resistance with change of temperature) can be had as little as ±
10 parts per million per degree Celsius.

Film-type resistors are made by depositing a thin layer of resistive material on a
ceramic or glass core, attaching axial leads, and encapsulating the assembly in an
insulating outer cover (see Figure R-23). Resistive materials presently used are
microcrystalline carbon, boron-carbon, various metallic oxides, and precious metals.
(The last two materials are used for higher operating temperature and lower noise
level.) For lower values of resistance a continuous film is applied to the core, the
thickness varying inversely with the resistance required. To get higher values the film
is deposited in a spiral pattern, coarse for intermediate values, fine for high values.
Obviously this introduces inductance effects at higher frequencies, but it is not as
serious as in wire-wound resistors and can be ignored up to about 10 megahertz.

Standard values for fixed resistors are given under *Preferred Values.*

In microminiature hybrid thick-film circuits resistors are deposited as films of
palladium-silver or palladium oxide-silver and glass by silkscreening, with a tolerance
as deposited of ±10 to ±15 percent. This can be adjusted to ±1 percent by trimming. In
thin-film circuits materials such as nichrome, tin oxide, tantalum-tantalum nitride, and
cermet chromium-silicon monoxide are deposited by vacuum evaporation, vapor
plating or sputtering, using masks, screens, or stencils.

In the fabrication of silicon monolithic integrated circuits using MOSFET's (metal-
oxide-silicon field-effect transistors), resistors are passive MOSFET's, these being
much smaller than conventional deposited resistors. The value of the resistor is
established by the amount of the fixed bias applied to the device (see Figure R-24).

For further details on microelectronic fabrication, see *Microelectronics.*

Figure R-23
Film-Type Resistor

Adjustable Resistors

The two basic types of adjustable resistor are the rheostat and potentiometer. The
rheostat (from two Greek words meaning "current setting") is a two-terminal device
in which one terminal is connected to a resistive element and the other to a contact
which is moveable, and slides along the resistive element to place a longer or shorter
section of it between the two terminals (see Figure R-25a). Connected in series with a
voltage source, it can indeed be used to adjust the current in the circuit.

The potentiometer (from two Latin words meaning "potential measuring device") is a
special type of rheostat with three terminals. Two of these are connected to the
opposite ends of the resistive element, which is in series with the voltage source. A
constant current therefore flows through the entire length of the resistive element. The

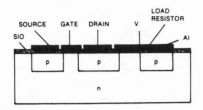

Figure R-24
(Same as Figure M-23a)

third terminal connects to a sliding contact which can be adjusted to touch the resistive element at any point. The device is consequently a voltage divider, since the voltage at the third terminal will be determined by the resistances between the sliding contact and each fixed terminal, multiplied by the current. As a laboratory instrument, the potentiometer is used to measure an unknown voltage by balancing it with a known potential difference, but the more common potentiometer is used for such purposes as a volume control on a radio or audio amplifier (see Figure R-25b).

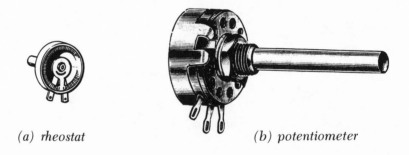

(a) *rheostat*   (b) *potentiometer*

Figure R-25
Variable Resistors

Both of these adjustable resistors are usually made in a circular form, in which the moveable contact is attached to a shaft that rotates through 300 degrees. The resistive element may be carbon composition, wire-wound or metallic film, as described above. The movable contact does not go all the way to the end of the resistive element. A small amount of residual resistance, called hop-off resistance, exists. This is provided to avoid accidentally burning the extreme end of the resistive element with a current too high for it. The wattage rating of an adjustable resistor is based on using the whole length of the resistance element, so the allowable wattage is reduced as this is shortened.

Composition adjustable-resistor elements may be formed on a paper-base phenolic laminate, or they may be hot molded integral with a molded base. The hot-molded type is more durable. The element may also be tapered to provide for a nonlinear change of resistance with shaft rotation. Potentiometers have four standard tapers: linear (which is no taper); "S" taper; CW log taper; and CCW log taper. The linear taper is the most commonly used; the others are used for some audio applications (tone controls, equalization, and the like). Potentiometers for tone controls may also have two fixed taps for frequency-sensitive circuit adjustments.

The ordinary potentiometer has less than one turn, but multi-turn potentiometers with up to 25 turns are available. These generally consist of a strip of resistive material and a sliding contact mounted on a lead screw that moves in a straight line as the screw is turned. A more advanced version has the contact rotate as it moves forward, the resistive element being in the form of a helix (e.g., a Beckman "Helipot"). Most multi-turn potentiometers have wire-wound elements. The advantage of this type of potentiometer is the greater resolution it gives, so that it can be used with a turns-counting dial for very precise settings (see Figure R-26).

Figure R-26
Ten-Turn Precision Potentiometer and Turns-Counting Dial

One other type of adjustable resistor is the tapped resistor (see Figure R-27). This type of resistor has a number of "taps," or connections. Different resistances can be connected into the circuit by using the taps. These can be slid along the resistive element to divide it into various resistances and then clamped in place by tightening the locking screw. Tapped resistors are used for voltage dividers in power supplies and the like.

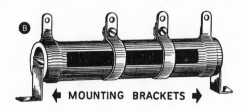

Figure R-27
Tapped Resistor

resistor-capacitance-transistor logic (RCTL)—Variation of resistor-transistor logic (q.v.), in which bypass capacitors are placed across the gate input resistors. This increases operating speed, but the noise immunity is poor.

resistor color code—Method of marking resistors with colored bands to indicate their values. *Color Coding.*

resistor-transistor logic (RTL)—Earliest IC logic family, noted for its economy and high speed-power product, but has the disadvantages of poor noise immunity and fan-out. *Microelectronics.*

resolution—1. The ability of an optical system or a photographic film to form distinguishable images of closely spaced points, lines, or objects. Also called resolving power.   2. Number of divisions on a scale: the larger the scale, the more divisions, so the finer the resolution of the measurement.

resolver—Electromechanical device for sensing a change in angular mechanical position and converting it to an electrical signal. Similar to a synchro (q.v.), but with many more poles, it has finer resolution and a pulse output, so that the angle of rotation is given by the number of pulses, which are fed to a computer from which correction signals are issued as necessary. Frequently used in inertial guidance systems.

resonance—In a series circuit containing inductance, capacitance, and resistance, resonance occurs at the frequency when the inductive reactance and capacitive reactance are equal. Since they are 180 degrees out of phase they cancel each other completely, and the amount of current flow is determined entirely by the resistance, so is at the maximum. In a parallel circuit under the same conditions the current at resonance will be the minimum. The resonant frequency is given by $f_r = 1/2\pi \sqrt{LC}$, where f_r = resonant frequency, in hertz; L = inductance, in henries; C = capacitance, in farads. L is given by $L = 1/4 \pi^2 f_r^2 C$, and C by $C = 1/4 \pi^2 f_r^2 L$. (See Figure R-28.)

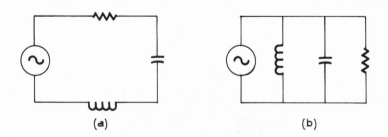

(a)   (b)

Figure R-28
Resonance

(a) series circuit, current maximum

(b) parallel circuit, current minimum

resonance bridge—In the resonance bridge the capacitor is tuned to resonance with the unknown inductor. This serves to measure its inductance. The resistive element is then balanced by R3 (at resonance a series-resonant circuit has no reactance) as in a Wheatstone Bridge (q.v.). (See Figure R-29.)

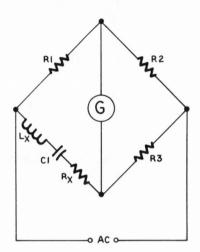

Figure R-29
Resonance Bridge

From *Complete Guide to Reading Schematic Diagrams*, First Edition, page 213. ©1972. Reprinted with permission of Parker Publishing Company, Inc.

resonant circuit—(See resonance.)

resonant frequency—(See resonance.)

resonant voltage step-up—In a series-resonant circuit at resonance (q.v.), the current in the circuit is limited only by the resistor. Nevertheless, it also flows through the inductor and the capacitor. The voltage across either of these is given by IX, or the current in amperes multiplied by the reactance in ohms. As explained under resonance, the reactances are of opposite polarity, so the voltages are of opposite phase; the net voltage around the circuit is still only the applied voltage. The voltage across the inductor or the capacitor is given also by the Q of the circuit multiplied by the applied voltage.

response curve—Graph that shows the variation with frequency of the output, measured in volts or decibels, of a device (amplifier, filter, microphone, etc.) to which an input of constant amplitude but varying frequency is being applied.

retentivity—(See remanence.)

RETMA—Abbreviation for the former Radio-Electronic-Television Manufacturers Association, now the Electronic Industries Association (EIA).

retrace blanking—Application of a negative potential to the control grid of a cathode-ray tube or picture tube at the end of each sweep, in order to cut off the electron beam until it is ready to start the next sweep, so avoiding the display of retrace lines on the screen.

reverberation—1. Time required for a sound to die away when suddenly interrupted. It varies with frequency and the characteristics and amounts of sound-absorbent materials present. *Electroacoustics.* 2. In sound recording, the echoing phenomenon used to make a sound rounder and mellower.

reverse bias—At a pn junction, reverse bias increases the potential barrier, reduces the forward component, and leaves only a small reverse current which does not depend on barrier height. *Semiconductors.*

r.f. amplifier—(See radio frequency amplification.)

r.f.i.—Radio-frequency interference (q.v.).

r.f. signal generator—Radio-frequency signal generator (q.v.).

rhenium—Silvery-white, extremely hard, very rare metal, with one of the highest melting points of the elements (3180°C). Used for contact points, high-temperature use. Symbol, Re; atomic number, 75; atomic weight, 186.2.

rheostat—Adjustable resistor used for adjustment of current or the varying of resistance in a circuit. It has two terminals instead of the three of a potentiometer. *Resistors.*

rhombic antenna—High-frequency linear antenna with the plan view of a diamond. *Antennas.*

RIAA curve—Equalization curve for phonograph records produced to the standards of the Recording Industry Association of America. *Electroacoustics.*

ribbon microphone—Microphone in which a corrugated strip of duralumin is suspended between the pole pieces of a permanent magnet. Sound waves make the strip vibrate, so that currents are induced in it by the magnetic field. These currents are led away by leads connected to the ends of the ribbon.

"rig"—Slang expression used by "hams" and "good buddies" to refer to their equipment.

rim drive—Phonograph turntable driven by small rubber pulley pressed against the inner surface of the rim. The pulley is mounted on a shaft driven by the motor via the speed reduction pulleys.

ring counter—Serial shift register in which the output of the last flip-flop is fed back to the input of the first, so that data are not only shifted but recirculated as well.

ringing—Transient oscillation that dies away in a few cycles, usually caused by a waveform with a steeply rising leading edge, or by self-resonant transformer or loudspeaker windings.

ripple—Residual a.c. voltage in the output of a power supply, consisting of a small a.c. component riding on the d.c. component. Usually ignored if it does not exceed a few millivolts.

rise time—Time in which the amplitude of the leading edge of a pulse increases from 10 percent of the pulse amplitude to 90 percent of the pulse amplitude. (See also pulse.)

r.m.s.—Abbreviation for root mean square (q.v.).

Rochelle-salt crystal—Sodium potassium tartrate tetrahydrate ($KNaC_4H_4O_6.4H_2O$), crystalline solid having a large piezoelectric effect (effective piezo modulus $H = 5.67$ coulombs/meter^2), crystals of which are used in phonograph pickups, microphones, ultrasonic generators, and the like.

roentgen—Unit of X or gamma radiation; the amount that will produce in one cubic centimeter of air an amount of ionization equal to one electrostatic unit of charge (a charge that will repel a similar charge at one centimeter distance with a force of one dyne). (See also rad and rem.)

rolloff—In a frequency-response curve, the portion where the response is no longer flat, but slopes downward.

ROM—Abbreviation for read-only memory (q.v.).

root mean square—The average value obtained by taking the square root of the arithmetic mean of the squares of the set values, given by

$$v_{rms} = \sqrt{\frac{v_1^2 + v_2^2 + \ldots v_n^2}{n}}$$

where $v_1, v_2 \ldots v_n$ are the values to be averaged, n is the number of values, and v_{rms} the root-mean-square value.

rotary switch—Selector or multiposition switch in which rotation of the control knob or actuator causes a sliding contact to be transferred from fixed contact to fixed contact, the fixed contacts being arranged in circles on one or more decks or sections. This type of switch is used for attenuators, channel tuners, and the like. Also called a wafer switch.

rotator—Device mounted on a mast to turn the antenna to point in a desired direction.

rotor—The moveable part of a variable capacitor, etc., or the revolving part of a motor or generator.

round off—Since it is physically impossible to write down an infinite decimal, the decimals after a certain number of places are dropped. "Rounding off" reduces the error involved in the approximation. If the first digit dropped is less than 5, the last digit retained is not altered. If the first digit dropped is 5 through 9, the last digit retained is increased by 1.

R-S flip-flop—Flip-flop with inputs S and R. A 1 at the S input results in a 1 at the output; a 1 at the R input results in a 0 at the output. Also called set-reset flip-flop. *Computer Hardware, Microelectronics.*

RTL—Abbreviation for resistor-transistor logic (q.v.).

ruby laser—Solid-state laser using triply ionized chromium (Cr^{3+}) ions in a ruby (Al_2O_3) rod, and excited by a xenon flashlamp that irradiates it with energy with a wavelength of about 550 nanometers. If the energy is sufficiently intense, laser oscillation will take place at 694.3 nanometers. *Lasers.*

Ruhmkorff coil—Transformer consisting of a primary coil wound on a soft-iron cylindrical core and a secondary coil wound around the primary. There is a very high turns ratio between the two windings. The primary has the smaller number of turns of comparatively thicker wire; the secondary has a very large number of turns of finer wire. A vibrator, or interruptor, operated by the magnetic field induced in the core by the primary winding, causes the d.c. from a battery to be interrupted in the manner of a buzzer, so that a very high pulsating voltage is induced in the secondary, which then can be used to demonstrate spark discharges across an air gap, behavior of Geissler tubes, and the like. The sharpness of the current interruption is enhanced by a capacitor. Also called an inductor coil.

rumble—Low frequency noise generated by the turntable, its associated pickup, and equalizer. Rumble must be at least 40 dB for monophonic records and 35 dB for stereophonic records below a reference level of 1.4 cm/s peak velocity at 100 Hz.

R-Y signal—Red minus brightness color difference signal in color television. When combined with the brightness (Y) signal it is then used to excite the red gun in the picture tube to display the color red. *Broadcasting.*

S—1. Symbol for siemens (q.v.).   2. Symbol for south.   3. Symbol for chemical element sulfur (q.v.).

s—Symbol for second (q.v.).

sal ammoniac—Ammonium chloride, the electrolyte used in Leclanché and dry cells (qq.v.).

sampling oscilloscope—Oscilloscope using a technique that displays the amplitude at one instant only of each cycle of a recurrent waveform. Each point, or sample, displayed is slightly later in time than the one next before it, so eventually samples will have been taken of every point on the waveform. In this way a very high frequency signal can be shown with a comparatively slow sweep. When all the samples are displayed close together they have the appearance of a continuous trace, as many as 1000 being taken. (See Figure S-1.)

sapphire—Transparent or translucent variety of corundum, or aluminum oxide (Al_2O_3). Used for jewel bearings in meter movements, phonograph styli, and the substrate in silicon-on-sapphire (SOS) devices (q.v.). *Microelectronics, Recording.*

satellite communication—Man-made vehicles in orbit around the Earth conveying

Figure S-1

Principle of Sampling Oscilloscope

From *Complete Guide to Electronic Test Equipment and Troubleshooting Techniques,* page 180.

information between points on Earth, such as television programs, telephone calls, etc. Synchronous satellites remain in fixed positions with respect to points on the Earth's surface; passive satellites are unpowered repeaters; active satellites usually obtain power from solar panels and receive, amplify, and reradiate signals from Earth. *Space Communications.*

saturable reactor—Magnetic-core reactor that becomes a very low impedance when the flux in the core reaches saturation. *Magnetic Amplifiers.*

saturation—1. Degree of chromatic purity, or narrowness of distribution of wavelengths in light.   2. Maximum density of magnetic flux that can be present in a magnetic material.   3. Maximum flow of current in an electron tube or transistor.

SAW—Abbreviation for surface acoustic wave (q.v.).

sawtooth generator—Oscillator with a sawtooth output waveform, obtained by charging a capacitor through a resistor and then discharging it almost instantly by shorting it with a transistor switch, etc.

sawtooth wave—Continuous periodic waveform of which each cycle consists of a triangle with its base on the baseline and the other two sides of uneven length, the longer one rising (or falling), the shorter returning to the baseline. (See Figure S-2.)

Figure S-2

Sawtooth Wave

Ibid, page 120.

S band—Microwave band with frequencies from 1.55 to 5.20 gigahertz. *Frequency Data.*

SCA—Abbreviation for Subsidiary Communications Authorization (q.v.).

scalar—Physical quantity that is completely described by its magnitude, such as weight, volume, density, etc., as distinct from quantities that have both magnitude and direction (e.g., force and velocity), which are called vectors.

scale—1. Series of marks along a line, at regular or graduated intervals, used in measuring or computing.   2. Any instrument marked in this manner.

scale factor—1. Number by which a meter dial reading, etc., must be multiplied, to give the proper value for the selected range (e.g., $\times 10, \times 100$, etc.).   2. Number by which a value given in a table must be multiplied to make it in proper proportion to other values.   3. Any number used in a similar way.

scaling circuit—Circuit that gives an output after a certain number of input pulses have been received.

scanning—1. Process of forming a television picture line by line and frame by frame. *Broadcasting.*   2. Systematically directing a radar beam so that it covers a given area.   3. Automatically tuning a receiver to a number of selected frequencies in rapid succession over and over again, but stopping at any one that is active.

scattering—Alteration in direction of a moving particle because of collision with another particle, in accordance with Coulomb's law (the force varies inversely as the square of the distance between the particles). Similar behavior of radiation incident upon matter resulting in the scattering of photons by electrons is called the Compton effect. *Nuclear Physics.*

schematic diagram—Diagram that shows the electrical connections of an electronic device, using graphic symbols and straight lines to represent the parts and their connections. *Graphic Symbols.*

Schering bridge—Bridge for measuring capacitance. (See Figure S-3.)

Schmitt trigger—Single-shot (monostable) multivibrator, in which any input signal over a certain amplitude results in an output pulse of constant amplitude as long as the input signal lasts. Used to produce stable trigger pulses from miscellaneous input signals. (See Figure M-40.)

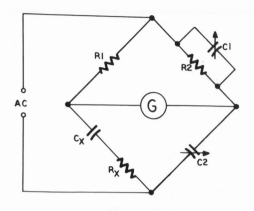

Figure S-3
Schering Bridge

From *Complete Guide to Reading Schematic Diagrams*, **First Edition,**
page 214.

Schottky barrier diode—Diode with a rectifying metal-to-semiconductor contact. Also called hot-electron diode or hot-carrier diode. *Semiconductors.*

scientific notation—Use for writing very large or very small numbers. The significant figures of the number are expressed as some number equal to or greater than one, but less than ten, multiplied by the appropriate power of ten. For example, the speed of light is 299,792,510 meters per second. Expressed in scientific notation it is 2.9979251×10^8 meter per second. Also, the mass of an electron at rest is 0.000 000 000 000 000 000 000 000 000 000 910 955 854 kilograms. Expressed in scientific notation it is $9.109\,558\,54 \times 10^{-31}$ kilograms.

scintillation counter—Radiation detector in which, when a charged particle passes through certain transparent solids or liquids, a flash of light is produced. This is converted photoelectrically into an electrical impulse and amplified in a photomultiplier tube until it is strong enough to drive a counter. *Nuclear Physics.*

scope—Colloquialism for oscilloscope (q.v.).

SCR—Abbreviation of silicon controlled rectifier (q.v.).

scramble—To modify speech before transmission in such a way that only the intended recipient, using a receiver that unscrambles the signal, can understand the message. Various methods of doing this involve such procedures as dividing the speech band into several smaller bands using band-pass filters, multiplying or dividing their frequencies by various factors, inverting some, etc., and reversing the procedure at the receiver to synthesize the original.

screen—Surface on which an image is projected, the most commonly used being the phosphor screen of the cathode-ray tube. *Electron Tubes.*

screen grid—Wire mesh or spiral electrode in an electron tube between the control grid and anode that acts as a shield to reduce the interelectrode capacitance between them to a negligible value. *Electron Tubes.*

S-curve—S-shaped display on oscilloscope screen when connected to check the alignment of a ratio detector. It should be symmetrical about the center frequency.

Se—Symbol for the chemical element selenium (q.v.).

search radar—Radar system designed to scan rapidly over a large area, using vertical and horizontal fan-shaped beams. *Radar.*

second—Fundamental unit of time, defined as the duration of 9 192 631 770 periods of the radiation corresponding to the transition between two hyperfine levels of the ground state of the cesium-133 atom. Symbol, s. *S.I. Units.*

secondary—1. Secondary cell, cell that is rechargeable. 2. Secondary emission, ejection of electrons from a solid that is bombarded by a stream of charged particles. 3. Secondary winding, output winding of a transformer.

Seebeck effect—Production of a voltage (and consequently a current) in a loop of material containing at least two dissimilar conductors when two junctions are at different temperatures. Used to measure temperature with great accuracy and sensitivity.

seismic wave—Elastic wave in a solid at an acoustic frequency. *Electroacoustics.*

selectivity—The ability to select one desired frequency and to discriminate against others.

selenium—Metalloid element, existing most commonly in a gray metallic form. Its electrical conductivity increases when light strikes it, and it is also able to convert light directly into electricity, so it is used in photoelectric devices. It has rectifying properties as well (see selenium rectifier). Symbol, Se; atomic weight, 78.96; atomic number, 34.

selenium rectifier—Semiconductor junction consisting of selenium and tin-cadmium alloy coatings on an iron or aluminum plate. Electrons flow more freely from the alloy coating to the selenium than in the opposite direction. Was used widely as a power-supply rectifier, but has been largely superseded by silicon rectifiers. *Semiconductors, Rectifiers and Filters.*

semiconductors—Materials with conductivities intermediate between those of conductors and insulators, that conduct better as their temperature increases. *Semiconductors.*

SEMICONDUCTORS

It is not necessary to delve very deeply into atomic physics to understand semiconductors. An atom consists of a nucleus with a positive charge (see *Nuclear Physics*) and a cloud of electrons around it. The electrons have negative charges, and there are usually exactly the right number to balance the positive charge on the nucleus.

Because of their negative charges the electrons repel each other, so the way they are arranged is a compromise between the attraction of the nucleus, the repulsion they have for each other, and the energy they possess due to their temperature. This last keeps them in a constant state of agitation, so that they have a continuous high-speed movement around the nucleus while maintaining their distances from each other.

The result may be visualized as a fuzzy ball made up of layers or shells of electrons. The smallest shell, closest to the nucleus, has the fewest electrons (two), and they are also the most tightly bound because the Coulomb attraction (electrostatic attraction between oppositely charged particles) of the nucleus is strongest in this shell. Going outwards from the nucleus, the number of electrons in each shell increases as its size increases. However, the number of electrons in the atom, which is the same as its atomic number, determines the number of shells occupied.

The number of electrons in the outermost shell is often fewer than it can hold. They also are not as tightly bound in it as those in shells closer to the nucleus. Consequently they can spread out more, so that these shells can be filled by fewer electrons.

The outermost shells of carbon, silicon, germanium, and other atoms in Group IV of the periodic table of the elements (see element) all have four electrons, although they could hold eight. These atoms arrange themselves in such a way that by sharing each of their four outer electrons with each of four neighboring atoms they all fill up their outer shells completely. In doing this they also combine into a symmetrical three-dimensional structure of small "cubes," all oriented the same way, called a crystal lattice.

This chemical binding together is called covalent bonding, and the electrons that do it are called valence electrons. In carbon the valence shell is the second one from the nucleus, so the Coulomb force is very strong and it is extremely difficult to dislodge a carbon valence electron from a covalent bond. This explains why diamonds, the crystalline form of carbon, are so hard. But in silicon the valence shell is number three, and in germanium it is number four, so the covalent bonds are not as strong because the valence electrons are farther from the nucleus.

As already mentioned, all electrons have energies which vary with the temperature. If an electron is heated to a higher temperature it absorbs more energy, or becomes more excited. If it receives enough energy (which may be from other radiant energy also, such as light) it may be able to break loose and become a conduction electron. Conduction electrons may be made to move easily by an external electric field, because they are free of the attractive force of the nucleus.

Electrons in isolated atoms can exist only in certain definite energy levels, because energy is absorbed or released in definite amounts, called quanta. (The amount of water in a tank would always be an exact number of gallons if it was removed or added in gallon amounts.) Electrons with lower energy occupy shells close to the nucleus. If an electron absorbs a quantum of energy from an outside source (e.g., ultraviolet light) it makes a "quantum jump" to a higher energy orbit.

When a number of atoms is associated in a crystal lattice the electrical fields of neighboring atoms affect them also, so the discrete energy levels of isolated atoms spread out into energy bands. However, the bands do not normally overlap. There are gaps between them, which cannot be occupied because the increases or decreases of energy are always in quanta.

Electrons in the valence band which are excited to a conduction band must therefore cross the energy gap separating these two levels. They are then free of the Coulomb force of the nucleus and available to travel under the influence of an external force field. If a conduction band of an element, such as germanium, gains more excited electrons, the germanium becomes more conductive. At room temperature the resistivity of a cubic centimeter of pure germanium is about 45 ohms, but at 150 degrees Celsius it has fallen to approximately one ohm. An electron that leaves the valence band and enters the conduction band takes away a negative charge from its parent atom, which therefore has an excess of one positive charge. As it is no longer neutral it is now a positive ion. It also has a "hole" in its valence band.

Because of this way of looking at it, it has become customary to think of conduction in semiconductors as consisting of two charge carriers: electrons (negative) and holes (positive). In a pure semiconductor, thermal energy excites electrons out of the valence band into a conduction band, at the same time creating holes. Since every conduction electron leaves behind a hole, conduction takes place by hole-electron pairs. However, since the electrons are no longer under the influence of nuclear attraction, they have a higher velocity in an electric field; the holes have a lower velocity in the opposite direction.

This is called intrinsic conduction, since it is an intrinsic, or a characteristic property of the pure semiconductor. As mentioned above, it can be greatly enhanced by raising the temperature of the material. However, adding certain impurities, in typical concentrations of one part in a million, will also increase conduction. This type of conduction is called extrinsic, since it is due to the addition of external impurities.

These impurities are substances in Group III or Group V of the periodic table of the elements. For example, antimony (Sb) in Group V has five electrons in its valence shell. When an antimony atom is introduced into the crystal lattice of a Group IV element like silicon, it readily forms covalent bonds with four adjacent silicon atoms, but it still has that fifth electron "hanging loose," as it were. This electron is only weakly attached and easily becomes a conduction electron. The antimony with which the silicon was "doped" has "donated" a negative charge carrier. Antimony, therefore, is a donor impurity, and because the donated charge carriers are negative, the antimony-doped silicon is called n-type. Other Group V dopants that could be used in the same way are arsenic and phosphorus.

Conversely, a Group III atom such as indium (In) has only three electrons in its valence shell. When it is introduced into the silicon crystal lattice it can readily form covalent bonds with three neighboring silicon atoms, but that still leaves it with one vacancy to fill. However, indium in a silicon crystal lattice has a very strong attraction for silicon valence electrons, so it wastes no time in stealing an electron from a nearby silicon-silicon covalent bond. The covalent bond that lost an electron now has a hole, which acts as if it was a positive charge, free to hop from covalent bond to covalent bond in the material. Indium, therefore, is an acceptor impurity, and because the holes vacated by the electrons it steals are positive charge carriers, the indium-doped silicon is called p-type. Other Group III dopants that could be used in the same way are aluminum and boron.

The semiconductors from Group IV are all elements (silicon, germanium, diamond, and gray tin). However, compound semiconductors also exist, in which one atom of an element from Group III is surrounded by four atoms of an element from Group V. This arrangement enables them to form a lattice structure similar to a Group IV element, except that the nearest-neighbor atoms are of different species. An example of this kind of semiconductor is gallium arsenide (GaAs).

If a Group IV semiconductor such as silicon is doped so that it consists of two regions, one n-type, the other p-type, the borderline between them is called a pn junction. The n region, as explained above, contains surplus conduction electrons and a few thermally generated holes. The more numerous electrons are called majority carriers; the holes are therefore minority carriers. In the p region the situation is reversed, with holes as majority carriers and electrons as minority carriers.

The majority carriers diffuse across the junction because like charges repel each other, and so they spread out as much as possible. A large number neutralize each other by electrons combining with holes. Each n-type atom that donated an electron is left with a positive charge, but it has to stay behind, of course, because it is fixed in the crystal lattice. The same applies to each acceptor atom in the p region that donated a hole; it is left behind with a negative charge. Consequently the n-type region has an overall positive charge and the p-type region an overall negative charge. This charge distribution stops the flow of any more majority carriers, but stimulates the flow of minority carriers, so the region close to the junction becomes depleted of carriers. Majority carriers are repelled from it; minority carriers are pulled together and neutralize each other. Another effect is that since the conduction electrons diffusing into the p region had a higher energy level than the valence holes, the energy bands of the p-type material have become higher while those of the n-type material have been lowered.

If an external positive potential is applied to the n-type material it becomes even more positive, and if an external negative potential is applied to the p-type material it becomes even more negative, hence the opposition to majority carrier flow is increased. A small minority-carrier current does flow, however. This application of potentials (positive to n-type, negative to p-type) is called reversed bias.

If the exernal potentials are changed so that the negative potential is applied to the n-type material its positive charge is greatly reduced. The positive potential applied to the p-type material also reduces its negative charge. Opposition to majority carrier flow is now negligible and current flows freely. The junction is said to be positively biased.

This behavior of the pn junction is therefore used in general-purpose diodes, and in the following special types.

Power rectifiers are usually made of silicon because it can withstand higher temperatures. Currents up to 1000 A and peak reverse voltages of 1000 V or more can be handled by the larger rectifiers.

Zener diodes, or reference diodes, make use of another pn junction characteristic. If the reverse-bias potential is made high enough, charge carriers can acquire enough energy to make hole-electron pairs by impact ionization. This is a nondestructive breakdown of the potential barrier of the junction, and because the creation of hole-electron pairs is cumulative it is called avalanche breakdown. The reverse-bias potential at which this will take place depends upon the impurity concentration and may be from a few hundred millivolts to several hundred volts. The breakdown potential is very sharp and well defined, allowing the device to be used as a reference in a regulated power supply, a protective device against overloads and transients, a clipper, or an arc suppressor in small circuit breakers, to mention only a few representative uses.

Varactor diodes are also reverse-biased pn junctions, with a specially graded impurity concentration at the junction. The width of the depletion, or barrier region, can then be altered by varying the bias. Since the barrier region is equivalent to the dielectric of a capacitor, with the p and n regions as the plates, a voltage-controlled capacitor results.

Tunnel diodes have an extremely narrow depletion, or barrier region. Under very low forward bias (0.3 to 0.4 V), electrons tunnel through the junction from the conduction band to an equal level in the valence band, producing a negative-resistance effect. At a higher forward bias the tunnel diode behaves like an ordinary diode. This diode is also called an Esaki diode, and the juxtaposition of conduction and valence bands at a very thin depletion region is brought about by heavy doping. As a stable, low-noise and highly efficient device, this diode has many applications in UHF.

The Schottky diode consists of a metal-semiconductor junction instead of a pn junction. Usually gold and n-type silicon are used. When this junction is forward-biased, electrons are injected into the metal. These electrons have a relatively high energy level, so they are called "hot carriers." The Schottky diode is therefore often called a hot-electron diode. Because the diode uses only majority carriers in conduction across the junction, which have a very short lifetime, the noise due to generation and recombination of hole-electron pairs is very low, making the Schottky diode suitable for UHF mixer and switching circuits.

Step-recovery diodes have a doping profile near the junction that causes minority carriers to be stored in this region when the diode is forward-biased. When reverse-bias is applied these minority carriers flow out as a reverse current. While this current flows the diode offers a very low impedance, but this ceases abruptly as soon as the stored charge has been used up. This characteristic can be used to initiate or terminate a fast pulse, or to generate harmonics. Other names for this device are snap-off diode and charge-storage diode, and it is used for microwave-frequency multiplication and switching.

Pin diodes have a layer of intrinsic, or lightly doped, semiconductor sandwiched between heavily doped p and n regions. Under forward bias of 50 mA, injection and storage of carriers in the i region reduce a typical diode's impedance to about one ohm, but when the bias is reversed the impedance increases to 8 kilohms. This allows the pin diode to be used as a d.c. controlled switch, or as a modulator, at microwave frequencies. Pin diodes are also used as high-voltage rectifiers, and when constructed with a very thin i region they can be operated as charge-storage diodes (see step-recovery diodes).

Gunn and Impatt (impact ionization avalanche transit time) diodes are used as microwave oscillators. The Gunn diode is not really a junction diode, but consists of a gallium arsenide or indium phosphide crystal sandwiched between two ohmic contacts. When proper voltage is applied across this device, electrons in the first conduction band are transferred into a higher band, where their effective mass increases about six times but their mobility decreases about sixty times. This causes them to bunch together into field domains as they travel from the cathode to the anode, and as each domain has a lowered mobility, current through the device is greatly reduced. On reaching the anode the field disappears and current increases sharply, but only momentarily, as a new field is formed at the cathode. The frequency of oscillation is the reciprocal of the time taken for the field domain to transit from the cathode to the anode. Another type of bulk-effect diode, working on the same principle but differently doped, is the limited space-charge accumulation (LSA) diode, in which the high-energy field domain is as wide as the space between the electrodes. Since transit time is nil, the frequency of oscillation is limited only by the rate of build up of successive fields.

The IMPATT diode's pn junction is reverse-biased to its avalanche threshold. When an a.c. voltage of sufficiently high frequency is super-imposed on the bias voltage the diode is driven in and out of the avalanche condition at the signal frequency. Electrons generated by field-induced impact ionization in the avalanche condition are still in transit, however, when the field reverses, producing the condition where the current and voltage are of opposite phase, which is a negative resistance effect, and harmonics at microwave frequencies are produced.

Photo diodes may be operated in a photovoltaic (solar cell) or a photoconductive mode. In the photovoltaic mode, light falling on the unbiased junction stimulates the production of hole-electron pairs. These charges separate at the junction, the electrons going to the n-type material, the holes to the p-type. A potential difference develops across the diode, depending upon the intensity of the light, up to 0.45 volt. In the photoconductive mode the diode is reverse-biased. Light falling on the junction allows current to flow in proportion to its intensity. An avalanche photodiode is biased just below the avalanche threshold. When light falls on the junction avalanche, multiplication (see Zener diode) takes place. This diode is therefore very sensitive to low levels of light and is suitable for uses similar to those for which photomultiplier tubes (q.v.) are used.

Light-emitting diodes (LED's) are mostly made of gallium compounds and emit radiation when hole-electron pairs recombine. The radiation wavelength ranges from 0.55 to 34 micrometers, or from green, via yellow and red light to far infrared. (See *Optoelectronics*.)

Injection lasers are also a type of diode (see *Lasers*).

For other semiconductors see *Transistors* and *Transistor Circuits*.

sensitivity—Strength of the signal in microvolts at the input of a receiver that is required to produce a specified audio power output.

separation—Degree to which either channel of a stereo system does not contain information that belongs in the other channel. Expressed as a ratio in decibels. The standard for a tape recording is not less than 40 dB from 100 Hz to 10 kHz.

series circuit—Circuit in which components are connected end to end, so that the current flows through them in a single path and has the same value at every point.

series regulator—An electron tube or transistor placed in series with the rectified voltage of a power supply. Its resistance varies in accordance with an error voltage applied to its control grid or base, this error voltage being derived from a comparator that compares the output voltage of the series regulator with a reference. The resistance of the series regulator changes so as to raise or lower its output voltage to maintain the required value. (See also regulation.)

series resonant circuit—Circuit with an inductor and capacitor in series. When their reactances are equal, which occurs at the resonant frequency, the amount of current flow is at its maximum, being limited only by the resistance of the circuit. The resonant frequency f_r is given by $1/2\pi\sqrt{LC}$, where $2\pi = 6.28$, L = inductance in henries, and C = capacitance in farads.

servomechanism—Automatic device used to correct the performance of a mechanism by means of an error-sensing feedback. The feedback compares the actual performance of the mechanism with the desired performance and generates an error signal, which actuates the mechanism so that the difference is eliminated. Examples of servomechanisms are automatic pilots, engine speed governors, and inertial navigators.

set-reset flip-flop—(See R-S flip-flop.)

shaded-pole motor—Small a.c. motor ($\leqslant 1/4$ h.p.) with a low-resistance short-circuited coil embedded in one side of each stator coil area. This coil shifts the flux of the area enclosed by the coil about 75 degrees in time, allowing an imperfect but effective rotating flux to be produced. Because the starting torque is small these motors are used mostly for fans and blowers. Full-load fixed speeds range from 1050 to 1550 r.p.m., and they are not reversible.

shadow mask—Thin perforated metal sheet that lies directly behind the phosphor screen of a three-gun color picture tube. Its 200,000 precisely aligned holes ensure that each electron beam strikes only its corresponding color dots on the screen. Now superseded by the Trinitron tube (q.v.). *Electron Tubes.*

shell—In quantum theory, electrons of atoms are thought of as occupying diffuse shells around the nucleus. The first shell (closest to the nucleus) is the K shell; the next, the L shell; and so forth. The maximum number of electrons that theoretically can occupy each of the first seven shells are 2, 8, 18, 32, 50, 72, and 98 respectively. However, only the heaviest elements have the first four shells fully occupied. It is their electronic configuration that determines the chemical properties of atoms.

SHF—Abbreviation for super-high frequency (q.v.).

shielding—Enclosing a circuit, conductor, or component in a grounded, low-resistance metal container, or placing a grounded metal plate between two components, provides isolation by short-circuiting the lines of force of any electric field within or without the shielded area. At radio frequencies similar shielding will prevent magnetic coupling, since the magnetic field induces a current in the shield, which in turn sets up an opposing magnetic field.

shift register—Register having the capability of serially shifting data from each stage of the register to the adjacent one. A shift register may be unidirectional (shifts one way only) or bidirectional (shifts both ways). Input and output may be serial (one bit at a time) or parallel (bits simultaneously to or from all stages). *Computer Hardware.*

Shockley diode—Four-layer (pnpn) switching device that operates like a thyratron. *Transistors.*

short circuit—Connection, often unintentional, that provides a low-resistance path between two points in a circuit or to ground. Depending upon the circumstances, a short circuit may stop the circuit operation, alter it, or have no effect. If it results in excessive current flow, damage to the circuit may result.

shorting switch—Type of switch in which shorting (bridging) takes place during contact transfer. Also called make-before-break switch (q.v.).

short wave—Nonstandard term used of wavelengths less than 187 meters (1605 kilohertz). *Frequency Data.*

shot noise—Noise consisting of separated impulses of high amplitude, generated in electron tubes and transistors, or externally by ignition systems, etc. It is different from hiss (overlapping impulses) caused by thermal agitation of molecules, since it is caused by fluctuations in cathode emissions or bias current.

shunt—(See parallel, 2.)

sidebands—New groups of frequencies above and below the frequency of a carrier that are formed when it is modulated. The group of frequencies higher than the carrier is called the upper sideband, that which is lower is called the lower sideband. The modulation (or information) in the signal is in the sidebands, so that the carrier can be suppressed if desired, without losing it—in fact, only one sideband need be transmitted. The carrier and its sidebands constitute a channel. *Modulation.*

siemens—Unit of electrical conductance, being the conductance of a conductor that has an electrical resistance of one ohm. Symbol, S. *S.I. Units.*

signal—Any detectable physical quality, quantity, or impulse that can be used to convey information from one point to another.

signal generator—Device for generating radio or audio-frequency signals. (See radio-frequency generator and audio-frequency oscillator.)

signal ground—Electrical connection to common reference point for signals, often via a capacitor. (See also ground.)

signal strength—Field strength in millivolts per meter (of receiving antenna effective length) of the transmission of a station. The value decreases with distance from the transmitter and with ground conductivity.

signal-to-noise ratio—Relationship of the strength of a useful signal to the strength of the background noise, expressed in decibels. *Radio Noise and Interference.*

signal tracer—Test instrument with loudspeaker or meter, singly or in combination, that enables the user to verify the presence of an input signal at following stages in a receiver or amplifier using r.f. and audio probes.

significant digits—In a decimal number, all nonzero digits are significant. All zeros between significant digits are significant. A zero following a nonzero digit may or may not be significant, depending upon whether it is accurate or whether its only function is to help place the decimal point in an approximate number. For example, 673,924 has six significant digits, but its approximation 674,000 has only three. However, if 674,000 was the exact number it would have six significant digits.

silicon—Nonmetallic chemical element comprising 27.7 percent of the Earth's crust (second only to oxygen); a hard, dark-gray solid with a metallic luster and the same crystalline formation as the diamond form of carbon. Used for semiconductor devices. Symbol, Si; atomic weight, 28.086; atomic number, 14. *Semiconductors.*

silicon controlled rectifier (SCR)—Four-layer diode with external electric connection to the gate (p_1 region), used as a switch. *Transistors.*

silicon dioxide (SiO_2)—Compound of the two most abundant elements in the Earth's crust, silicon dioxide, or silica, is the main constituent of more than 95 percent of known rocks. When silicon is heated to red heat in the presence of oxygen, a layer of silicon dioxide is formed on its surface. It is the best of all insulators, with a resistivity greater than 10^{19} ohms per centimeter. *Microelectronics.*

silicone—Any of the family of polymer compounds of silicon, oxygen, and the methyl and phenyl organic groups, which together form long polysiloxane chain molecules; used for synthetic rubber, resins, water-repellent textiles, etc.

silver—Brilliant white-colored metal, with the highest-known electrical and thermal conductivity, used for electrical contacts, printed circuit boards, photography, and glass mirrors. Symbol, Ag; atomic weight, 107.886; atomic number, 47.

sine—1. Trigonometric function; defined for a right triangle, the sine of either of the acute angles is the length of the side opposite the angle divided by the length of the hypotenuse. Abbreviated sin.   2. Hyperbolic function: hyperbolic sine of x = (θ in radians) = ($\epsilon^x - \epsilon^{-x}$)/2. Abbreviated sinh. (See Figure S-4.)

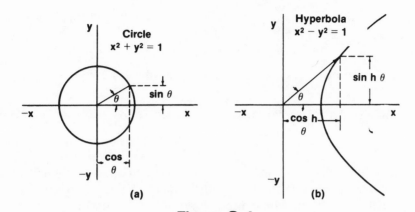

Figure S-4

Illustration Showing a Comparison of (a) trigonometric and (b) hyperbolic functions

sine wave—Waveform of a single cycle of pure alternating current or voltage in which the amplitude at each instant is proportional to the sine of the angle as it changes from zero to 360 degrees (one cycle). (See Figure S-5.)

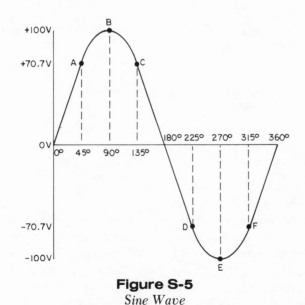

Figure S-5
Sine Wave

From *Complete Guide to Electronic Test Equipment and Troubleshooting Techniques*, page 228.

single crystal—Any solid throughout which an orderly three-dimensional arrangement ("lattice") of the atoms, ions, or molecules is repeated without interruption.

single-pole, single-throw switch (spst)—Two terminal switch that opens or closes a circuit, such as an electric light wall switch.

single-shot multivibrator—(See monostable multivibrator.)

single sideband—AM transmission with one sideband and carrier suppressed (A3J), or one sideband and reduced carrier (A3A).

single sweep—On an oscilloscope with triggered sweep, a single-sweep mode of operation may be provided. When this mode is selected, the next trigger pulse initiates the sweep, and a single trace is presented on the screen. After this sweep is complete, the sweep generator is "locked out" until reset. When reset, another single sweep takes place on the next trigger, and so on. This mode is used when the signal to be displayed is not repetitive or varies in amplitude, time or shape, when a conventional repetitive display may produce an unstable presentation.

sink—1. Component or connection into which current flows. 2. Means of dissipating heat: heat sink.

sinusoidal wave—Wave, the profile of which is given by a sine function (see sine wave).

S.I. Units—Units of measure of the Système International d'Unités (S.I.), or International System of Units ("metric system"). *S.I. Units.*

S.I. UNITS

The Metric System

Three major features characterize the metric system: decimalization, rational prefixes, and basing fundamental units on invariable natural phenomena, instead of such arbitrary standards as the length of Pharaoh's arm or the volume of a jar that rings with a certain pitch when struck. But although the metric system was proposed in 1670 as a reasonable, uniform system to replace the welter of medieval weights and measures, the vast inertia of ancient systems permeating industry and commerce and involving containers, measures, tools, and machines, not to mention popular resistance to change—partly religious and entirely psychological—prevented any general acceptance of the metric system.

However, in 1789, with the fall of the Bastille, the first of a series of political upheavals began that was to facilitate the adoption of the metric system by one country after another. The French revolution was followed by the conquests of Napoleon that spread the system to most of Europe. The Japanese adopted it in the wake of the Meiji restoration (1868), the Russians after their revolution in 1917, and Latin America and China similarly. Curiously enough, the United States, whose own revolution was the original spark for the others, retained the English system, even attempting to bring its standards into closer harmony with the British standards.

Meanwhile, the rapid development of the sciences resulted in a proliferation of subsystems to serve particular disciplines. Scientists also began to realize that the original eighteenth-century standards were not accurate enough for twentieth-century use. In the 1960's a series of international conferences on weights and measures began which resulted in the modernization of the metric system and its revised title "Le Système International d'Unités" (S.I. Units) or, in English, "The International System of Units." The United States's participation in this activity was through the National Bureau of Standards.

Scientists now use S.I. units as a matter of course. In the U.S., S.I. units are used in electronics, optometry, photography, chemical industries, and the medical profession. Many foreign imports require metric tools to work on them, and acceptance of U.S. exports requires more and more that they be designed to metric dimensions.

UNITS OF MEASURE

SI is a simplifed system in that it is founded upon seven units of measure called base units. Two additional units, generally referred to as supplementary units, complete the foundation from which units for measuring other physical quantities are derived mathematically.

SI BASE UNITS

Unit Name	Plural Form	Pronunciation	Symbol	Quantity
ampere	amperes	am′ pār	A	electric current
candela	candelas	kăn de′ lə	cd	luminous intensity
kelvin	kelvins	kĕl′ vin	K	thermodynamic temperature*
kilogram	kilograms	kil′ō grăm	kg	mass
meter**	meter	mē′tᵊr	m	length
mole	moles	mōl	mol	amount of substance
second	seconds	sek ′ənd	s	time

*Degree Celsius (°C) accepted (°C = K − 273.15). Plural form is degrees Celsius.
**Also spelled metre.

SI SUPPLEMENTARY UNITS

Name	Plural	Pronunciation	Symbol	Quantity
radian*	radian	rā′ dē ən	rad	plane angle
steradian	steradian	stərā′ dē ən	sr	solid angle

*Use of degree, minute, and second is acceptable.

DERIVED units most commonly used have been given special names and symbols, thus reducing lengthy mathematical expression. These units, plus base and supplementary units, when combined, form other derived units without special names. Algebraic functions of multiplication and division are used in the new unit created.

SI DERIVED UNITS WITH SPECIAL NAMES

Unit Name	Plural Form	Pronunciation	Symbol	Quantity	Formula
becquerel	becquerels	be 'krel	Bq	radioactivity	s^{-1}
coulomb	coulombs	koo'lom	C	electric charge	$A \cdot s$
farad	farads	far' ad	F	electric capacitance	C/V
gray	grays	grā	Gy	absorbed dose	J/kg
henry	henries	hen' re	H	inductance	Wb/A
hertz	hertz	hᵊrts'	Hz	frequency	1/s or s^{-1}
joule	joules	jowl	J	energy	$N \cdot m$
lumen	lumens	loo' men	lm	luminous flux	cd.sr
lux	lux	luks	lx	illuminance	$m^{-2} \cdot cd \cdot sr$
newton	newtons	n(y)u' tᵊn	N	force or weight	$m \cdot kg \cdot s^{-2}$
ohm	ohms	ōm	Ω	electric resistance	V/A
pascal	pascals	pas'kᵊl	Pa	pressure or stress	N/m^2
siemens	siemens	sē' mᵊns	S	conductance	A/V
tesla	teslas	tes' lă	T	magnetic flux density	Wb/m^2
volt	volts	vōlt'	V	electric potential	W/A
watt	watts	wät'	W	power	J/s
weber	webers	web'ǝr	Wb	magnetic flux	$V \cdot s$

SOME DERIVED UNITS WITHOUT SPECIAL NAMES

Unit Name	Symbol	Quantity
square meter	m^2	area
cubic meter	m^3	volume
meter per second	m/s	velocity (linear)
radian per second	rad/s	angular velocity
meter per second squared	m/s^2	acceleration (linear)
radian per second squared	rad/s^2	angular acceleration
newton meter	$N \cdot m$	moment of force (torque)
kilogram per cubic meter	kg/m^3	density
joule per kelvin	J/K	entropy
watt per square meter	W/m^2	thermal flux density

NON-SI units used in specialized fields and those of practical importance will remain in use internationally.

NON-SI UNITS MOST COMMONLY USED WITH SI

Name	Plural	Symbol	Value in SI	
minute	minutes	min	1 min	$= 60\ s$
hour	hours	h	1h	$= 3\ 600\ s$
day	days	d	1d	$= 86\ 400\ s$
degree	degrees	°	1°	$= (\pi/180)\ rad$
minute	minutes	'	1'	$= (\pi/10\ 800)\ rad$
second	seconds	''	1''	$= (\pi/648\ 000)\ rad$
liter	liters	l	1l	$= 10^{-3}\ m^3$
metric ton	metric tons	t	1t	$= 10^3\ kg$
bar	bar or bars	bar	1 bar	$= 10^5 Pa$

PREFIXES

SI utilizes a series of prefixes based on the powers of 10 to accommodate increases or decreases in magnitude of the quantity relative to a given unit. Manipulation of the decimal point through multiplication or division and application of the prefix to the unit defines a new unit and quantity that is more convenient and practical to use.

PREFIXES OF SI UNITS

Name	Pronunciation	Symbols	Amount	Multiples and Submultiples	Definition
exa	ex' a	E	1 000 000 000 000 000 000	10^{18}	one million million million times
peta	pet' a	P	1 000 000 000 000 000	10^{15}	one thousand million million times
tera	ter' a	T	1 000 000 000 000	10^{12}	one million million times
giga	ji' ga	G	1 000 000 000	10^9	one thousand million times
mega	meg' a	M	1 000 000	10^6	one million times
kilo	kil' o	k	1 000	10^3	one thousand times
hecto	hek' to	h*	100	10^2	one hundred times
deka	dek' a	da*	10	10	ten times
deci	des i	d*	0.1	10^{-1}	one tenth of
centi	sen' ti	c*	0.01	10^{-2}	one hundredth of
milli	mil' i	m	0.001	10^{-3}	one thousandth of
micro	mi' kro	μ	0.000 001	10^{-6}	one millionth of
nano	nan' o	n	0.000 000 001	10^{-9}	one thousandth millionth of
pico	pe' co	p	0.000 000 000 001	10^{-12}	one millionth millionth of
femto	fem' to	f	0.000 000 000 000 001	10^{-15}	one thousandth millionth millionth of
atto	at' to	a	0.000 000 000 000 000 001	10^{-18}	one millionth millionth millionth of

*These prefixes should generally be avoided except for measurement of area and volume, and for nontechnical uses of centimeter.

PREFIXES FOR MASS

Although the name of the base unit for mass (kilogram) contains the name of the prefix *kilo,* names of the multiples and submultiples of mass are formed by adding the prefixes to the word *gram.*

Examples:

Correct	Incorrect
megagram (Mg)	kilokilogram (kkg)
milligram (mg)	microkilogram (μkg)

Spelling, Capitalization, and Symbols

1. The foregoing tables provide the correct spelling and pluralization of units. Prefixes are never used alone, and no space or hyphen is used in attaching a prefix to a unit.

2. Unit and prefix names are not capitalized unless they occur at the beginning of a sentence. (Exception: The name of the unit of the practical temperature scale, Celsius, is always capitalized.) An abbreviation for a unit derived from a proper name is capitalized. Those base units and derived units with special names that have uppercase symbols are derived from proper names.

3. Symbols of units and prefixes are printed in an upright (Roman) typeface regardless of the typeface used for accompanying text. The uppercase and lowercase lettering of symbols must be as shown in the tables. When using characters of the Greek alphabet, it is preferable to use a typed symbol. If not available, the word, and words of a derived unit of which it is a part, should be spelled out. A lowercase "u" must not be used as a symbol for the prefix micro. Except for drawings, hand-drawn or template characters should be avoided in formal publications.

Punctuation and Spacing

1. Number Grouping. When it is desirable to facilitate reading of numbers having five or more digits, the digits should be placed in groups of three, separated by a space instead of a comma, counting both to the left and right from the decimal point. DO NOT use commas since many countries recognize the comma as a decimal sign. In numbers of four digits, the space is not recommended, unless four-digit numbers are grouped in a column containing numbers with five or more digits.

Examples:

Correct	Incorrect
4 720 525	4,720,525
0.528 75	0.52875
6875 or 6 875	6,875.

Exception: On engineering drawings and documents, neither spaces nor commas are permitted in numerical expressions.

2. Symbol Spacing. Use a space between the numerical value and the unit symbol.

Examples:

Correct	Incorrect
2 A	2A
4 mm	4mm

3. Decimal Signs. The period (.) is used as a decimal point. No space is left before or after a decimal point. Commas are not to be used as decimal signs.

Example:
Correct	Incorrect
2.5	2. 5 or 2,5

4. Periods. Do not use periods following symbols unless they occur at the end of a sentence.

Examples:
Correct	Incorrect
A	A.
km/h	km/h.

Format and Usage

1. Numeric Values. Numbers smaller than one are shown with a zero preceding the decimal point. DO NOT use fractions.

Example:
Correct	Incorrect
0.25 kg	.25 kg or 1/4 kg

2. Symbol Forms. Symbols are used in the singular form. A plural form may indicate a new symbol.

Examples:
Correct	Incorrect
5 kg	5 kgs
ms (millisecond)	ms for meters is read as milliseconds

3. Prefixes. Only one prefix is used in forming a multiple of a unit. A prefix name or symbol should not be used alone in writing or speech.

Examples:
Correct	Incorrect
nm (nanometer)	mμm (millimicrometer)
pF (picofarad)	μμF (micromicrofarad)
kilogram	kilo

4. Prefix Selection. Prefixes that give numerical values of 0.1 through 1000 should be used unless this would result in mixing of prefixes in text.

Examples: 0.003 94 m can be written 3.94 mm
1401 Pa can be written 1.401 kPa

Exception: Dimensions of length on engineering drawings are expressed in millimeters. In engineering practice, the use of hecto, deka, deci, and centi should be avoided where possible.

5. Prefixes in Denominators. The denominator should be the basic unit without a prefix. The only exception is the kilogram (kg) which itself is a base SI unit.

Example:
Correct	Incorrect
km/s	m/ms

6. Multiples. When a compound unit is formed by the multiplication of two or more units, use a raised dot (·) to define the function. DO NOT use ×. An acceptable alternate, for typed or machine-printed text when a raised dot is not practical, is to leave a space between the unit symbols. The prefixed unit (if any) should appear first.

Examples:

Correct	Incorrect
m·K (meter kelvin)	mK (millikelvin)
mN·m (millinewton meter)	N·mm (newton millimeter)

7. Quotients. Compound units formed by division of quantities use the slash or solidus (/) to define the function. To avoid ambiguity, no more than one solidus in a combination should be used unless parentheses are inserted.

Examples:

Correct	Incorrect
m/s	$\dfrac{m}{s}$ or m÷s

8. Mixing of Symbols and Words. Avoid mixtures of symbols, units, and words.

Examples:

Correct	Incorrect
meter per second	meter/second
m/s	meter/s
12.75 m	12 m 750 mm

9. Mathematical Operations. Only symbols should be used when showing mathematical operations.

Example:

Correct	Incorrect
$N = kg·m/s^2$	$N = kilogram × meter per second^2$

10. Tolerances. In single-spaced copy, tolerances are placed on the same line. If both plus and minus tolerances are the same, show both signs and the value together.

Examples: 43.8 +0.4 −0.2   or (limit dim) 43.6–44.2
48.5 ±0.2   or (limit dim) 48.3–48.7

On engineering drawings, unilateral and bilateral tolerances are shown as follows:

$$32.00 \quad {}^{0}_{-0.02} \qquad 32.00 \quad {}^{+0.25}_{-0.10}$$

11. Limit Dimensioning. The high limit (maximum value) is placed above the low limit (minimum value). When expressed in a single line, the low limit precedes the high limit and both are separated by a dash. The high and low limits contain the same number of decimal places.

Examples: 3.9
3.3   or   3.3–3.9

12. Dual Systems of Measure. When both S.I. units and customary units are required to be shown in text, the S.I. units are shown first, followed by the customary units in parentheses.

Examples: 62 mm (2.44 in.)
              1 kg (2.205 lb)

Note

When the number 1 and the lowercase *l* are the same on a typewriter, confusion can result when 1 liter is typed 1 *l*. Spelling out the word liter is preferred.

Definitions

Base Units

ampere (electric current): The ampere is that constant current which, if maintained in two straight parallel conductors of infinite length and of negligible cross section, and placed 1 meter apart in a vacuum, would produce, between these conductors, a force equal to 2×10^{-7} newtons per meter of length.

candela (luminous intensity): The candela is the luminous intensity, in a perpendicular direction, of a surface of 1/600 000 of a square meter of a blackbody at the temperature of freezing platinum (2045 K) under a pressure of 101 325 pascals.

kelvin (thermodynamic temperature): The kelvin is the fraction 1/273.15 of the thermodynamic temperature of the triple point of water. The unit kelvin and its symbol K should be used to express an interval or difference of temperature.

In addition to the thermodynamic temperature expressed in kelvins, the Celsius (formerly centigrade) temperature will also be used. Celsius temperature is expressed in degrees Celsius (°C). The Celsius scale is related directly to the kelvin scale; thus, a temperature interval may be expressed in kelvins or degrees Celsius as follows:

One degree Celsius equals one kelvin (exact)

A Celsius temperature (t) is related to a kelvin temperature (T) as follows: $T = 273.15 + t$ (exact)

kilogram (mass): The standard for the unit of mass (not force), the kilogram is a cylinder of platinum-iridium alloy kept by the International Bureau of Weights and Measures in Paris. A duplicate in the custody of the National Bureau of Standards serves as the mass standard for the United States. This is the only base unit defined by an artifact, and is the only base unit having a prefix.

meter (length): The meter is defined as 1 650 763.73 wavelengths in a vacuum of the orange-red line of the spectrum of krypton-86.

mole (amount of substance): The mole is the amount of substance of a system that contains as many elementary entities as there are atoms in 0.012 kilogram of carbon 12.

When the mole is used, the elementary entities must be specified and may be atoms, molecules, ions, electrons, other particles, or specified groups of such particles.

second (time): The second is the duration of 9 192 631 770 periods of the radiation corresponding to the transition between two hyperfine levels of the ground state of the cesium-133 atom.

Supplementary Units

radian (plane angle): The radian is the plane angle between two radii of a circle which subtends, on the circumference, an arc equal in length to the radius.

steradian (solid angle): The steradian is the solid angle which, having its vertex in the center of a sphere, subtends an area on the surface of the sphere equal to that of a square with sides of length equal to the radius of the sphere.

S.I. Derived Units with Special Names

becquerel (radioactive activity): The unit of radioactivity equal to one disintegration per second.

coulomb (electric charge): The coulomb is the quantity of electric charge that is transferred each second by an electric current of 1 ampere.

farad (electric capacitance): The farad is the electrical capacitance that exists between two conductors when the transfer of an electric charge of 1 coulomb from one conductor to the other changes the potential difference between them by 1 volt.

gray (absorbed dose): The unit of absorbed dose of radiation equal to one joule absorbed by one kilogram.

henry (inductance): The henry is the electrical inductance of a closed circuit in which an electromotive force of 1 volt is produced when the electric current that traverses the circuit varies uniformly at the rate of 1 ampere per second.

hertz (frequency): The hertz is the unit of frequency of a periodic phenomenon of which the period is 1 second.

joule (energy, work, quantity of heat): The joule is the work done or the energy expended when a force of 1 newton moves the point of application a distance of 1 meter in the direction of that force.

lumen (luminous flux): The lumen is the luminous flux emitted in a unit solid angle by an isotropic point source having a luminous intensity of 1 candela.

lux (illuminance): The lux is the unit of illumination of 1 lumen per square meter.

newton (force): The newton is the force which, when applied to a body having a mass of 1 kilogram, causes an acceleration of 1 meter per second per second in the direction of application of the force.

ohm (electric resistance): The ohm is the electrical resistance betwen two points on a conductor which does not contain any source of electromotive force when a constant potential difference of 1 volt maintained between those points results in a current of 1 ampere in the conductor.

pascal (pressure or stress): The pascal is the pressure or stress that arises when a force of 1 newton is applied uniformly over an area of 1 square meter.

siemens (conductance): The siemens is the electrical conductance of a conductor that has an electrical resistance of 1 ohm.

tesla (magnetic flux density): The tesla is the unit of magnetic flux density equal to 1 weber of magnetic flux per square meter.

volt (electric potential): The volt is the potential difference that exists between two points on a conductor carrying an unvarying electric current of 1 ampere when the power dissipated between these points is equal to 1 watt.

watt (power): The watt is the power used when work is done or energy is expended at the rate of 1 joule per second.

weber (magnetic flux): The weber is the unit of the magnetic flux which, linking a circuit of one turn, produces in it an electromotive force of 1 volt as it is reduced to zero at a uniform rate in 1 second.

Non S.I. Unit with Special Name

bar (pressure): The bar is a multiple of pascals used in certain applications for fluid pressure. 1 bar = 10^5 pascals (approximately one atmosphere).

skew—Time delay or offset between any two signals.

skin effect—Tendency of high-frequency alternating currents to crowd toward the surface of a conductor.

skip—1. High-frequency (3-30 megahertz) radio waves travel along the Earth's surface and also are reflected from the ionosphere. The ground wave has a shorter range, limited by the Earth's curvature. The sky wave may return to Earth beyond the ground wave zone, resulting in a dead or skip zone between them. *Electromagnetic-*

Wave Propagation. 2. Computer instruction to bypass the following instruction or group of instructions if some prerequisite condition is met. *Computer Software.*

sky wave—(See skip.)

slave—Any device that operates on command from another device, but does not do so independently.

slew rate—Rate at which output can be driven from limit to limit over the dynamic range.

slice—Thin circular wafer of single-crystal silicon cut from a cylinder in microelectronics fabrication. *Microelectronics.*

slider—Sliding contact, as in a potentiometer or rheostat.

slide switch—Switch in which contact is made by a sliding element instead of by a toggle mechanism (q.v.).

slide wire—Bare resistance wire with sliding contact used in laboratory potentiometer (see potentiometer, (1)).

slotted section—Section of waveguide with a longitudinal slot through which the tip of a probe is inserted. The probe is mounted on a carriage that allows it to be moved along the slot so that it can be positioned in accordance with the reading of a standing-wave ratio meter. Used for testing the impedance of a microwave component by measuring the amplitude of standing waves reflected by it.

slow-scan television—Television system in which the horizontal scanning rate is slower than that used in broadcasting; used to transmit facsimiles of printed material, photographs, etc.

slow-wave structure—Structure, such as a helix, in a traveling-wave tube that reduces the velocity of the r.f. wave so that it is approximately the same as that of the electron beam. *Electron Tubes, Masers.*

slug tuning—Varying the inductance of a coil by adjusting the length of the magnetic core that is within the coil. The magnetic core is a threaded "slug" of powdered iron in an insulating binder and is screwed in or out to change the core permeability, hence this method is also called permeability tuning.

small-signal characteristics—Performance of an amplifier when operated on the linear portion of the characteristic curve of the electron tube or transistor.

S-meter—Signal-strength indicator, often built-in in a communications receiver, to compare signal strengths of received signals, and calibrated in steps in the S-scale. This scale has 6-dB steps from 0 to 9, S9 being the value corresponding to a signal strength of 100 μV.

smoothing filter—Low-pass filter used in a power supply. *Rectifiers and Filters.*

snap-off diode—Charge-storage diode. *Semiconductors.*

snow—Receiver noise in a television set, which in the absence of a strong signal produces a speckled appearance on the screen. A monochrome receiver has white snow, a color receiver has colored snow, or "confetti."

socket—Female connector or receptacle for a plug, light bulb, electron tube, etc., designed to hold the component plugged into it as well as to make electrical connection.

sodium—Very soft, silvery-white, alkali metal that is lighter than water and can be cut with a knife. Extremely active chemically, unites with the oxygen in the air and reacts with water, so has to be kept immersed in kerosene or naphtha. Used in sodium vapor lamps and as a heat exchanger in some nuclear reactors. Melting point, 98 degrees Celsius. Symbol, Na; atomic weight, 22.9898; atomic number, 11.

software—In computer technology, nonhardware items, principally programs and programming routines. *Computer Software.*

soft x-rays—X-rays of low energy (not exceeding 100,000 electron volts) used in medical diagnosis, and with wavelengths overlapping those of the ultraviolet region of the electromagnetic spectrum (q.v.).

solar cell—Photovoltaic device. (See photoelectric effect.)

solder—Alloy used to bond metals together. Ordinary soft solder commonly consists of 60 percent lead and 40 percent tin and melts at about 190°C (373°F). It secures attachment by dissolving a small amount of the copper, or other metal being soldered, so that a fusion alloy is formed between them. Soldering flux is required to remove the oxide film from the metal, which would otherwise prevent this intermetallic solvent action. For electrical work the flux should be rosin, which is highly active at soldering temperature, but inactive at ordinary temperatures. Chloride ("acid") and organic fluxes remain corrosive, and also are electrically conductive, which may result in leakage.

solenoid—Uniformly wound coil in the form of a cylinder, with its length much greater than its diameter. When d.c. flows through the coil the magnetic field so created draws a core or plunger of iron into the solenoid. This device is used to actuate switches and other devices, the most well-known being the solenoid that connects the battery to the self-starter of an automobile when the ignition switch is operated.

solid state—State of matter which is neither liquid nor vapor, in which substances are classed as insulators, conductors, or semiconductors. *Semiconductors.*

sonar—Acronym derived from "sound navigation and ranging," a system for determining the bearing and distance of underwater objects by ultrasonic beams and echoes sent out and received by transducers, which may be magnetostrictive, electrostrictive, or piezoelectric. The most widely used today is the electrostrictive, employing barium titanate or lead zirconate in slices sandwiched between metal plates.

SOS—1. Distress signal used in radio telegraphy (formerly CQD).   2. Abbreviation for silicon-on-sapphire, or silicon-on-spinel, a microelectronic process in which MOS devices are formed on an insulating substrate (sapphire or spinel) by heteroepitaxial growth. *Microelectronics.*

sound—Transmission of mechanical energy as wave motion in a medium. *Electroacoustics.*

sound takeoff—Circuit that separates the sound carrier from the video portion of the television signal and routes it to the sound section.

sound track—Magnetic or optical band on the edge of a motion-picture film that carries the sound recording.

south pole—The end of a bar magnet suspended in the Earth's magnetic field that points toward the south magnetic pole, and the pole of any magnet into which the lines of force of its magnetic field are assumed to enter.

space charge—In an electron tube, the cloud of electrons formed around the cathode by thermionic emission. *Electron Tubes.*

space communication—Microwave communication between planet and spacecraft, spacecraft and spacecraft, planet and satellite. *Space Communications.*

SPACE COMMUNICATIONS

At the present time practically all space communication is between the Earth and the spacecraft. Communication between different spacecraft has been limited to certain programs such as Apollo, Skylab, and Project Viking. Consequently most space radio paths to date have also had to pass through the Earth's atmosphere.

The atmosphere will only allow waves of certain frequencies to pass through it. Figure S-6 shows that "windows" exist for some frequencies, though not all can be used. For instance, the window for visible light can be used only when the sky is clear and not obscured by clouds. But below 50 gigahertz the atmosphere is transparent to radio waves down to about 10 megahertz, below which transmission is severely affected by the ionosphere. Frequencies above 10 gigahertz are also attenuated by rain. As a result, satellites at first were using frequencies between 3.7 and 4.2 gigahertz only, and their Earth stations, with much higher power, were operating between 5.925 and 6.425 gigahertz. (These are still the frequencies used by the geostationary television satellites.)

The introduction of synchronous satellites in geostationary orbits at an altitude of 35,900 kilometers (22,300 miles) has resulted in considerable progress. At that height a satellite travels at the same angular velocity as the surface of the rotating Earth, and thus remains constantly over the same point. By using parabolic reflectors, which beam the signal in a narrow cone, the low power of the satellite transmitter is optimized, since it is directed only towards the vicinity of the Earth station. The Earth station benefits as well by not having to track the satellite. Since the area on the Earth's surface that receives the signal is relatively small, it is possible to have many satellites on the same frequency without interfering with each other. Also, one satellite can beam its transmission to two Earth stations at once, these stations being a hundred miles or more apart, so that if heavy rain is affecting one it is not likely to be affecting the other. Thus higher frequencies can be used which require smaller antennas, which allow them to be focussed into sharper beams. Noise effects are also reduced by using FM.

While the Earth station has no trouble in obtaining all the power it needs, the satellite has a serious problem. At present the most widely used source is the solar cell. This is a thin wafer of p-type silicon which has had its surface changed to n-type by diffusion with a suitable impurity. Solar radiation falling on the cell is changed directly into electrical energy. An array of solar cells that is adequate for the satellite's power requirements is large and very expensive.

Most of the rest of the circuitry is solid state, but the output amplifier is a traveling-wave electron tube (see *Electron Tubes*). Typically, this requires an input power of 12 watts to provide an output of four watts. To receive and amplify this very low-power signal,

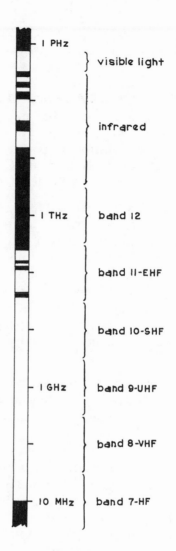

Figure S-6
Windows into Space

the Earth station usually makes use of a parametric amplifier cooled to a temperature of around four kelvins, which gives it a noise temperature of some 65 kelvins.

Noise temperature is approximately the blackbody radio temperature of the source. All bodies with temperatures above absolute zero have it. That of the Earth is 254 kelvins. The satellite's antenna will "see" only a part of the Earth, however, so it will receive only a part of the Earth's noise contribution. However, this is much less than the satellite receiver noise, since satellite receivers are relatively noisy.

The problems involved in the use of satellites close to Earth become multiplied many times when the spacecraft is a probe such as Voyagers 1 and 2, which followed Pioneers 10 and 11 to the major planets. The transmission path is in the billions of kilometers, and the precision required to keep the probe's and Earth's antennas aligned over such distances is formidable. From the probe's point of view Earth at that range subtends an angle of about two seconds of arc, so that it would be a telescopic object.

Because the probe's signal at this range cannot help but be very weak, the background noise tends to bury it. This is overcome by digitizing the information (converting it to digital format). The pulses can then be enhanced (built up), while the noise, being random, is not enhanced, allowing the pulses to stand out. The strengthened pulses can then be reconverted to analog format if necessary (e.g., video information in a televised picture).

Although laser communication between Earth and space is restricted by atmospheric attenuation, laser communications between craft in space are feasible and have the advantages of exceptionally high antenna gain, very high radiated power, and compact transmitting optics. Space laser signals could, of course, be received and transmitted by a synchronous satellite, using a microwave link to an Earth station. Such systems have been explored and are still in a growth stage.

spade lug—Forked, flat metal terminal soldered or crimped to the end of a lead to facilitate attachment to a binding post.

spaghetti—Insulating tubing of plastic or varnished cloth used to slip over bare wires that are too close to other exposed conductors.

spdt—Abbreviation for single-pole, double-throw (switch). (See Figure S-7.)

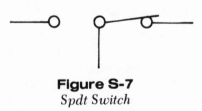

Figure S-7
Spdt Switch

speaker—(See loudspeaker.)

spectrum—(See electromagnetic spectrum.)

spectrum analyzer—Type of oscilloscope that displays a graph of amplitude against frequency instead of the usual amplitude against time. Useful for studying the energy distribution of an electrical signal as a function of the frequencies within that signal (e.g., a modulated carrier and its sidebands).

spherical aberration—Blurring of the image projected by an optical lens because not all the light rays from a point on the optical axis of the lens meet at the same image point. Rays passing through the lens close to its center are focussed farther away than rays that pass through the lens near to its rim. When the light rays come from a point not on the optical axis the small blurred spot is elongated into a comet shape, hence this is called coma.

spider—Light flexible corrugated disk, or similar device, that keeps the voice coil of a dynamic loudspeaker centered within the pole piece of the permanent magnet.

spike—Transient impulse of short duration but high amplitude.

spinel—Any of a group of oxide minerals with the general formula AB_2O_4, where A can be magnesium, iron, zinc, manganese or nickel, and B may be aluminum, chromium, or iron. (See SOS.)

sporadic-E ionization—Scattered patches of relatively dense ionization at the E layer level that raise the critical frequency to a value perhaps twice that which is returned from any of the regular layers by normal reflection.

spot—Small illuminated disk on cathode-ray tube screen where the sharply focussed electron beam strikes it, if the beam is stationary. When the beam is moving the spot becomes a trace due to the persistence of the phosphor.

spst—Abbreviation for single-pole, single-throw (q.v.).

sputtering—Deposition of material in film form using high-voltage glow discharge.

square wave—Waveform consisting of a voltage that is constant in amplitude but reverses its polarity at regular intervals. The time taken to reverse polarity is negligible in comparison to the duration of the positive and negative excursions. A square wave may also be thought of as consisting of the addition of a fundamental sine wave and an infinite number of its odd harmonics. (See Figure S-8.)

squelch circuit—Circuit that eliminates background noise in a receiver, that becomes very loud in the absence of a signal, by reducing sensitivity in accordance with the setting of a squelch control. Sensitivity is automatically restored when a signal is received.

squirrel-cage induction motor—Widely used type of a.c. motor with a rotor winding consisting of a cylindrical assembly of heavy copper conductors joined to metal disks at each end, similar to the revolving part of a squirrel cage, and embedded in slots in

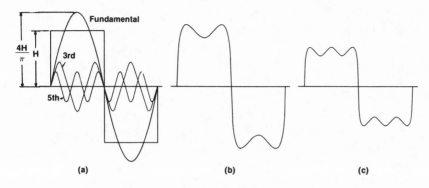

Figure S-8

Composition or a square wave, showing the first three harmonics, and the approximation to a square wave as the first two and the first three odd-harmonics are combined

the iron armature, parallel to the rotor shaft. Squirrel-cage motors are available commercially from fractional horsepower to 12,000 horsepower (larger on special order), and high starting-torque models are manufactured for operating loaded compressors, hoists, elevators, etc.

stability—Condition whereby a slight disturbance does not produce too disrupting an effect on a circuit or system, usually achieved by some type of feedback. *Feedback.*

stable platform—Platform carrying three accelerometers in an aircraft, misile or space vehicle, that measure acceleration in three axes (two for a ship). The resultant signals are converted by a computer to distance and direction traveled. The platform is stabilized by three rate-gyroscopes that detect pitch, roll or yaw, and produce signal voltages, which when amplified drive servomotors to rotate the platform to its original position.

stack—In computer programming, a memory organization to handle subroutine calls and returns on a last-in, first-out basis. *Computer Hardware.*

stage—Each amplifier, when two or more are cascaded.

staggered tuning—Tuning two or more stages of an amplifier to slightly different frequencies so as to obtain a substantially uniform response over a wide bandwidth. Commonly used in television i.f. amplifiers.

stamper—In phonograph disk recording, a metal negative from which phonograph records are stamped. *Recording.*

standard cell—A saturated standard cell consists of a glass tube in the form of a letter H. A pool of mercury in the bottom of each limb makes contact with a metal electrode inserted through the glass. One mercury pool is amalgamated with a 10-percent cadmium: it is the anode. The other has a layer of mercurous sulfate resting on it: it is the cathode. Both anode and cathode are covered with a layer of cadmium sulfate crystals to ensure that the cadmium sulfate electrolyte remains saturated. The cell must be maintained at 20.00 ± 0.02 degrees Celsius and, provided it is not jarred, tilted or otherwise mishandled, its voltage will never vary by more than 5 parts per million from the standard value, which is 1.01864 volts. If it is mishandled it takes three weeks to restabilize; therefore a more rugged version is available. This is called an unsaturated standard cell. It requires only a few hours to recover after being moved. Its nominal value is 1.019 volts ± 0.002 percent, and it may be kept at temperatures ranging from 22 to 25 degrees Celsius. These cells are used to standardize laboratory potentiometers. They are ruined by drawing any current from them (even the small current drawn by a voltmeter). (See Figure S-9.)

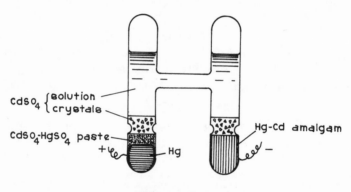

Figure S-9
Saturated Standard Cell

standard deviation—Measure of the variability (dispersion or spread) of any set of numerical values about their arithmetic mean. It is specifically defined as the square root of the arithmetic mean of the squared deviations.

standing waves—When a transmission line is not terminated with a load equal to its characteristic impedance, power is reflected back along the line toward the source. Interaction between the outgoing and reflected power results in current loops (maxima) and nodes (minima) called standing waves. The ratio between the value of the loops and the nodes is called the standing wave ratio (SWR). *Transmission Lines.*

star connection—(See wye-connected system.)

static—Atmospheric noise. *Radio Noise and Interference.*

static electricity—Electricity in motionless charges.

stator—The fixed part of any device that also has a moving part, or rotor.

step function—Signal such as a square wave or pulse with a virtually instantaneous risetime, used for testing transient or frequency response.

stepping relay—Multicontact switch in which the wiper is moved on from one contact to the next by a relay.

steradian—Cone-shaped solid angle with its vertex at the center of a sphere and its base enclosing an area on the surface of the sphere equal to the square of the radius of the sphere. (See also radian measure.)

stereo—Colloquialism for stereophonic equipment that attempts to give a three-dimensional effect by the use of two channels. *Electroacoustics, Recording.*

stimulated emission—Release of energy from an excited atom by artificial means. *Lasers, Masers.*

stitch bonding—(See ball bonding.)

stopband filter—(See band elimination filter.)

storage—1. Placing data in computer memory; the memory itself. *Computer Hardware.* 2. Type of cathode-ray tube that produces a display of controllable duration. *Electron Tubes.*

storage battery—Secondary battery. (See battery.)

strain gage—Device for measuring the changes in distances between points in solid bodies that occur when the body is deformed. The resistance strain gage consists of a flat foil produced by a printed-circuit technique in the form of a grid on a plastic backing. It is firmly bonded to the surface on which the strain is to be measured and connected to a current source and a meter. The resistance of the foil increases when it is stretched, so the meter shows a decrease of current that is proportional to the increase of distance between opposite ends of the foil pattern. Conversely, a decrease in the distance is reflected by an increase of current. The deflections of the meter pointer are read against a dial scale calibrated for strain, or the change in current is displayed on a digital readout calibrated in appropriate units.

strain insulator—Egg-shaped insulator used to break guy wires of antenna masts into nonresonant lengths.

stranded wire—Wire made by twisting together a number (usually 7) of smaller wires, to obtain greater flexibility than can be gotten from solid wire.

stratosphere—Nearly isothermal layer of the atmosphere (q.v.) above the troposphere.

stray capacitance—Capacitance due to the wires and components of a circuit. Although small, the effect may be considerable at radio frequencies.

strobing—In synchronous data communication, the transfer of data from the computer data bus into the accumulator, for example, by the controller. The data is read in during a very short time interval compared with the length of time it is on the bus.

stroboscopic disk—Paper or cardboard disk printed with concentric circular bands divided into various numbers of black and white segments. When placed on a revolving phonograph turntable and illuminated by a strobe light flashing at a known rate one of the bands appears to stand still, indicating the r.p.m. of the turntable.

stub—Short length of transmission line attached to a transmission line to match its impedance to the antenna. *Transmission Lines.*

stylus—Pointed device used to cut or play phonograph records. *Recording.*

subassembly—Subordinate section of a main assembly fabricated separately for convenience.

subatomic particles—Elementary particles, such as electrons and neutrons, that are believed to be the fundamental units of matter and energy. *Nuclear Physics.*

subcarrier—Lower-frequency carrier used to modulate a main carrier and often itself modulated to carry information, for instance in television and FM stereo. *Broadcasting.*

subroutine—In computer programming, particular operations that are frequently used are made into subroutines that can be called as a whole instead of copying out the same sequence of instructions each time.

subscript—Figure or character that is written below and to the right of a symbol to distinguish it from others, as n_1, n_2, n_3, where 1, 2 and 3 are subscripts. Also called subindices.

Subsidiary Communications Authorization (SCA)—Authorization to an FM broadcasting station to transmit in the 53-75 kilohertz portion of the stereo channel (20-75 kHz for monophonic channels). *Broadcasting*.

substrate—Single body of material on or in which circuit elements are fabricated in the manufacture of integrated circuits. *Microelectronics*.

subtractive process—In the fabrication of a printed circuit, the process in which unwanted copper is removed to leave the desired conductor pattern. *Printed Circuits*.

sulfur—Nonmetallic element, tasteless, odorless, brittle solid, pale yellow in color, poor conductor of electricity. Symbol, S; atomic weight, 32.064; atomic number, 16.

sulfuric acid—Dense, colorless, oily, corrosive liquid, sometimes called oil of vitriol, with the chemical formula H_2SO_4. It has a very strong affinity to water, so that adding water to sulfuric acid is highly dangerous. Even adding the concentrated acid to water heats the water. Diluted with pure water, the acid is widely used for the electrolyte in lead-acid storage batteries, which should have a specific gravity of 1.260 if used in an automobile.

summing point—In an amplifier such as an operational amplifier, in which a feedback network returns a portion of the output to the input, the point in the input where the feedback signal is applied is the summing point. *Operational Amplifiers*.

superconductivity—State of matter into which many materials enter on being cooled to exceedingly low temperatures, in which they have negligible resistance.

superheterodyne reception—Combination of local oscillator and an incoming carrier frequency, giving a beat (or heterodyne) frequency that is the difference between the original combining frequencies. This difference frequency, called the intermediate frequency (i.f.), retains the modulation of the original signal and is subsequently demodulated to recover the audio information after passing through the i.f. amplifier, whose high gain and selectivity are enhanced by having to handle only one frequency.

super-high frequency (SHF)—Frequency range from 3 to 30 gigahertz, in band number 10 (centimetric waves).

superscript—Figure or character that is written above and to the right of a symbol, usually an algebraic exponent, as in x^2, etc., where 2 is a superscript denoting x to the power of 2, or "squared."

supply reel—In a tape deck, the reel from which the tape is being unwound. *Recording.*

suppressed carrier—(See sidebands.)

suppressor grid—In an electron tube, a grid placed close to the anode to capture secondary electrons emitted by it. *Electron Tubes.*

surface wave—Seismic wave (q.v.) traveling on the surface of a substrate.

susceptance—Reciprocal of reactance, given by:

$B = X/R^2 + X^2$, or, when there is no resistance,
$B = 1/X$, where B = susceptance in siemens,
R = resistance in ohms,
X = reactance in ohms.

sweep—The horizontal movement of the electron beam in a cathode-ray tube. *Electron Tubes.*

swinging choke—Employed in a choke-input power supply filter, a swinging choke has the required critical inductance with the bleeder load only, and about the optimum inductance value at full load. *Rectifiers and Filters.*

switch—Device for making and breaking connections, and thereby closing and opening circuits.

SWR—Abbreviation for standing wave ratio. (See standing waves.)

sync—Abbreviation for synchronizing signal, etc. (q.v.).

synchronizing signal—Pulses that keep one circuit in step with another, as in television. *Broadcasting.*

synchrotron—Particle accelerator. *Nuclear Physics.*

T—1. Symbol for prefix tera- (q.v.). 2. Symbol for tesla (q.v.). 3. In a schematic diagram, the class letter for a transformer.

t—Symbol for metric ton (1000 kg).

Ta—Symbol for tantalum (q.v.).

take-up reel—On a tape recorder, the reel that winds on the tape.

tandem office—Intermediary exchange between end offices in a telephone system. *Telephony.*

tank circuit—Parallel-resonant circuit that couples the output stage of a transmitter to the antenna. (See Figure T-1.)

tantalum—Very hard, silver-gray metal, with high density, high melting point (2996°C), and excellent resistance to all acids except hydrofluoric acid. Used in electrolytic capacitors and as getters and electrodes in electron tubes. Symbol, Ta; atomic weight, 180.948; atomic number, 73.

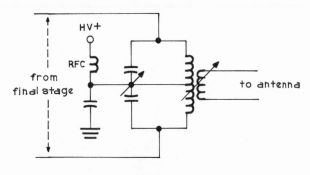

Figure T-1
Tank Circuit

tantalum capacitor—Electrolytic capacitor made with tantalum foil. *Capacitors.*

tape—(See magnetic tape recording.)

tape deck—Tape transport and head assembly used for magnetic tape recording or playback. *Recording.*

target—1. In radar, an object that produces an echo, or reflection of the transmitted pulse. *Radar.*   2. In an electron tube, a structure at which a beam of electrons is directed. *Electron Tubes.*

taut-band suspension—Meter movement suspended on thin metal strips held taut by springs, that twist as the movement turns, thereby applying restoring torque. Taut-band suspension eliminates friction error.

Tchebychev filter—(See Chebishev filter.)

technician—Engineering assistant, graduate of a technical institution or school, but not having an engineering degree. Expert in the practical side of laboratory or manufacturing processes. Generally considered by engineers to be the backbone of the organization. (Many engineers were technicians first.)

telecommunications—General term for the transmission of any information by any means between entities that cannot communicate directly.

telegraphy—Method of sending coded messages over wires by means of electrical impulses. Also, the on-off keying of a radio carrier wave, using a code. (See Morse Code.)

telemetry—Measurement at a distance, transmitting by electronic means (wire, radio, etc.) data from a transducer at a remote site, and then, after receipt, converting the data to a display, readout, or printout of some kind.

telephone—Device for voice communication by wire or radio between parties at a distance from each other. *Telephone.*

TELEPHONE

The basic principles of the telephone have not changed since Alexander Graham Bell's patent was granted in 1876. The voice makes the air vibrate, and it in turn causes a diaphragm to vibrate. The diaphragm modulates an electric current so that an electrical signal with fluctuation corresponding to the vibrations is produced. This signal is transmitted along a wire (or through space in a radio system) to the receiver. Here it flows in a coil wound around a magnet. The fluctuations in the current cause variations in the magnet's field, so that its attraction for an adjacent steel diaphragm varies in accordance with the electrical signal. This makes the diaphragm vibrate, moving the air, and reproducing the original sound.

The transmitter, or microphone, into which the user speaks, was not invented by Bell, but by Thomas Alva Edison, who had been commissioned to improve Bell's "speaking telegraph." This device is still in use in modern instruments. It consists of a metal diaphragm that presses against a pinch of shiny carbon granules in a metal cup. The diaphragm and cup are insulated from each other, so a current passing from one to the other has to flow through the carbon granules. The resistance offered by them varies in accordance with the pressure of the diaphragm, so that as this vibrates in response to sound waves in the air, the current also varies. Since the current is an input of power that is modulated by the pressure variations of the diaphragm, the carbon microphone is an amplifier as well, with a current output that is more powerful than the voice input.

As already mentioned, the receiver, or earphone, consists of a magnet, a coil assembly and a diaphragm, which operate on the same principle as the original Bell induction instrument. Both the receiver and transmitter are mounted in the handset of a modern telephone, so that the user can hold them in the proper positions with respect to mouth and ear, with one hand. Their four wires are contained in a coiled retractile cable that connects the handset to the base.

Up to this point the modern instrument is fundamentally the same as its predecessors for over a hundred years. However, it was realized soon after its invention that unlike the telegraph, which had a single terminal in each town, there were many telephone terminals and each had to be connected to any of the others at a moment's notice. Obviously, the lines had to be brought together at a common point where connections could be made. The first telephone exchange, which had a manual switchboard serving 21 subscribers, went into operation at New Haven, Connecticut in 1878. When the subscriber wanted to make a call he had to wind a crank handle, which sent a signal to the exchange to alert the operator, who would then "ring" the other party. Each telephone also had its own battery.

This all changed with the introduction of fully automatic exchanges, the first of which was installed in Omaha in 1921. The subscriber now dialed the number of the party being called; therefore the dial assembly became the principal feature of the base.

The dial assembly sends an addressing signal consisting of a series of from 1 to 10 pulses representing the corresponding numerical digits 1 to 9 and 0. The pulses are breaks in the continuous current on the line, at a rate of 10 pulses per second. The dial assembly is therefore a spring-driven rotary device that opens and closes a switch a number of times according to how far it has been turned before being released.

The dial is being replaced on modern telephones by a set of push buttons which, when pressed by the subscriber, cause transmission to the exchange of combinations of pairs of audio-frequency tones, as follows:

Pushbutton Number	Audio-Frequency Tones (hertz)	
0	941	1209
1	697	1209
2	697	1336
3	697	1477
4	770	1209
5	770	1336
6	770	1477
7	852	1209
8	852	1336
9	852	1477

At the exchange these tones are interpreted as if they were dial pulses, so either type of telephone is compatible if the exchange has this capability.

The desired connection is obtained from the number dialed. The standard seven-digit code allocates four digits for the individual subscriber in any one exchange, and three digits for the exchange itself. The modern automatic switching system is described under *Telephony*.

The other principal device in the base of the telephone is the ringer assembly. This is similar to a doorbell, consisting of a ringer coil which actuates a clapper that strikes two gongs as it vibrates between them. When another party dials the number of this telephone, the apparatus in the exchange causes the ringer assembly to ring by using a high-voltage low-frequency signal (about 90 to 100 volts at 16-2/3 to 25 hertz). As long as the handset is resting in its proper position on the base, the ringing signal will be connected by the switch hook to the ringer. If, however, it is not on the base, the connection to the ringer is open; the exchange senses this and sends a busy signal back to the calling party.

The switch hook gets its name from the old pedestal telephone, where it was a hook-operated switch that held the earpiece when it was "hung up." Many telephone expressions date from this era, and are still used, although some are not strictly applicable with modern equipment. A telephone is described as being "on hook" or "off hook" although there no longer is a hook.

When the subscriber answers the telephone it goes from on hook to off hook, and the hook switch disconnects the ringer and connects the handset to the line. At the exchange the subscriber's line is disconnected from the ringing signal and connected to the other party's line. Originally this would be a d.c. connection, but subscriber carrier systems are presently used to provide service to many customers, since by using carriers with different frequencies many people can use the same line at the same time.

Telephones that transmit a picture as well as a voice have been planned since 1927, but although introduced in Germany in the 1930's by the German Post Office, they have been slow in development in the U.S. The reason for this is obvious. U.S. telephone companies are commercial operations, and the expense of a videotelephone service requiring a bandwidth equivalent to the demands of more than 300 long-distance telephone connections (about one megahertz) did not find encouraging acceptance by the public.

In spite of this, the American Telephone and Telegraph Company did introduce the Picturephone service in Chicago and Pittsburgh in 1971, but a year later only about 100 were in regular use by subscribers. This set has a viewing screen 5 by 5.5 inches, which displays a 250-line picture at 30 frames a second, using interlaced scanning. The specially designed camera tube has a mosaic of tiny photosensitive semiconductor elements.

Efforts are now being made to develop a system that digitizes the picture signal so that it can be transmitted at a lower frequency and reconstituted at the receiving subscriber's set. Even if this system might still be too expensive for the general public it may have value, in conjunction with a Touch-Tone telephone, as a remote computer terminal.

telephony—Communications system to facilitate telephones anywhere to be connected with each other. *Telephony.*

TELEPHONY

Telephony is the name for the system that enables people to converse by telephone over any distance. It is a part of the broader field of telecommunications, which includes telegraphy and data transmission among other nonverbal communications over a distance.

A telephone subscriber is usually connected by way of a "loop" (two- or four-wire line) to a local telephone exchange, called an "end office," or "central office." There are five classes of switching centers, as shown in Figure T-2, and the end office is the fifth or lowest class. It is responsible for interconnecting customers' lines directly in

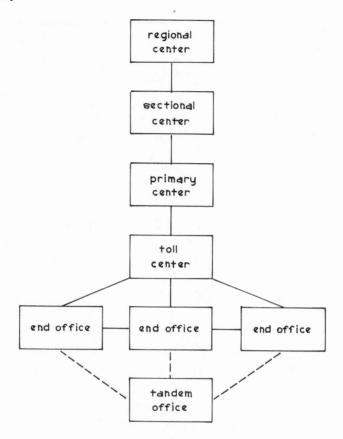

Figure T-2
Simplified Telephone System

the area served by it. End offices in the same city can also be connected to each other. In some densely populated exchange areas the end offices may have an exchange of their own, called a "tandem office." These end offices comprise the area in which the subscriber can make local calls.

When a subscriber wishes to make a call to a number in another area, his end office connects him to a class four office, called a toll center. His call is now a toll call, or long-distance call, for which an extra charge is made. In heavily populated areas there may be several toll centers grouped under a class three office, called a primary center. Above the primary center are two other levels, the sectional center (class two) and the regional center (class one).

These centers are all interconnected by connection media called trunks. A trunk may be a cable, coaxial line, or microwave link. The selection of which to use is done by a computer in accordance with traffic intensity.

Direct distance dialing is generally used to make connections. The addressing signal to a party in the U.S. made from a country outside North America might be 1-714-679-4590, where 1 is the country code, 714 the area code, and 679-4590 the local code. The country code is not used when the call originates in the same country, nor the area code when in the same area. (However, in the U.S., where commercial telephone service is used, there are variations in this system. For instance, some telephone companies do not offer direct dialing to other countries, and some use a 1-prefix to connect to other toll centers in the same area.)

The local exchange office (end office) to which a telephone is connected must make the connection between the subscriber's telephone and another subscriber, or with an outgoing trunk. The connection is made by the switching system in the exchange. The type of switching used depends upon the sophistication of the system, and could be by stepping switches, rotary switches, panel switches, crossbar switches, or semiconductor devices. Generally speaking, the switching network consists of connecting devices that may be in either of two states, with a low impedance in one state and a very high impedance in the other. These devices are called crosspoints, regardless of what type of switch is used. Figure T-3 shows a single-stage rectangular-coordinate switching matrix that connects any of N inlets with any of M outlets. This is called full availability. Other networks may also be constructed with limited availability, in which some lines do not connect. Small stages like that in Figure T-3, but with up to 200 inlets and outlets, are combined into larger networks. While in theory a 200-inlet, 200-outlet, network would require 40,000 cross-points, in fact the use of secondary switches between stages reduces the actual number to 15,200. The maximum number of subscribers that can be connected to one exchange is 10,000, using all possible combinations of the four digits in the number (i.e., 0000 through 9999). In modern systems the dial pulses are stored, and then the appropriate crosspoint switches, which are pnpn devices, are turned on by the stored signal. The seven digits allow nine million telephones to be assigned to each area code, and there can be up to a thousand area codes for each geographical zone. The actual number of telephones in the U.S. as of January 1, 1978, was 162,072,146, so even with Canada and Mexico included there is still plenty of room for more.

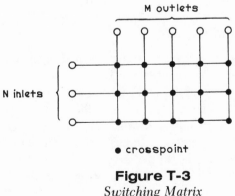

Figure T-3
Switching Matrix

Modern electronic telephone exchanges contain highly sophisticated computing elements that perform most of the functions of a human operator. They can monitor thousands of telephone lines, determine which are demanding service, provide dial tone, remember the digits of each number as it is dialed, translate the number to identify the central office and line of the called party, transmit the consequent billing between offices, send ringing current to the called party, set up the required connections, monitor the call during its progress, and disconnect the phones when the parties hang up. They also time and bill toll calls, make repeated attempts to find alternate routes for calls if the most direct path is unavailable, and perform self-tests and print out trouble-shooting information. New developments that are coming into general use will add additional capabilities. These include automatic transfer of the call to another number, calling the user back when a busy line becomes available, and other services in response to dialed codes.

As mentioned previously, there are various paths that can be used for transmitting telephone messages. The major wire lines employed are open wire (bare wires mounted on poles), multipair cables (on poles, buried in the ground, in underground conduits or submarine cables), and coaxial cables.

A single open-wire pair may be used for carrier telephony up to a frequency of 150 kilohertz. This can provide 12 voice channels, each requiring four kilohertz, frequency multiplexed using single-sideband suppressed carrier (SSSC). Transmission in one direction uses frequencies from 40 through 88 kilohertz, in the other direction frequencies from 100 through 148 kilohertz.

Cable systems allow the use of frequencies up to 500 kilohertz. This enables 24 voice channels to be used on each pair of wires. Transmission in one direction is between 40 and 140 kilohertz; in the other between 164 and 264 kilohertz. Coaxial cables transmit signals ranging into megahertz frequencies, so that as many as 600 voice channels may be frequency-multiplexed on one cable.

Microwave links account for about one-half the trunk mileage in the U.S. In these links the groups of frequency-multiplexed messages given above are used to modulate carriers of 6 and 11 gigahertz for transmission up to 250 miles, while carriers of 6 and 4 gigahertz are used for longer distances up to 4,000 miles. Microwave repeaters are located 20 to 30 miles apart so that the signal can be repowered periodically, and as much of the noise filtered out as possible.

The two last-named frequencies are also used for up and down links for Intelsat satellites, operated by a consortium of more than 100 nations. Other satellites used for telephony between countries, especially across the Atlantic and Pacific Oceans, are the Comstar and Marisat (for use by ships) satellites. Many others are in the development stage. Domestic communications satellites also exist, or soon will, such as Canada's Anik B. Synchronous satellites with highly directional antennas can use the same frequencies without interference, and therefore, since there is lots of room up there, could provide hundreds of millions of telephone circuits, far more than are likely to be needed in the near future.

teleprinter—Printing device that prints like a typewriter in response to a keyboard somewhere else, to which it is connected by landline.

television (TV)—Conversion of a scene in motion with its accompanying sounds into an electrical signal, transmission of the signal, and its reconversion into image and sound at the receiver. *Broadcasting.*

tera-—Prefix of S.I. units meaning one million million times (10^{12}); symbol, T.

terminal—1. Either end of an electrical circuit, or a connector at either end.   2. Device, such as a keyboard, cathode-ray tube, etc., for the input or output of data to or from a computer. *Computer Hardware.*

termination—Load at the output end of a transmission line. *Transmission Lines.*

tesla (T)—Unit of magnetic flux density, equal to one weber of magnetic flux per square meter.

Tesla coil—High-frequency induction coil, similar to the Ruhmkorff coil (q.v.), but with an air core and without interrupter and capacitor.

test equipment—Devices such as voltmeters, oscilloscopes and the like, used to evaluate the performance of electronic equipment.

test leads—Color-coded, insulated flexible leads for making temporary connections between test equipment and a unit being tested.

test pattern—Still picture transmitted by a television station or obtained from a test instrument (e.g., a TV analyst) that is displayed on the screen of a receiver so that adjustments may be made to improve linearity, focus, contrast, etc. (See Figure T-4.)

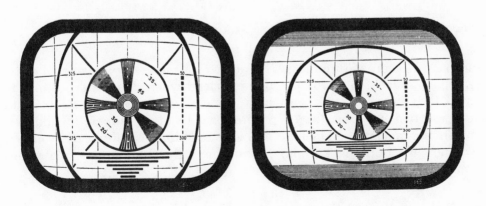

Figure T-4
Test Patterns Showing Effect of Misadjusted Height Control

test point—Point indicated in a schematic diagram at which to connect a test instrument to verify a specific voltage, waveform, etc. Sometimes an actual connection (e.g., a jack) on a panel, for the convenience of the technician.

test probe—Sharp metal point, or hook, with insulated handle on the end of a test lead, for making temporary contact with some part of an electronic circuit to obtain a voltage reading, waveform, etc. The sharp-pointed type is also called a test prod or, colloquially, a "toad stabber." (Some test probes contain electronic circuitry: see probe, demodulator probe, high-voltage probe.)

test record—Phonograph record used to test high-fidelity amplifiers, speakers, and record players.

TE waves—Transverse-electric, or H waves, a mode of propagation of energy in a waveguide in which the electric vector is always perpendicular to the direction of propagation. *Waveguides and Resonators.*

thermal agitation—Random kinetic energy of atoms and molecules at any temperature above absolute zero. Its amplitude increases with temperature and generates noise in the antenna and first stage of a receiver. *Radio Noise and Interference.*

thermistor—Electrical resistance element made of the oxides of manganese and nickel, the resistance of which varies with temperature. Used for temperature measurement and temperature compensation (a component which might overheat from excessive current may be protected by a thermistor located to sense a rise in its temperature, if the thermistor is also in series with the current).

thermocompression bonding—(See ball bonding.)

thermocouple—Temperature-measuring device consisting of two wires of different metals joined at each end. One junction is placed where the temperature is to be measured; the other is kept at a constant or reference temperature. A voltage appears between the two junctions that is proportional to their temperature difference. Combinations of metals used include antimony and bismuth, copper and iron, copper and constantan (copper-nickel alloy), and (for high temperatures) platinum and rhodium.

thermopile—Number of thermocouples (q.v.) connected in series for greater sensitivity and to give an average of several readings.

thermostat—Device for maintaining the temperature of an enclosed area essentially constant by controlling the flow of fuel or current to the heating or cooling equipment. Most thermostats consist of a bimetallic strip or spiral of two dissimilar metals, that expand or contract by different amounts when the temperature changes. This causes the strip to bend so as to open or close a switch, and so supply or shut off current to a solenoid-operated valve or contactor.

thick-film process—Method of manufacturing integrated circuits by screen deposition of thick films; usually only passive elements are made this way.

thin-film process—Method of manufacturing integrated circuits by depositing thin layers of materials to perform electrical functions; usually evaporation or sputtering is used.

thorium—Radioactive silvery-white metal, used as breeder reactor fuel; commercial photoelectric cells for measuring ultraviolet radiation (with wavelengths between 3.75×10^{-7} and 2.00×10^{-7} meters); and as an additive to magnesium and magnesium alloys to improve their high-temperature strength, and to tungsten used in some electron-tube filaments (*Electron Tubes*). Symbol, Th; atomic weight, 232.088; atomic number, 90.

three-phase circuit—Three wires carrying alternating currents which differ in phase by 120 degrees. A fourth (ground) wire may also be used. (See delta connection and wye connection.)

thru hole connection—Conductive material used to make electrical and mechanical connection from the conductive pattern on one side to the conductive pattern on the opposite side of a printed circuit board. *Printed Circuits.*

thyratron—Gas-filled, three-element, hot-cathode tube in which the grid controls only the starting of a current, thus providing a triggering effect, used for converting a.c. to d.c. or vice versa. *Electron Tubes.*

thyristor—Transistor have three pn junctions, that is the solid-state counterpart of the thyratron. The silicon-controlled rectifier (SCR) is the most common form of thyristor. *Transistors.*

tickler coil—Coil that couples the output of the triode in a regenerative detector to the input to give positive feedback. (See regenerative detector.)

time—(See second.)

time constant—Time in seconds required for capacitor charging through a resistor to reach 63 percent of the appied voltage, given by:

T = CR

where   T = time constant in seconds
        C = capacitance of capacitor in farads
        R = resistance of resistor in ohms

(See Figure T-5.)

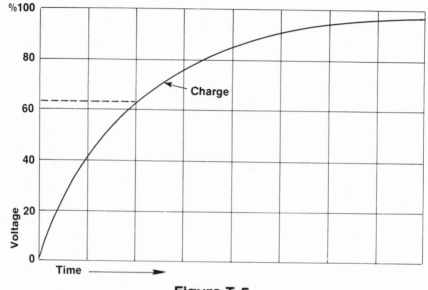

Figure T-5
Time-Constant Curve

time-mark generator—Test instrument that generates precisely spaced pulses. When connected to the input of an oscilloscope they appear as time marks, the display resembling a comb, which are used to calibrate the horizontal sweep.

time sharing—Process in which one computer can perform several tasks simultaneously. The computer accepts inputs, processes them, and delivers outputs to each user in turn, but at such high speed that it seems each program is being handled continuously, instead of in equal time segments in strict rotation. *Computer Hardware.*

tip jack—Small jack (q.v.).

titanium—Silvery-gray, lightweight, high-strength, low-corrosion construction metal, used in alloy form for high-performance aircraft, missile, and spacecraft parts. Symbol, Ti; atomic weight, 47.90; atomic number, 22.

TM waves—Transverse-magnetic, or E waves, a mode of propagation of energy in a waveguide in which the magnetic vector is always perpendicular to the direction of propagation. *Waveguides and Resonators.*

toggle switch—Switch in which the contacts are opened or closed by a toggle mechanism (force-amplifying leverage device) actuated by the external level. The most common type of switch. (See Figure T-6.)

Figure T-6
Toggle Switch

tone arm—On a record player, the moveable arm that holds the pickup. *Recording.*

tone control—Resistance-capacitance filter, usually variable, to adjust an amplifier's response to bass or treble audio frequencies. *Electroacoustics.*

toroid—Magnetic core in the form of a ring, upon which one or more windings are wound to produce a doughnut-shaped inductor or transformer.

torr—Unit of pressure equal to the pressure of a column of mercury one millimeter high (standard atmospheric pressure is equivalent to 760 millimeters of mercury), or 1.33322×10^2 pascals.

touch-tone dialing—Push-button telephone dialing. *Telephone.*

T-pad—Type of attenuator. *Attenuators.*

transconductance—In an electron tube, the change of plate current divided by the change in grid voltage that causes the plate-current change. Also called mutual conductance. Symbol, gm. The unit of transconductance is the siemens (q.v.).

transducer—Device that converts input energy into output energy, usually of a different kind, but bearing a known relation to the input as, for example, a microphone, which converts sound energy into analogous electrical energy.

transformer—Device that transfers electrical energy from one circuit to another, in most cases without direct connection, and often raising or lowering the voltage in the process. *Transformers.*

TRANSFORMERS

When an alternating voltage is applied to an inductance, an e.m.f. is induced by the varying magnetic field accompanying the flow of alternating current. If a second coil is brought into the same field, a similar e.m.f. will be induced in this coil as well. If a circuit is connected to the terminals of the second coil a current will flow in it also. The coils are said to be coupled, and together they constitute a transformer. The input coil is called the primary coil or winding, the output coil the secondary.

The usefulness of the transformer lies in the fact that electrical energy can be transferred from one circuit to another without direct connection, and in the process can readily be changed from one voltage level to another. For instance, a doorbell requires a supply of 10 volts to operate, but the house supply of 120 volts is more convenient than providing 10-volt batteries, so a transformer is used to change the house voltage to the lower level.

The magnetic field generated by the current in the primary may be greatly concentrated by providing a core of magnetic material on which both coils are wound. This increases the inductance of the coils, so that a smaller number of turns may be used. A closed core having a continuous magnetic path also ensures that practically all of the field set up by the current in the primary coil will cut the turns of the secondary coil. However, the core introduces a power loss because of hysteresis and eddy currents, so this type of construction is practical only at power and audio frequencies. For radio frequencies air-core transformers are generally used.

For a given alternating magnetic field the voltage induced is proportional to the number of turns in the coil. Since the primary and secondary windings are in the same field, the voltages induced will be proportional to the number of turns on each coil. The induced voltage in the primary coil is practically the same, though opposite to, the applied voltage, so:

$$E_s = \frac{n_s}{n_p} E_p \tag{1}$$

where E_p = induced voltage in the primary, = input voltage
E_s = induced voltage in the secondary
n_p = number of turns in the primary
n_s = number of turns in the secondary

The ratio $\frac{n_s}{n_p}$ is called the turns ratio of the transformer. Since the applied voltage and the induced voltage in the primary are equal and opposite, no current would flow at all in the primary if there were no losses, and no load on the secondary. In fact, there are losses, as already mentioned, and a small current does flow in the amount necessary to supply these losses and to overcome the resistance of the primary wiring. This current, which is so small it can be neglected for most purposes, is called the magnetizing current, or exciting current (I_{ex}).

However, when current is drawn from the secondary by any load connected to it, the secondary current sets up a magnetic field of its own in the core. This reduces the strength of the original field because it is of opposite phase. But the original field must always be maintained; otherwise the induced voltage will not equal the applied voltage, so the current in the primary increases by the amount necessary to intensify the primary field so that the effect of the secondary field is completely canceled, and the primary field maintained at the original level.

This means that whatever power is taken from the secondary is made up by power taken by the primary from the source. Allowing for losses, the power output is given by:

$$P_o = nP_i \tag{2}$$

where P_o = power output from secondary
P_i = power input to primary
n = efficiency factor

The efficiency factor is the full-load efficiency of the transformer, and decreases with either higher or lower outputs. For small power transformers such as are used in radio receivers and transmitters, $n = 0.60$ through 0.90 (60 to 90 percent).

Since the voltage in the primary is given by:

$$E_p = \frac{P_i}{I_p} \tag{3}$$

where E_p = induced voltage in primary input voltage
P_i = input power to primary
I_p = primary current

and the voltage in the secondary is given by:

$$E_s = \frac{P_o}{I_s} \tag{4}$$

where E_s = induced voltage in secondary, = input voltage
P_o = output power from secondary
I_s = secondary current

substituting (3) and (4) for E_p and E_s in (1) gives:

$$\frac{P_o}{I_s} = \frac{n_s}{n_p} \left(\frac{P_i}{I_p} \right)$$

Ignoring all losses, and assuming 100 percent efficiency so that $P_o = P_i$, allows this expression to be simplifed to:

$$I_p = \frac{n_s}{n_p} I_s \tag{5}$$

Comparing (5) with (1) shows that when the secondary voltage is higher than the primary voltage, the secondary current will be lower than the primary current, and vice versa. For example, if n_s is 2800, n_p is 400 and E_p is 115 volts, the secondary voltage is given by:

$$E_s = \frac{2800}{400} \times 115 = 805 \text{ volts}$$

Assuming I_p is 1.4 amperes, and transposing the terms to get the secondary current:

$$I_s = \frac{400}{2800} \times 1.4 = 0.2 \text{ ampere}.$$

The primary power is $115 \times 1.4 = 161$ watts, and the secondary power is $805 \times 0.2 = 161$ watts, which agrees with formula (2), ignoring the efficiency factor.

These calculations are near enough for most practical needs, but something must be said about the various losses touched on so far.

Hysteresis and eddy currents are core losses. Hysteresis is the flux that remains in the core when the magnetizing force returns to zero. If the voltage applied to the primary is in the form of a sine wave, the flux density increases as the voltage increases to a peak, and in a perfect magnetic material would return to zero as the voltage fell back to its starting point. Then as the voltage rose to the opposite peak the flux would increase again, but with the reverse polarity, and so on. But even the best magnetic materials cannot quite do this. Some of the magnetic domains are still in the former alignment at the moment the magnetizing force changes polarity, so that additional power must be used to turn them around. This power is therefore wasted. The amount lost will depend on the material used. Best core materials require low coercive force, defined as the magnetizing force that must be applied in a direction opposite the residual flux in order to reduce it to zero. High-performance core materials, such as alloys of iron with nickel, silicon, molybdenum, cobalt or other substances, require a coercive force of between 0.16 and 40 amperes per meter (0.002 and 0.5 oersteds). (For contrast, silmanal, an alloy used for permanent magnets, has a coercivity of 4.8×10^5 amperes per meter (6000 oersteds)).

Eddy currents are currents induced in the core by the magnetic fields of the windings. To minimize the energy lost in this way transformer cores are made of thin sheets called laminations, which are coated with shellac, or some other nonconductive material, and stacked as shown in Figure T-7. Although this does not entirely eliminate eddy-currents losses, because some current paths are still present, it does reduce eddy-current losses to a low value.

Both hysteresis and eddy-current losses increase with frequency, hence even laminated iron cores are useless at radio frequencies. However, transformers for frequencies up to one megahertz are made with ferrite cores for use in the telephone carrier range, and i.f. transformers with cores of powered iron alloy in a binder are used for frequencies as high as 40-50 megahertz (television i.f.). The latter are tuned transformers, which means they are made for certain frequencies, and therefore can handle higher frequencies as long as they are confined to a limited bandwidth. Pulse transformers are also of this type.

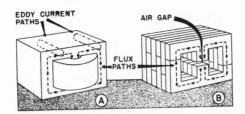

Figure T-7

*If the core of a transformer is solid as at A, there are many eddy
current paths in it; these can be reduced by making a laminated
core as at B.*

Another loss is due to the power lost in overcoming the d.c. resistance of the windings
and is called copper loss, or I²R loss (from $P = I^2R$). A typical winding of 3000 turns of
AWG #28 wire might total 2000 feet. From the Copper Wire Table *(Properties of
Materials)*, this would have a d.c. resistance of 129.8 ohms. Assuming a current of 0.5
ampere, the power dissipated in this winding would be $(0.5)^2 \times 129.8 = 32.45$ watts.

In a practical transformer some of the flux is not common to both windings. It escapes,
and travels through the air surrounding the core, and is therefore called leakage flux. If
the magnetic material in the core is near its saturation point, where it cannot hold much
more flux, the leakage flux increases rapidly if the coil current increases further, so that
a much greater amount of power is wasted. Transformers should usually be operated
well below the saturation level to minimize this, but even so there will always be some.

Leakage flux acts in the same way as flux about any coil that is not coupled to another
coil. It generates a voltage in the coil by self induction. Consequently there are small
amounts of leakage inductance in both windings, but *not* common to them. These
offer an equivalent amount of inductive reactance to the current flowing in the
windings, a reactance called leakage reactance. Since it is no different from other
inductive reactance, its value increases with frequency.

There is also capacitance between the turns of each winding, which has the effect of
placing a capacitor in parallel with the winding. At low frequencies the shunting effect
is negligible, because the reactance is high, but the reactance decreases as the
frequency increases. Furthermore, there will be some frequency at which the
inductive and capacitive reactances are equal, so that the winding is self resonant. This
will generally be of importance only at radio frequencies, however.

These various parameters of the transformer are shown schematically in the
equivalent network in Figure T-8.

Another important characteristic of transformers is the way the impedance of a fixed
load can be transformed to any desired value, within practical limits. A case in point is
the use of an audio output transformer to couple the output of the final stage of an
amplifier to a loudspeaker. For maximum power transfer, the load impedance should
equal the output impedance of the amplifier. For instance, an electron-tube audio-
frequency amplifier might require a load of 5000 ohms for optimum performance, yet
has to drive a loudspeaker with an impedance of only 8 ohms (quite a common

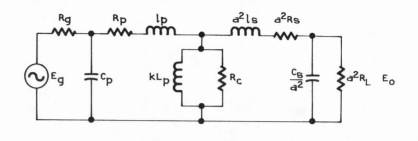

a	= turns ratio = N_p/N_s	
C_p	= primary shunt capacitance*	
C_s	= secondary shunt capacitance*	
E_g	= generator voltage (rms)	
E_o	= output voltage = aE_g in iron-core transformer	
k	= coefficient of coupling (ignore in iron-core transformer)	

L_p	= primary inductance
l_p	= primary leakage inductance
l_s	= secondary leakage inductance
R_c	= core-loss shunt resistance
R_g	= generator impedance
R_L	= load impedance
R_p	= primary winding resistance
R_s	= secondary winding resistance

*Negligible at low frequencies.

Figure T-8
Equivalent Network of a Transformer

impedance for a loudspeaker). To design or select a suitable transformer the following formula is used, although it is an approximation that ignores all the losses:

$$N = \sqrt{\frac{Z_s}{Z_p}} \tag{6}$$

where N = required turns ratio, secondary to primary
Zₛ = impedance of load connected to secondary
Zₚ = impedance required by source

Substituting the given values for Z_s and Z_p

$$N = \sqrt{\frac{8}{5000}} = 0.04, \text{ or } 1/25$$

The primary must therefore have 25 times as many turns as the secondary.

The transformer principle also works with only one winding, as in the autotransformer, of which the Variac is probably the best known. This transformer is used for adjusting the power line voltage to some required value, and works best when the primary and secondary voltages are not very different. Unlike other transformers, it does not provide any isolation between the primary and secondary circuits.

As mentioned already, r.f. transformers are generally air-core transformers. The permeability of air is thousands of times less than that of most core materials, so inductance is also much less, although there are no core losses. On the other hand, distributed capacitance has a much greater shunting effect, although this is reduced as much as possible by universal winding, in which the turns are crisscrossed.

In most cases the secondary of the transformer, at least, is tuned by a variable capacitor to form a resonant circuit. This results in a resonant voltage step-up across the secondary, which is independent of the turns ratio.

Leakage flux is very considerable. Instead of being only 2 to 5 percent of the total magnetic flux, it may even exceed the mutual flux. However, the air core itself cannot saturate.

This makes the mathematical analysis of air-core transformers quite different from that of iron-core types. Iron-core transformers have a mutual flux of 95 to 98 percent. This is called their coefficient of coupling, or K. For air-core transformers K is only 0.5 to 1.0 percent.

The coupling coefficient K can be calculated from

$$K = \frac{M}{\sqrt{L_pL_s}} \tag{7}$$

where   M = the mutual inductance value (see below)
L_p = the primary inductance in henries
L_s = the secondary inductance in henries.

The mutual inductance M is given by

$$M = \frac{L_A - L_o}{4} \tag{8}$$

where L_A = total inductance of L_p and L_s with fields aiding
L_o = total inductance of L_p and L_s with fields opposing.

R.f. transformers with powdered-iron or ferrite cores behave much more like iron-core transformers, except that distributed capacitance has a more serious shunting effect. However, these are usually tuned to a frequency by a variable capacitor or by varying the core permeability (by moving the core) to obtain the resonant voltage step-up already mentioned (see Figure T-9).

Figure T-9
Variable Inductor Using a Powdered-Iron Slug as a Core.
Typical type used for television i-f stage.

Many transformers are enclosed in metal shields which are either conductive or magnetic. R.f. transformers are enclosed in conductive shields of copper or aluminum. Currents induced in the shield oppose the changing flux field and cancel it at external points. L.f transformers have shields of magnetic material that provide an easy path for flux immediately surrounding the transformer, so only a negligible external field can exist. Any shield must be at a certain minimum distance from the transformer for optimal operation.

transient—Abrupt change of voltage of short duration, such as a brief pulse caused by the operation of a switch, etc.

transistor—Solid-state device for amplifying, controlling, or generating electrical signals. *Transistors.*

TRANSISTORS

In the article on *Semiconductors* it is stated (a) that with no bias the electrostatic field in the vicinity of a pn junction aids the flow of minority carriers across it but opposes the flow of majority carriers; (b) that forward bias across the junction encourages the flow of majority carriers while discouraging the flow of minority carriers; and (c) that reverse bias encourages the flow of minority carriers across the junction, while discouraging the flow of majority carriers. These characteristics of a pn junction are what make a junction transistor possible.

In a *junction transistor* there are in effect two pn-junction diodes in series, but with opposite polarity. That is to say, they are arranged in a pn-np sequence, or an np-pn sequence. However, instead of being separate diodes they are combined in a pnp or npn configuration. The included n or p region is also made very thin.

Figure T-10 shows an npn transistor connected so that the left-hand "diode" is forward biased and the right-hand "diode" is reverse biased. Majority carriers flow readily across the left-hand junction, but not across the one on the right. In the n region on the left, electrons are majority carriers, so they are "injected" by the forward bias into the p region. But in the p region electrons are minority carriers, and because this region is very thin they soon find themselves within the electrostatic field of the other, reverse-biased junction. This results in the electrons being swept through this junction into the right-hand n region.

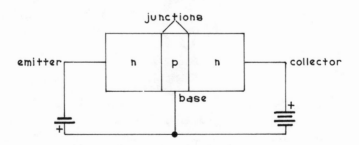

Figure T-10

Transistor Connected for Forward Bias on the Emitter-Base Junction and Reverse Bias on the Collector-Base Junction

In a pnp transistor the action is similar, but the majority carriers injected in the base are holes, and the biasing voltages are the other way round.

For reasons that are now obvious the region that provides the majority carriers is called the emitter, and the region that receives most of them is called the collector. A small percentage of carriers do not reach the second junction, however, as they encounter majority carriers in the base and disappear by recombination. This small base current, added to the collector current, equals the emitter current, or:

$$I_e = I_c + I_b \qquad (1)$$

I_c can be expressed as a fraction of I_e. For instance, if I_c is 95 percent of I_e, it can be given as 0.95 of I_e, assuming I_e to be 1.00. The value 0.95 is called alpha (α) when the base is grounded.

If the value of I_c is 0.95 of I_e, the value of I_b must be 0.05 to satisfy (1). The ratio between these two values is called beta (β) when the emitter is grounded, given by:

$$\beta = I_c/I_b \qquad (2)$$

Beta is called the transport factor of the transistor.

The foregoing description of transistor operation should have made it clear that the collector current is always a fixed percentage of the emitter current. The emitter current, however, varies according to the bias on the pn junction between it and the base. Consequently, variations in this bias produce corresponding variations in the collector current.

If the emitter is grounded, as in Figure T-11, variations in the bias must be brought about by variations in the base voltage. These variations make the base-emitter junction act like a variable resistance, which results in a nonlinear collector current. In other words, the collector current is not a linear function of the base voltage. However, as already mentioned, there is a fixed relationship between the base current and collector current, so the collector current is a reasonable facsimile of the input if the input signal is a current.

Figure T-11
Collector Current I_C Is a Linear Function of Base Current I_B

The signal source of a transistor, however, often happens to be a voltage source rather than a current source. If the variations in base voltage are small the distortion in collector current will be negligible, but if they are not, a linearizing resistor R_g must be inserted in series with the voltage source. This fixed resistance, which is large compared to the transistor's input resistance, absorbs most of the voltage variations, while the consequent current variations are common to R_g and the junction. The larger R_g is made with respect to the junction resistance, the greater the linearity, but the less the amplification, since the two resistances are dividing the voltage in accordance with their magnitude.

Most transistors today are made of silicon or germanium. Silicon transistors are more easily manufactured, using the technology described in the article on *Microelectronics*, but germanium is used for microwave transistors, since carrier mobility is higher.

Small-signal transistors are available for frequencies as high as three gigahertz. They rarely have power output capabilities greater than a few hundred milliwatts and are mostly used in the common-emitter connection (see *Transistor Circuits*). This is also the connection used for *switching transistors,* which are operated either saturated or cutoff instead of around an operating point midway between.

Data sheets for small signal applications of junction transistors often give values for hybrid parameters. The parameters most often used are as follows:

Common Base	Common Emitter	Common Collector	Definition
h_{11}, h_{11b}, or h_{ib}	h_{11e}, or h_{ie}	h_{11c}, or h_{1c}	Input impedance with output shortcircuited
h_{12}, h_{12b}, or h_{rb}	h_{12e} or h_{re}	h_{12c}, or h_{rc}	Reverse open-circuit voltage amplification factor
h_{21}, h_{21b}, or h_{fb}	h_{21e}, or h_{fe}	h_{21c}, or h_{fc}	Forward short-circuit current amplification factor
h_{22}, h_{22b}, or h_{ob}	h_{22e}, or h_{oe}	h_{22c}, or h_{oc}	Output admittance with open-circuit input

Power transistors are those with a power output capability higher than about one watt. They may be used as either switches or amplifiers, and range from cheap germanium-alloy devices for low-frequency power amplifiers and slow switches to silicon r.-f. power transistors delivering 50 watts at 150 megahertz, and up to 2 watts at 2 gigahertz. Since they generate much heat they have to be mounted on heat sinks and may require fans for additional cooling.

In contrast to the junction transistor, which is a bipolar (two-charge carrier) device, the *field-effect transistor* is a unipolar transistor in which operation is by electrons or holes, but not both in the same device. The first FET's were junction-type devices, in which control is exercised by a reverse-biased pn junction. Its depletion layer extends into a conducting channel and removes carriers from it. The JFET is now overshadowed by the more recently developed insulated-gate field-effect transistor (IGFET), often called metal-oxide-semiconductor field-effect transistor (MOSFET), because control is by means of the electric field from a metal electrode, extending through a thin insulator of silicon dioxide into the semiconductor substrate. As shown in Figure T-12, a substrate of n-type silicon has had two p-type regions formed in it by diffusing boron into it. The part of the substrate between these regions is a channel, typically 25 micrometers long, over which is a layer of silicon dioxide about 0.1

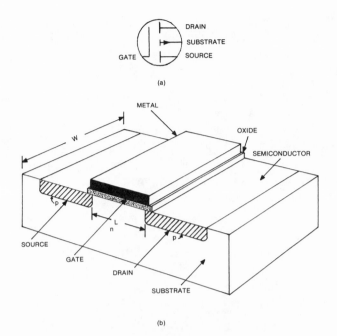

Figure T-12

(a) *P-channel enchancement-type MOSFET, with single gate,*
active substrate (envelope symbol may be omitted if no
confusion will be caused);

(b) *P-channel enhancement-type MOSFET structure.*

micrometer thick. A layer of aluminum is deposited on top of the oxide, and similar aluminum electrodes are formed over the two p-type regions.

When a negative voltage is applied to the control electrode (called the gate in this device) an electric field extends from it through the insulator into the n-type substrate. At ordinary temperatures free electrons and holes are appearing by thermal excitation and disappearing by recombination everywhere in the substrate. The field repels the electrons from a thin ($\approx$ 5 nanometers) surface layer so that it is populated only by holes. Since this is characteristic of p-type silicon, the layer is called an inversion layer. There is now a continuous conductive path, or p-channel, between the two p regions. If a potential difference is applied between them, hole current will flow. When the negative voltage on the control electrode is removed and the field disappears, electrons return to the channel and the hole current ceases.

It makes no difference which way the potential is applied to the device, because hole current can flow just as well one way as the other. The p region from which the holes emerge is called the source, and that into which they go is called the drain. For holes to be attracted to the drain it must be negative with respect to the source.

A similar device may also be built using a p-type substrate and n-type source and drain regions. In this case the gate voltage is positive and holes are repelled, so the inversion layer is an n-channel, with conduction by electrons.

Both devices are called enhancement-type FET's, because conductivity is enhanced in proportion to gate voltage. A p-channel enhancement type has an n-type substrate, and an n-channel enhancement type has a p-type substrate. Both draw no current in the absence of gate voltage, so are the first choice of designers of logic circuits.

The opposite of the enhancement device is the depletion device shown in Figure T-13. In this transistor, current flows all the time in the absence of a gate voltage, because source, drain, and the channel between them are all of the same type. When a gate voltage is applied the electric field repels the carriers from the channel so that an inversion layer is created. The deeper this goes the more deeply the channel is inverted, until eventually there are no majority carriers available and conduction ceases. The gate potential required to bring this about is called the pinch-off voltage.

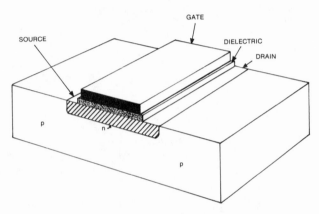

Figure T-13
Depletion-Mode MOSFET

MOSFET's offer certain advantages in comparison to junction transistors. They have high input impedance (10^9 - 10^{15} ohms), so they are voltage devices (like electron tubes) instead of current devices. They are thermally stable, because current through them decreases with increasing temperature, instead of increasing as in junction transistors. They are easier to fabricate and well suited for production in large-scale integration by simple means. Their bilateral symmetry makes them useful as bilateral switches.

On the other hand, conduction is in the surface layer of the channel, and therefore carrier mobility is lower than in a junction transistor, where conduction is through the bulk of the device. The gate construction (metal-dielectric-semiconductor) of a MOSFET is actually a small capacitor, which takes several nanoseconds to charge, thereby adding to the time the MOSFET takes to switch. These two factors make the MOSFET slower than the bipolar junction transistor, although subsequent developments in gate fabrication, such as ion implantation, which gives more sharply defined doped regions, have notably improved performance.

Phototransistors are bipolar transistors in which photons irradiate the base region through a transparent window. This causes generation of hole-electron pairs, so that control is by means of an optical signal on the base, which therefore does not require a lead. Most commercial phototransistors have only emitter and collector leads and are silicon npn devices.

Unijunction transistors have only one junction. As shown in Figure T-14, the UJT consists of a base of n-type silicon with two ohmic contacts called base-1 and base-2. A p-region approximately halfway along the base is called the emitter. When base-1 is grounded and a positive supply voltage is connected to base-2, the base region, which has a fairly high resistance, acts as a voltage divider, so that the base region in the vicinity of the emitter region is at roughly half the supply voltage. As this is positive, the pn junction between the two regions will be reverse biased if there is little or no external voltage on the emitter. But if a higher positive voltage is applied to the emitter so that the junction is forward biased, holes are injected into the base region. These are attracted toward base-1, which is at zero potential and therefore negative with respect to the emitter. The resistance of this portion of the base then drops to a low level, so the UJT acts like a switch.

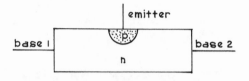

Figure T-14
Unijunction Transistor

A *diac* has a pnp structure like a transistor, but no lead is connected to the base region. It is therefore two diodes connected to share one n region. Either diode avalanches with a suitably high applied voltage, but as they are polarized oppositely they can pass an alternating current. The major application of a diac is in conjunction with a triac in a.c. phase-control circuits (see light-dimmer switch).

Adding another junction to a pnp device makes it a pnpn device, or *four-layer diode* (Schockley diode), which acts very like a gas discharge tube. In the forward-blocking (off) state, the center junction J2, in Figure T-15, is reverse biased, and a small reverse current flows. Under forward bias, electrons are injected at junction J1 into the base region p_1, and these electrons diffuse across to be collected at J2 with current gain α_1, as in a junction transistor. However, the regions p_1-n_2-p_2 also form a junction transistor, and the electrons collected at J2 provide base current for this transistor.

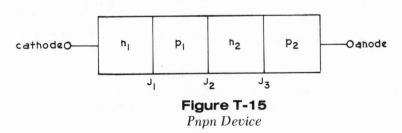

Figure T-15
Pnpn Device

Similarly, holes injected at J3 are collected by J2 with current gain α_2, and provide base current for the transistor n_1-p_1-n_2.

When both "transistors" are saturated the device has a very low impedance, but to get them into that state it must be triggered by momentarily raising the forward voltage until sufficient avalanche current flows across J2.

The *silicon controlled rectifier* (SCR) is a well-known type of four-layer diode with an external connection to the p_1 region. A potential momentarily applied to this gate forward-biases J1, causing electrons to be injected into p_1, and initiates turnon as in a Shockley diode. It is turned off by disconnecting the load or momentarily shorting the anode to the cathode.

A *triac* is a modified pnpn device, having five regions and capable of switching on for either polarity of an applied voltage. It is therefore the a.-c. equivalent of an SCR. Triacs are frequently used in a.c. power control circuits (see diac above).

transistor circuits—Circuits in which transistors are the active devices. *Transistor Circuits*.

TRANSISTOR CIRCUITS

Junction Transistor Circuits

Small-Signal Amplifiers

Small-signal amplifiers are generally of the common-emitter configuration shown in Figure T-16a. Bipolar transistors may be cascaded in much the same way as electron tubes. Common-base and common-collector circuits (Figure T-16b and T-16c) may also be used. Coupling between stages may be direct, R-C, or transformer. Except in common-collector circuits the input impedance is equal to or less than the output impedance, so care in impedance matching is required. The three types of coupling are shown in Figure T-17.

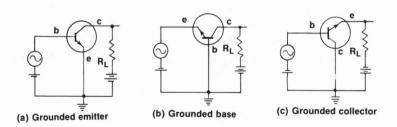

(a) Grounded emitter   (b) Grounded base   (c) Grounded collector

Figure T-16
Small-Signal Amplifier Configurations
(c=collector, e=emitter, b=base, R_L=load resistance)

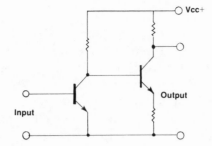

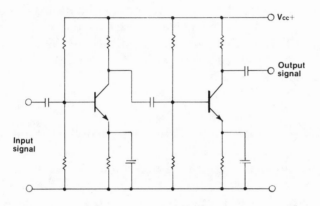

(a) direct-coupled n-p-n amplifier stages

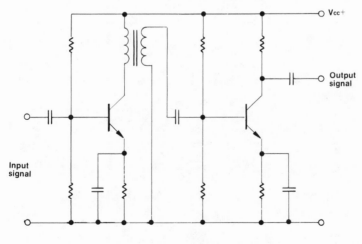

(b) R-C coupled stages

(c) transformer-coupled stages

Figure T-17
Coupling Between Stages

Large-Signal Operation

Bipolar transistors have two power limitations. The most important is rise in temperature because it can have a run-away effect. As the temperature increases the collector current increases, leading to a further rise in temperature, a further current increase and so on, until the transistor is destroyed. The other limitation is the maximum voltage that can be applied between collector and base. Class-A transistor amplifiers can have efficiencies up to about 47 percent, but it is best to use Class-B transistor amplifiers, since the output can approach three times the total dissipated power, which is equivalent to six times the allowable dissipation for each unit. A Class-B amplifier also consumes negligible power in the absence of an input signal. A Class-B amplifier can be constructed using complementary symmetry as in Figure P-21. This does away with the need for a phase inverter or a push-pull output transformer. The pnp transistor amplifies the negative part of the input signal, and the npn transistor amplifies the positive part. The amplified positive and negative parts are combined in the output, where output currents of both transistors flow through the common load resistance.

Oscillators

Feedback oscillators operate by returning a portion of the output signal to the input to sustain oscillation by positive feedback. L-C oscillators, such as the Hartley and Colpitts oscillators shown in Figure T-18, determine the frequency of oscillation by an inductance-capacitance network. R-C oscillators, such as the blocking oscillator and astable multivibrator shown in Figure T-19, determine frequency by a resistance-capacitance network.

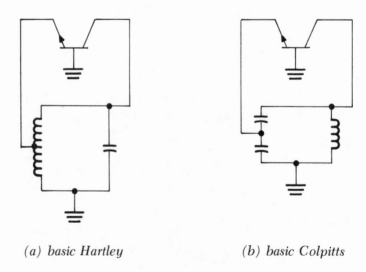

(a) basic Hartley   (b) basic Colpitts

Figure T-18
L-C Oscillators

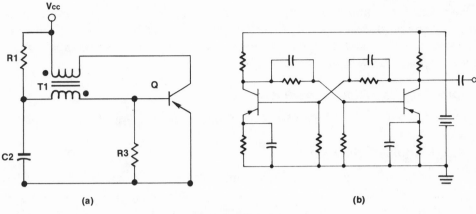

(a)   **(b)**

Figure T-19

(a) blocking oscillator

(b) astable multivibrator (same as Figure A-34)

Some transistors, such as point-contact or hook-collector types, which exhibit negative resistance when the junction is biased above its breakdown voltage, may also be used in oscillator circuits. An example is shown of the basic circuit in Figure T-20.

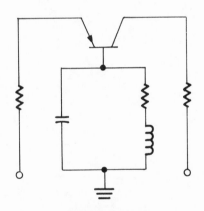

Figure T-20

Negative-Resistance Oscillator

Video-Frequency Amplifiers

Video-frequency amplifiers are required to handle a wider range of frequencies, from d.c. to four megahertz or so. Low-frequency compensation is achieved by bypassing for high-frequencies part of the collector load via a low-value capacitor, while high-frequency compensation is performed by the use of peaking coils, as in Figure P-11.

I.-f. Amplifiers

I.-f. amplifiers may have series-resonant or parallel-resonant coupling between stages.

The parallel-resonant coupling, as shown in Figure I-7, is more commonly used, and may be single tuned or double tuned. The input to the transistor is taken from a tap on the secondary of the i.-f. transformer so that the low input impedance of the transistor will not load the winding excessively.

Logic Circuits   (See Microelectronics.)

Field-Effect Transistor Circuits

Small-Signal Amplifiers

The three basic circuits for MOSFET amplifiers are shown in Figure T-21. They are the same for JFET's. The common-source configuration is the usual arrangement, since it provides very high input impedance, moderate output impedance, and a voltage gain greater than unity.

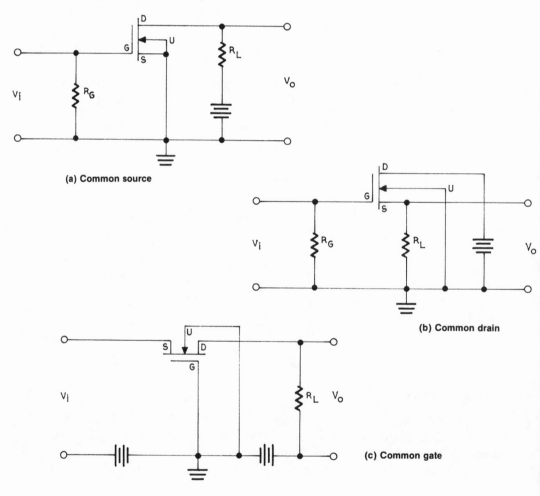

(a) Common source

(b) Common drain

(c) Common gate

Figure T-21
Basic MOSFET Small-Signal Amplifier Configurations

Typical MOSFET Circuits

MOSFET's are used mostly in integrated circuits (see *Micro-Electronics*) such as operational amplifiers and so on, and are less often used in circuits with discrete components. A few typical MOSFET circuits are shown in Figure T-22.

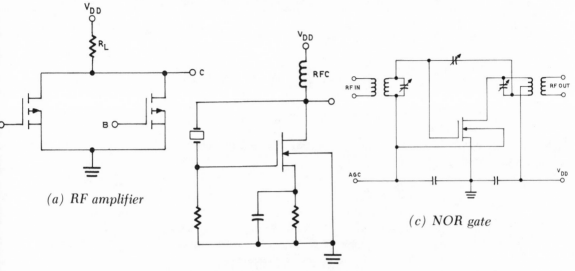

(a) RF amplifier

(b) Pierce crystal oscillator (same as Figure P-17)

(c) NOR gate

Figure T-22
Typical MOSFET Circuits

transmission line—Conducting path for guiding electrical signals. *Transmission Lines.*

TRANSMISSION LINES

Whether it is hundreds of feet up on a steel tower, or merely mounted on a hand-held walkie-talkie, the antenna can never be in exactly the same place as the transmitter. A means must therefore be provided for conveying the power generated in the transmitter to the antenna so it can be radiated out into space.

But while the antenna is designed for maximum radiation, the transmission line must radiate as little as possible, so that all the energy of the transmitter reaches the antenna. A single wire will not do because it radiates a wave. However, if another conductor alongside the first conductor carries the same wave, but 180 degrees out of phase, the radiation is canceled out.

There are three major types of transmission lines: the open two-wire line, the coaxial line, and the waveguide. Waveguides are covered under *Waveguides and Resonators*, so only the two-wire line and the coaxial line are discussed here.

The open two-wire line shown in Figure T-23 consists of two conductors running side by side, with insulating spacers to keep them separated and parallel. Such lines use wire of 12 or 14 gage, or metal tubing with a diameter of 1/4 to 1/2 inches, with

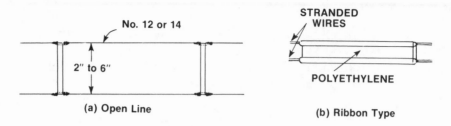

(a) Open Line

(b) Ribbon Type

Figure T-23
Two-Wire Transmission Lines

(a) open line

(b) ribbon type

spacing varying from two to six inches. The characteristic impedance of this type of line is given by

$$Z_o = 276 \log \frac{b}{a} \tag{1}$$

where Z_o = characteristic impedance (explained below)
   b   = center-to-center distance between conductors
   a   = radius of conductor (same units as b)

This type of parallel-conductor line is said to be air insulated. There is also a "ribbon" type of transmission line with stranded conductors embedded in polyethylene. This may be obtained shielded or unshielded. The former has a braided copper sheath surrounding the two conductors. Ribbon lines are suitable for low-power transmitters (under 100 watts) and usually have a characteristic impedance of 300 ohms.

However, the most widely-used line today is the coaxial cable. It is similar to the open two-wire line except that one conductor is inside the other. The smaller inner conductor is surrounded by a tubular outer conductor. The dielectric insulation between the two conductors can consist of (a) air, with spacers at intervals along the line, (b) gas, with spacers along the line, or (c) a solid dielectric. Air and gas-insulated coaxial lines are much more expensive and difficult to install than the more generally used solid-dielectric type, but have lower losses.

A solid-dielectric coaxial line has either a solid or stranded copper inner conductor, surrounded by a polyethylene dielectric covered by a braided copper shield, and with a waterproof polyvinyl or teflon outer jacket. (Sometimes an armor coat of fiberglass or metal braid is used as well.) This is shown in Figure T-24. These cables are usually designated by RG numbers (RG is a military prefix denoting radio-frequency cables), of which the most widely used is RG-59/U, with a nominal characteristic impedance of 75 ohms. This cable has a diameter of approximately 1/4 inch, is fairly flexible, and can withstand voltages up to 2300 volts. This cable is also used to couple television and FM antennas to receivers. Since both antennas and receivers have 300-ohm impedances a 300 to 75-ohm matching transformer is required at each end of the line.

The characteristic impedance of a transmission line has been mentioned several times. It is an important characteristic. It is involved with the question of standing waves.

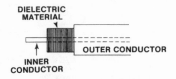

Figure T-24
Solid-Dielectric Coaxial Line

Standing waves occur when the impedance of the transmission line is not equal to the impedance of the antenna. When this is the case, part of the energy is reflected from the antenna back along the line. But if the characteristic impedance of the line matches that of the antenna, the antenna seems like another length of transmission line to the r.-f. wave, and all its energy passes into the antenna. A transmission line in that state is said to be nonresonant, or flat. When there is a mismatch, and standing waves appear, it is said to be resonant.

The characteristic impedance of the line arises from the distributed inductances and capacitances along its length. It does not vary with frequency and is independent of line length. It can therefore be matched permanently to that of the antenna. If the antenna has a different impedance, a match can be obtained by using a matching transformer as mentioned above, by connecting the line to the antenna at some point where the impedance is right (the impedance of an antenna varies along its length, being greatest at the ends and least in the center), or by the use of matching stubs.

For a center-fed half-wave antenna, the two points of connnection can be moved apart, as shown in Figure T-25. This increases the impedance of the line so that it matches that of the antenna at those points. For a 600-ohm line the dimension C is given by:

$$C = \frac{118}{f} \tag{2}$$

where f is the frequency for which the antenna is to be used, in megahertz. The dimension E is given by:

$$E = \frac{148}{f} \tag{3}$$

The antenna-length formula is given in the article on *Antennas*, and the dimensions of the open two-wire line can be calculated from equation (1) above. This type of matching is called a delta match.

Figure T-26 shows the method of coupling a nonresonant line to a center-fed antenna through a quarter-wave linear transformer, or matching section. The free end of the matching section Y is open, while the other end is connected to a low-impedance point (current loop) on the antenna. This would be the center of a half-wave dipole of the proper length, from the formula given in the article on *Antennas*. The matching section is made a quarter-wave long, using the formula:

$$\text{Length (in feet)} = \frac{240}{f}$$

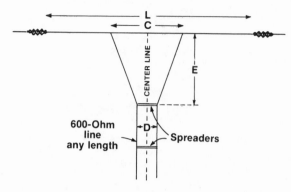

Figure T-25
Delta Match

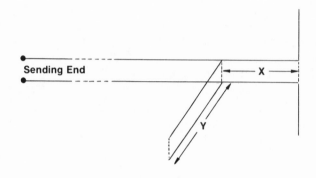

Figure T-26
Matching Section

where f = frequency in megahertz. In actual practice the line should be made a little longer than necessary, and then the system is adjusted to resonance by clipping small pieces from the end of the stub.

The distance of the line tap is found by trial and error. After selecting a trial point and attaching the line, power is applied to the transmitter and the transmission line is checked for standing waves. This is done by measuring the current in, or voltage along, the wires. At any one position along the line the currents in the two wires should be identical. Readings taken at quarter-wave intervals will indicate whether or not standing waves are present. The line is then reattached in different positions until the smallest standing-wave ratio is obtained (SWR = V_{max}/V_{min} or I_{max}/I_{min}). Some final trimming of the free ends of the stub may then be necessary to eliminate the remaining standing waves.

At the transmitter end of the transmission line suitable coupling is required to transform the input impedance of the line into a value of impedance that will load the transmitter properly. Since this may be several thousand ohms, far higher than the line, some type of transformer coupling is used, where the line is connected across only a few turns of the output tank coil, or is connected to a link coil inductively coupled to the tank coil. This circuitry is provided in all commercially available transmitters, which generally use 75-ohm coaxial transmission lines.

In conclusion, it should be remembered that there is no difference in principle between coaxial cables and open two-wire lines. However, coaxial cables are unbalanced, since the outer conductor also acts as a shield, and is grounded. Two-wire lines are balanced, each side being identical. Different types of transmission line are compared in the following table:

Type of Line	Characteristic Impedance (Ω)	Velocity Factor*	Attenuation in dB per 100 ft. at 28 MHZ**
Open two-wire line	400-600	0.975	0.10
Air-insulated coaxial line	50-120	0.850	0.55
Solid dielectric coaxial: RG-58/U	53	0.660	1.90
Solid dielectric coaxial: RG-59/U	73	0.660	1.80
Twin lead	300	0.820	0.84

*Ratio of velocity of wave along line to velocity in free space.
**Assuming no standing waves.

transmitter—1. Equipment for generating and radiating radio signals. 2. Telephone microphone. *Telephone.*

transponder—Electronic device that automatically transmits a signal in response to the receipt of a signal, as in a microwave repeater or a satellite.

trap—A circuit tuned to a certain frequency which it is desired to remove; for instance, a sound trap in a video amplifier to prevent the sound portion of the television signal from reaching the picture tube.

trapezoidal pattern—Oscilloscope display when scope is connected to evaluate modulation index. Modulation index is given by $\dfrac{H_1 + H_2}{H_1 - H_2} \times 100$. *Modulation.* (See Figure T-27.)

traveling-wave tube—Electron tube used for amplification or generation of microwaves. *Electron Tubes.*

treble—Higher audio frequencies, usually above the sound made by the middle C on the piano keyboard (261.63 Hz). *Electroacoustics.*

t.r.f.—Abbreviation for tuned radio frequency (q.v.).

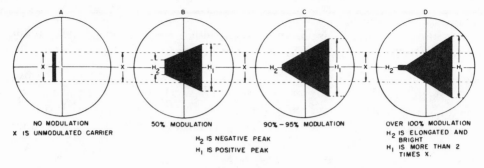

Figure T-27
Trapezoidal Patterns

triac—Modified pnpn device capable of switching on for either polarity of an applied voltage: the a.c. equivalent of an SCR. (See also light-dimming switch.) *Transistors.*

triaxial cable—Coaxial cable with second dielectric sheath and braiding to give a transmission line with three concentric conductors.

triboelectricity—Static electricity produced by friction.

Trinitron tube—Color picture tube that superseded the shadow-mask type. The shadow mask is replaced by a metal grille having vertical slits from the top to the bottom of the screen. The three electron beams pass through the slits to the color phosphors, which are in the form of vertical stripes aligned with the slits. A major portion of the electron gun is common to all three beams. The advantages are improved brightness and focus. *Electron Tubes.*

triode—Three-electrode electron tube, containing a cathode, grid, and anode. *Electron Tubes.*

tritium—Radioactive isotope of hydrogen, emitting negative beta particles with an energy of 19,000 electron volts and having a half-life of 12.5 years. Used for self-luminous paint for dial markings in watches and other instruments, in which a phosphor is made to glow by the beta particles emitted by tritium. Symbol, T or ^{3}H; atomic weight (approximately), 3; atomic number, 1.

troposphere—Lowest region of the atmosphere. (See atmosphere.)

troubleshooting—Locating and correcting malfunctions in electronic equipment.

trunk—Circuit between two telephone exchanges. *Telephony.*

truth table—Table that shows the output state of a logic gate for every possible combination of inputs. *Computer Hardware.*

TTL—Abbreviation for transistor-transistor logic. Also abbreviated T²L. *Microelectronics.*

tube—Electron tube (q.v.).

tube shield—Cylindrical metal cover placed over an electron tube to shield it.

tube socket—Receptacle for the base pins of an electron tube, which provides support and connections to the circuit.

tube tester—Instrument for ascertaining the condition of electron tubes.

tuned-grid, tuned-plate oscillator—Oscillator with resonant circuits in both input and output. Feedback is via plate-grid capacitance. (See Figure T-28.)

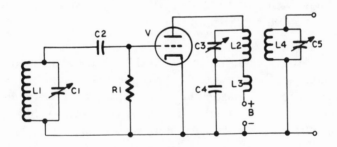

Figure T-28
Tuned-Grid, Tuned-Plate Oscillator

From *Complete Guide to Reading Schematic Diagrams*, First Edition, page 83.

tuned radio-frequency receiver—Type of radio receiver that preceded the super-heterodyne receiver. It consisted usually of three cascaded r.f. amplifiers with ganged tuning capacitors, a detector, and an audio amplifier.

tuner—The first stage or stages of a television or radio receiver that separates a particular channel or frequency from the other frequencies picked up by the antenna. A tuner may be a separate component or built in.

tungsten—Nickel-white to grayish metal. Its melting point is the highest of all metals (3410°C), it has the highest tensile strength at temperatures over 1650°C, and the lowest coefficient of thermal expansion (linear). Ordinarily it is brittle, but ductile at high temperature, so that it can be drawn into very fine wire for filaments. Also called wolfram. Symbol, W; atomic weight, 183.85; atomic number, 74.

tunnel diode—Semiconductor device that operates by means of the tunnel effect of electrons. Also called Esaki diode. *Semiconductors.*

turns ratio—Ratio of number of turns in the primary winding of a transformer to number of turns in the secondary winding. *Transformers.*

turntable—Rotating platform that carries a phonograph record. *Recording.*

turret tuner—Type of television-receiver tuner in which the coils required for each channel are mounted on strips around a drum, or turret, which is rotated so that the contacts on the strips carrying the coils for the selected channel make contact with stationary contacts in the tuner enclosure.

TV—Abbreviation for television (q.v.).

tweeter—In a high-fidelity loudspeaker system, the small speaker that handles frequencies above 8 kilohertz. *Recording.*

twin lead—The most popular, cheapest, and least efficient lead-in line used for television receivers, usually consisting of two parallel conductors embedded in the opposite edges of a ribbon of plastic material, with a characteristic impedance of 300 ohms as a general rule. Improved types of twin lead are shielded like a coaxial line, and give far better results, as does ordinary coax with suitable matching transformers.

two-phase circuit—Two wires carrying alternating currents which differ in phase by 90 degrees, with a third wire which is the common return.

TWT—Abbreviation for traveling-wave tube (q.v.).

U—1. Symbol for uranium (q.v.).   2. In a schematic diagram, the class letter for an integrated circuit.

UHF—Abbreviation for ultrahigh frequency (q.v.).

UJT—Abbreviation for unijunction transistor (q.v.).

ultor—High-voltage anode in cathode-ray tube or picture tube. *Electron Tubes.*

ultrahigh frequency—Frequency range from 300 to 3000 megahertz, in band number 9 (decimetric waves).

ultrasonic bonding—Method of attaching input and output connections to a microelectronic device by using ultrasonic (20 to 60 kilohertz) vibration to make the metals bond together. *Microelectronics.*

ultraviolet radiation—That portion of the electromagnetic spectrum (q.v.) from 370 to 10 nanometers. It is subdivided into near (370-300 nm), far (330-200 nm), and extreme 200-10 nm). *Optoelectronics.*

unbalanced circuit—Circuit in which two sides are electrically unlike, one of them often at ground or zero potential.

unblanking—Removal of negative potential from control grid of cathode-ray tube to allow electron beam to turn on.

unijunction transistor—Semiconductor device that can switch from nonconducting to conducting with the application of the requisite voltage. *Transistors.*

unipolar device—Semiconductor device, such as a field-effect transistor, in which only one type of charge-carrier is present—either electrons or holes, but not both.

Universal Time (UT)—Universal time scale based on the mean angle of rotation of the Earth about its axis in relation to the Sun. Also known as Greenwich Mean Time (GMT). *Frequency Data.*

universal winding—Coil winding method in which the turns are crisscrossed to reduce distributed capacitance. *Transformers.*

uranium—Silvery-white radioactive metal, heaviest naturally occurring element, easily worked, important nuclear fuel (one pound of uranium yields as much energy as three million pounds of coal). Symbol, U; atomic weight, 238.03; atomic number, 92.

V—1. Symbol for the chemical element vanadium (q.v.). 2. Symbol for volt (q.v.). 3. Symbol, with various subscripts, for power supply; e.g.,

V_{CC} = Collector power supply
V_{DD} = drain power supply
V_{EE} = emitter power supply
V_{SS} = source power supply

4. In a schematic diagram, the class letter for an electron tube.

v—Nonstandard symbol for volt.

VA—Symbol for volt-ampere (q.v.).

Vac, VAC, V.A.C., v.a.c.—Symbol for volt, alternating current.

vacuum—Any region of space devoid of matter; in common usage, the closest approach attainable to a true vacuum. The pressure in a vacuum until recently was measured in torrs (1 torr = 1 millimeter of mercury), but is now given in pascals (1 pascal = 1 newton per square meter = 7.5×10^{-3} torr). The vacuum in a picture tube is about 10^{-6} torr, which is about the same as interplanetary space. Pressures far lower ($\sim 10^{-12}$) have been achieved in the laboratory.

vacuum tube—(See electron tube.)

vacuum-tube voltmeter—Voltmeter, or multimeter, that consists of a bridge circuit with electron (vacuum) tubes in two of the arms. A very sensitive device with an extremely high input impedance (negligible loading effect). Comparable solid-state voltmeters achieve this by use of field-effect transistors. (See Figure V-1.)

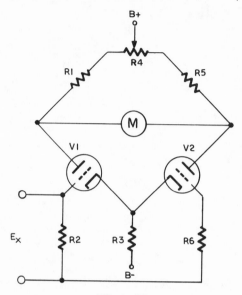

Figure V-1
Principle of Vacuum-Tube Voltmeter

From *Complete Guide to Electronic Test Equipoment and Trouble-shooting Techniques*, page 98.

valence—Property of an element that determines the number of other atoms with which an atom of the element can combine. This is according to the number of valence electrons occupying the atom's outermost shell. These have a range of energies called the valence band, which determines whether the element is a conductor, insulator, or semiconductor. *Semiconductors.*

valve—Original term for electron tube, still used by the British.

Van Allen radiation belts—Two doughnut-shaped zones of charged particles trapped in the Earth's magnetic field, in the equatorial plane, at altitudes exceeding several hundred miles. The particles arise from cosmic rays, the Sun, and high-altitude nuclear explosions.

Van de Graaff accelerator—Type of particle accelerator. *Nuclear Physics.*

VAR—Symbol for volt-ampere reactive (q.v.).

varactor diode—Semiconductor device whose capacitance varies with the applied voltage. *Semiconductors.*

variable capacitor—Capacitor whose capacitance is adjustable. *Capacitors.*

variable-frequency oscillator (VFO)—Oscillator whose frequency is adjusted by tuning the inductance or capacitance in the resonant circuit.

variable inductance—Coil whose inductance can be varied, usually by adjusting the position of its magnetic core.

variable reluctance—Type of phonograph pickup in which movements of the stylus as it follows the sound track produce changes in the magnetic field of an electromagnet between whose poles part of the stylus is positioned, so that analogous electrical currents are induced in the windings of the electromagnet. *Recording.*

variable resistor—Resistor whose resistance may be adjusted by a sliding contact. *Resistors.*

Variac—Type of variable autotransformer (q.v.). *Transformers.*

vector—Physical quantity that requires both magnitude and direction for its description. Velocity is a vector, since it is speed in a certain direction, but speed alone is not a vector. (See also scalar.)

velocity—Quantity that designates how fast and in which direction a point is moving. (See also vector.)

velocity microphone—Type of microphone, such as a ribbon microphone, which responds to the velocity of moving air molecules instead of to their pressure. *Recording.*

vertical amplifier—Section of oscilloscope that amplifies the input signal when the scope is operated in the usual Y-T mode. (See oscilloscope.)

vertical blanking—In the television composite-signal waveform, the interval between fields when the signal amplitude is above the blanking level in order to cut off the electron beam in the picture tube to avoid vertical retrace lines on the screen. *Broadcasting.*

vertical-deflection electrodes—In a cathode-ray tube with electrostatic deflection, the pair of plates that move the electron beam vertically. *Electron Tubes.*

vertical polarization—Polarization of field radiated by a vertical antenna. *Antennas, Electromagnetic-Wave Propagation.*

vertical synchronizing pulse—One of a set of six pulses transmitted between each of the fields of a television broadcast to enable the receiver's vertical deflection section to keep in step with the transmitter. *Broadcasting.*

very-high frequency (VHF)—Frequency range from 30 to 300 megahertz, in band number 8 (metric waves).

very-low frequency (VLF)—Frequency range from 3 to 30 kilohertz in band number 4 (myriametric waves).

vestigial sideband—Amplitude-modulated signal in which one sideband has been largely suppressed. *Broadcasting.*

VFO—Abbreviation for variable-frequency oscillator (q.v.).

VHF—Abbreviation for very-high frequency (q.v.).

vibrating-reed frequency meter—Frequency meter used to monitor 60 or 400-hertz power. A set of reeds visible through a window in the face of the meter that respond to, say, 58, 59, 60, 61 and 62 hertz, are influenced by the field set up by the 60-hertz current flowing in a coil, and that reed closest to the frequency vibrates, so that it appears to expand vertically. (See Figure V-2.)

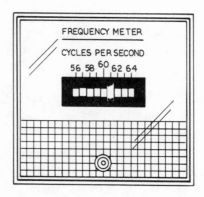

Figure V-2
Vibrating-Reed Frequency Meter

Ibid, page 46.

vibrator—Electromechanical device used for chopping a flow of direct current to convert it to a.c., as in a chopper or inverter (qq.v.).

video—1. Television picture signal. 2. Any wideband signal similar to 1. *Broadcasting, Radar.*

vidicon—Type of television camera tube. *Electron Tubes.*

volatile memory—Random-access memory that requires constant refreshing to retain data. Also called dynamic memory or storage. *Computer Hardware, Microelectronics.*

volt (V)—S.I. derived unit of electric potential, the potential difference that exists between two points on a conductor carrying an unvarying current of one ampere when the power dissipated between those points is equal to one watt. *S.I. Units.*

voltage amplifier—Amplifier for voltage, not designed for power amplification. *Transistor Circuits.*

voltage-controlled oscillator—Oscillator whose frequency is varied by varying the bias voltage to the active device (e.g., a varactor) in the frequency-generating circuit.

voltage divider—Two or more resistors in series across the output power supply, that divide the output voltage according to their resistance values. (See Figure V-3.)

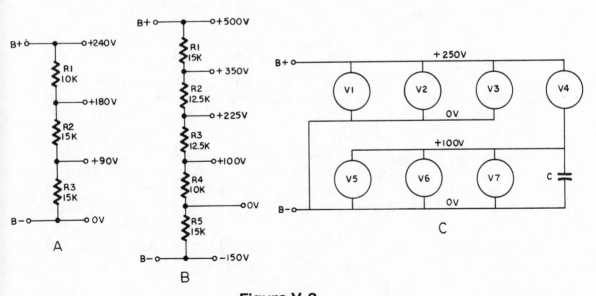

Figure V-3
Voltage Divider

From *Complete Guide to Reading Schematic Diagrams*, First Edition, page 164.

voltage doubler—Rectifier that has an output approximately twice the peak value of the input. *Rectifiers and Filters.*

voltage reference—Any device, such as a Zener diode, gas tube or standard cell, that provides a fixed voltage with which another voltage is compared, in a regulated power supply, etc. *Rectifiers and Filters.*

voltage regulation—Maintenance of output of a power supply at a desired value regardless of load or powerline variations. *Rectifiers and Filters.*

voltage standing-wave ratio (VSWR)—Ratio of standing-wave maximum amplitude to standing-wave minimum amplitude. *Transmission Lines.*

voltage tripler—Rectifier that has an output approximately three times the peak value of the input. *Rectifiers and Filters.*

voltaic cell—(See battery.)

volt, alternating current (V.a.c.)—Effective, or root-mean-square value, equal to $0.707 \times$ the peak value of the voltage. Often written Vrms.

volt-ampere (VA)—In an a.c. circuit, a measure of apparent power, given by:

$$VA = E I$$

where   E is the potential in volts
       I is the current in amperes
       VA is apparent power in volt-amperes

volt-ampere reactive (VAR)—Unit of the apparent power in an a.c. circuit which is delivered to the circuit during one part of the cycle but returned to the source during another part of the cycle.

voltmeter—Instrument that measures voltages of d.c. and a.c. on a scale graduated in volts. Simple voltmeters use a permanent-magnet moving-coil movement (q.v.), but since they draw current from the circuit they load it to some extent, so are less accurate than vacuum-tube voltmeters (q.v.) and other types which do not.

volt-ohm-milliameter (VOM)—Voltmeter adapted by shunt and series resistors to measure resistance and current as well. Also called a multimeter (q.v.).

volume control—Potentiometer for adjusting the input to the audio amplifier of a radio, television set, etc., so as to set the sound to the desired level.

volume unit (VU)—Unit of audio-frequency power, measured on a volume-unit meter, or VU meter.

VSWR—Abbreviation for voltage standing-wave ratio (q.v.).

VTVM—Abbreviation for vacuum-tube voltmeter (q.v.).

W—1. Symbol for watt (q.v.). 2. Symbol for chemical element tungsten (q.v.). 3. In schematic diagrams, class letter for cable, transmission line, wire, etc.

wafer—Thin slice cut from a cylinder of single-crystal silicon, upon which a large number of integrated circuits are formed. The wafer is then separated into the individual chips. *Microelectronics.*

wafer switch—(See rotary switch.)

walkie-talkie—Two-way portable radio communication set.

watt (W)—S.I. derived unit of power; the power used when work is done or energy is expended at the rate of one joule per second. *S.I. Units.*

wattmeter—Meter for measuring 60-hertz power, consisting of a pivoted potential coil mounted between two current coils. The potential coil attempts to align itself with the magnetic field created when the meter is connected to the power source, and a pointer attached to it indicates the power on a scale marked in watts.

waveguide—Hollow metal tube that channels radio waves to or from an antenna. *Waveguides and Resonators.*

WAVEGUIDES AND RESONATORS

Radio-frequency energy is propagated through a waveguide as an electromagnetic field. The waveguide is therefore a kind of pipe which confines the field and causes it to travel in the desired direction. The energy is injected at one end and propagated through the waveguide by reflections against its inner walls.

Waveguides are used to transmit microwave frequencies and are rectangular, circular, or elliptical. Of the three, the rectangular cross-section type is most commonly used. At microwave frequencies the losses in open two-wire or coaxial transmission lines make them unsuitable. These losses are radiation loss, dielectric loss, and copper loss (due to skin effect). Since standard waveguides have no dielectric and no center conductor, and cannot radiate, they do not suffer from these losses. They are also much more rugged.

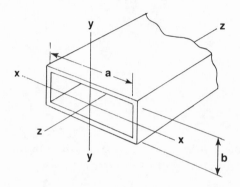

Figure W-1
Dimensions of Rectangular Waveguide

The dimensions of a rectangular waveguide, as shown in Figure W-1, determine the lowest frequency for which it can be used. The wavelength, λ_{co}, corresponding to the cutoff frequency is equal to twice the inside width of the guide, or $\lambda_{co} = 2a$.

Higher frequencies, however, can be transmitted. The distance b is not critical with regard to frequency, but does determine the voltage level at which the waveguide arcs over. As a general rule, a is about 0.7 times the wavelength in air, and b is from 0.2 to 0.5 times the wavelength in air.

The electromagnetic field in a rectangular waveguide consists of an electric field and a magnetic field at right angles to each other as when radiating in space. If the electric field lies in transverse planes in which the lines of force (E lines) are parallel to the Y axis and perpendicular to the Z axis, the waveguide is said to be operating in the TE (transverse electric) mode. If the magnetic field is oriented this way, the waveguide is operating in the TM (transverse-magnetic) mode. The TE mode is the one most commonly used.

The expression TE has subscript numbers to indicate the number of half-wave variations of the transverse field in the wide and narrow dimensions of the guide. The most often-used mode is TE$_{1.0}$, which means that the electric field has one half-wave variation in the wide dimension and none in the narrow dimension, as shown in Figure W-2. This mode is also called the dominant mode and is the one having the lowest cutoff frequency for a given size of guide.

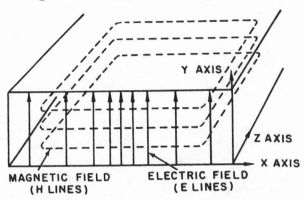

Y AXIS

Z AXIS

X AXIS

MAGNETIC FIELD
(H LINES)

ELECTRIC FIELD
(E LINES)

Figure W-2

In the TE$_{1.0}$ mode there is one intensification of the electric field (E lines) in the wide dimension of the waveguide, none in the narrow dimension. One intensification corresponds to one peak of a sine wave, or a half-wave variation.

There are three principal ways in which energy can be put into or removed from a waveguide. The first is by placing a small loop of wire so that it cuts or couples the magnetic field lines as in a simple transformer. The second is by providing an "antenna" or probe that can be placed parallel to the electric field lines. The third method is to link or contact the fields inside the guide by external fields through the use of slots or holes in the wall.

At low and medium radio frequencies resonant circuits usually consist of coils and capacitors. At microwave frequencies these components would have to be reduced to impractically small physical dimensions, so instead of LC resonant circuits cavity resonators are used. These are hollow metal boxes whose dimensions determine the resonant frequency.

A cavity resonator may be square, cylindrical, or spherical. The resonant frequency depends upon the dimensions of the cavity and the mode of oscillation of the waves (comparable to the transmission modes in a waveguide). For the lowest modes the resonant wavelength (λ_0) is given in Figure W-3.

Compared to ordinary resonant circuits, cavity resonators have extremely high Qs. A value of 1000 or more is readily obtainable. Energy can be coupled in or out in the same way as for a waveguide. A waveguide can also be connected directly to a cavity, provided a matching diaphragm is used. This is a plate with an aperture of the proper size mounted crosswise in the junction of the waveguide with the cavity. An example of the use of a waveguide-cavity input to a microwave receiver is shown in Figure W-4.

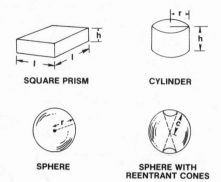

Figure W-3

Various Forms of Cavity Resonatiors. The resonant wavelength for the lowest mode of each is given by:

Square prism	1.41l
Cylinder	2.61r
Sphere	2.28r
Sphere with reentrant cones	4r

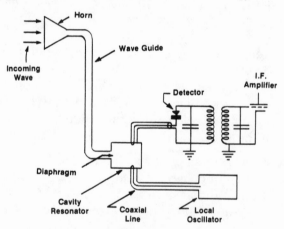

Figure W-4

Microwave Receiver Input
(Illustrating use of waveguide and cavity resonator)

wavelength—Distance in meters, centimeters, millimeters, etc., between corresponding points of two consecutive waves, given by:

$$\lambda = \frac{c}{f}$$

where λ = wavelength in meters
   c = speed of light ($\sim 3 \times 10^8$ m/s)
   f = frequency in hertz

waves—1. Traveling undulating field of electric and magnetic force (radio waves). 2. Alternating current in a conductor. 3. Vibrations in a solid (seismic waves). 4. Traveling zones of compression and decompression in air or other medium (sound waves). *Electroacoustics, Electromagnetic-Wave Propagation.* (See also sine wave, square wave, sawtooth wave.)

weber (Wb)—S.I. unit of magnetic flux which, linking a circuit of one turn, produces in it an electromotive force of one volt as it is reduced to zero at a uniform rate in one second. *S.I. Units.*

wet cell—Cell with liquid electrolyte. (See also battery.)

Wheatstone bridge—Resistance-measuring bridge. (See Figure W-5.)

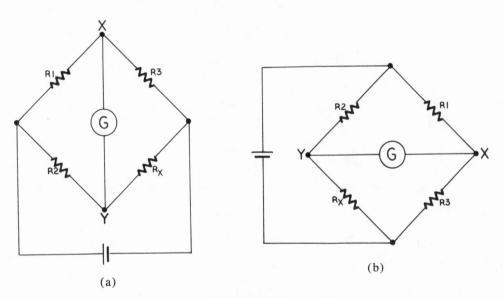

(a)

(b)

Figure W-5
Wheatstone Bridge

From *Complete Guide to Reading Schematic Diagrams*, First Edition, page 209.

white noise—Noise with the same average power density all across its spectrum, having constant energy per hertz bandwidth.

wideband—Said of an amplifier or filter that has a broad passband.

Wien bridge—Bridge for measuring frequency or capacitance. (See Figure W-6.)

winding—The manner in which a conductor is wound to form a coil. *Coil Data*

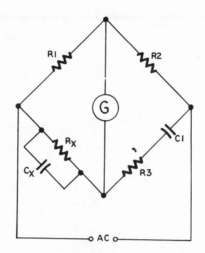

Figure W-6
Wien Bridge
Ibid, page 214.

wiper—Moveable contact in a rotary switch, potentiometer, etc.

wire—Metallic thread or slender rod. *Properties of Materials.*

wireless—Radio (British), as in wireless telegraphy (W/T), etc.

wire recording—Type of magnetic recording that uses a steel wire instead of magnetic tape.

wirewound resistor—Resistor made of resistance wire wound on an insulating rod. *Resistors.*

woofer—Low-frequency or bass component of a high-fidelity speaker system. *Recording.*

word—In a computer terminology, four bytes (q.v.), or 32 bits.

wow—Distortion caused by variations of speed of the turntable. *Recording.*

WVDC—Abbreviation for working voltage, direct current, or the maximum safe operating voltage for a capacitor. *Capacitors.*

WWV, WWVH—Call signs of National Bureau of Standards radio stations at Fort Collins, Colorado, and Maui, Hawaii. *Frequency Data.*

wye-connected system—Three-phase system used in power distribution lines. (See Figure W-7.)

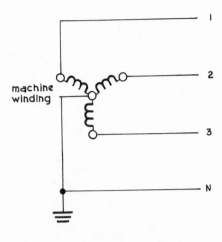

Figure W-7
Wye-Connected System

X—1. Symbol for reactance (q.v.).   2. In schematic diagrams, class letter for socket or holder.   3. In a three-dimensional reference frame, one of the three axes (see axis).

X-band—Microwave-frequency band from 5.20 to 10.90 gigahertz. *Frequency Data.*

xenon—Heavy and extremely rare gas, colorless, odorless, and tasteless, used in flash lamps for photography, ruby lasers, etc. Symbol, Xe; atomic weight, 131.30; atomic number, 54.

X-rays—Radiation similar to ultraviolet, but with shorter wavelength. *Nuclear Physics.*

Y—1. Symbol for admittance (q.v.).   2. In schematic diagrams, class letter for a piezoelectric or quartz crystal unit, a magnetostriction oscillator, or a tuning fork resonator.   3. In a three-dimensional reference frame, one of the three axes (see axis).   4. Symbol for chemical element yttrium (q.v.).

Yagi-Uda antenna—Multielement antenna, also called a Yagi antenna. *Antennas.*

Y-connected system—(See wye-connected system.)

yoke—Unit containing the vertical and horizontal deflection coils that is located on the neck of a television picture tube.

Y-signal—In color television, the brightness signal. *Broadcasting.*

yttrium—Silvery metal of the rare-earth group. Its compounds are used for the red phosphor in color-television tubes, in lasers, and microwave devices (yttrium-iron garnets, or "yig"). Symbol, Y; atomic weight, 88.905; atomic number, 39.

Z—1. Symbol for impedance.　2. In a three-dimensional reference frame, one of the three axes (see axis).　3. In a schematic diagram, class letter for various microwave devices.

zener diode—Silicon pn junction diode, whose breakdown voltage characteristics make it suitable for use as a voltage reference and as a clipper. *Semiconductors.*

zinc—Bluish-silver metal, a little more abundant than copper, used for zinc-carbon cells (ordinary flashlight batteries), galvanizing iron, and making brass (copper 50-60%, zinc 50-40%). Symbol, Zn; atomic weight, 65.37; atomic number, 30.

zirconium—White, soft metal, used for cladding fuel rods in nuclear reactors because it is highly transparent to neutrons, and as a getter in electron tubes. Symbol, Zr; atomic weight, 91.22; atomic number, 40.